S0-BLM-146

Secular Solar and Geomagnetic Variations in the Last 10,000 Years

ASTRONOMY

NATO ASI Series

Advanced Science Institutes Series

*A Series presenting the results of activities sponsored by the NATO Science Committee,
which aims at the dissemination of advanced scientific and technological knowledge,
with a view to strengthening links between scientific communities.*

The Series is published by an international board of publishers in conjunction with
the NATO Scientific Affairs Division

A	Life Sciences	Plenum Publishing Corporation
B	Physics	London and New York
C	Mathematical	Kluwer Academic Publishers
	and Physical Sciences	Dordrecht, Boston and London
D	Behavioural and Social Sciences	
E	Applied Sciences	
F	Computer and Systems Sciences	Springer-Verlag
G	Ecological Sciences	Berlin, Heidelberg, New York, London,
H	Cell Biology	Paris and Tokyo

Series C: Mathematical and Physical Sciences - Vol. 236

Secular Solar and Geomagnetic Variations in the Last 10,000 Years

edited by

F. R. Stephenson

and

A. W. Wolfendale

Physics Department,
University of Durham, Durham, England

Kluwer Academic Publishers

Dordrecht / Boston / London

Published in cooperation with NATO Scientific Affairs Division

o 305-6508

ASTRONOMY

Proceedings of the NATO Advanced Research Workshop on
Secular Solar and Geomagnetic Variations in the Last 10,000 Years
Durham, England
6–10 April 1987

Library of Congress Cataloging in Publication Data

Secular solar and geomagnetic variations in the last 10,000 years :
   proceedings of the NATO advanced research workshop held in Durham,
   England, April 6-10, 1987 / edited by F.R. Stephenson and A.W.
   Wolfendale.
        p.   cm. -- (NATO ASI series. Series C, Mathematical and
   physical sciences ; vol. 236)
     Includes index.
     ISBN 9027727554
     1. Solar oscillations--History--Congresses. 2. Magnetism,
   Terrestrial--Secular variation--History--Congresses.
   3. Paleoclimatology--Congresses.   I. Stephenson, F. Richard
   (Francis Richard), 1941-   . II. Wolfendale, A. W. III. Series:
   NATO ASI series. Series C, Mathematical and physical sciences ; no.
   236.
   QB539.083S43 1988
   523.7--dc19   88-12646
                                                        CIP

ISBN 90-277-2755-4

Published by Kluwer Academic Publishers,
P.O. Box 17, 3300 AA Dordrecht, The Netherlands.

Kluwer Academic Publishers incorporates the publishing programmes of
D. Reidel, Martinus Nijhoff, Dr W. Junk, and MTP Press.

Sold and distributed in the U.S.A. and Canada
by Kluwer Academic Publishers,
101 Philip Drive, Norwell, MA 02061, U.S.A.

In all other countries, sold and distributed
by Kluwer Academic Publishers Group,
P.O. Box 322, 3300 AH Dordrecht, The Netherlands.

All Rights Reserved
© 1988 by Kluwer Academic Publishers
No part of the material protected by this copyright notice may be reproduced or utilized
in any form or by any means, electronic or mechanical, including photocopying, recording
or by any information storage and retrieval system, without written permission from the
copyright owner.

Printed in The Netherlands

QB539
O83
N371
1988
ASTR

TABLE OF CONTENTS

PREFACE

Solar and geomagnetic variability are of considerable interest for scientists of many different persuasions and indeed one has the distinct impression that for the sun at least, there is direct relevance for mankind in general as the interrelation between solar and terrestrial phenomena is starting to be appreciated.

From the vast time scale of interest in the variability field, attention was confined to the last 10,000 years in a NATO Advanced Research Workshop held from April 6 - 10, 1987 in Durham, England, and the present publication comprises the lectures given there.

Such a Workshop was very timely in view of the impressive new data available from ^{14}C analysis in dated tree rings and ^{10}Be in polar ice cores, from natural palaeomagnetic records in lacustrine sediments and from archaeomagnetic material. Also to be mentioned are new studies of historical accounts of naked-eye sunspots and aurorae. All the data have contributed to improvements in understanding the relative variations of solar properties, the geomagnetic field and climate and it is hoped that this volume will convey the flavour of these advances in knowledge.

A feature of the Workshop was the lively discussions which followed so many of the papers. There were several instances of healthy disagreement and this is reflected in the opposing views presented in a number of the papers published here.

We were fortunate at the Workshop in having with us the doyen of British Astronomy, Professor Sir William McCrea FRS. His evening discourse on Cosmology was so interesting and thought provoking that we urged him to produce a manuscript for publication. The resulting paper is included as the finale to this volume.

Many members of the Physics Department and staff of St. Mary's College, where participants were accommodated, helped with the organisation and we are most grateful to them. Our particular thanks to our Departmental Superintendent, Mr. C.F. Cleveland, for his many contributions, to Ms. Margaret Chipchase for her considerable efforts in retyping the manuscripts and to Mrs. Pauline Russell for her art work. We also acknowledge the painstaking work of Mr. Kevin Yau who helped with Editorial duties.

F.R. Stephenson
A.W. Wolfendale

Durham, March 4, 1988.

NATO ADVANCED RESEARCH WORKSHOP
SECULAR SOLAR AND GEOMAGNETIC VARIATIONS IN THE LAST 10'000 YEARS

UNIVERSITY OF DURHAM. U.K.   APRIL 6-10, 1987

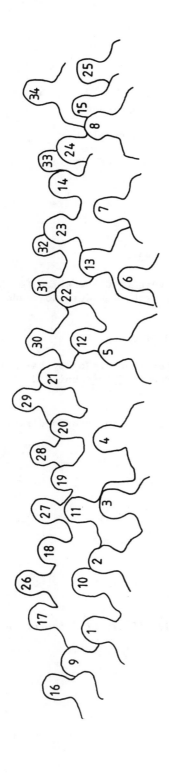

1. P.E. Damon
2. S. Cecchini
3. M. Chipchase
4. M. Galli
5. T.M.L. Wigley
6. J. Feynman
7. K.L. Verosub
8. F.R. Stephenson

9. N.-A. Mörner
10. R.L. Hanna
11. A.W. Wolfendale
12. K.K.C. Yau
13. S.K. Runcorn
14. N.O. Weiss
15. C.A. Jones

16. G. Bonino
17. M. Stuiver
18. J.A. Eddy
19. Sir William McCrea
20. G.M. Raisbeck
21. A.D. Wittmann
22. R. Jennings
23. D.G. McCartan
24. Y.K. Tulunay
25. G. Cini Castagnoli

26. C.F. Cleveland
27. A. Blinov
28. S.P. Papamarinopoulos
29. R.L. DuBois
30. J. Beer
31. D.H. Tarling
32. G.C. Reid
33. U. Siegenthaler
34. D.R. Soderblom

Institute Director: F.R. Stephenson

     Co-director: A.W. Wolfendale

                           Department of Physics
                           University of Durham
                           Durham, U.K.

Organizing Committee: J.A. Eddy

                           University Corporation for Atmospheric
                           Research, Boulder, Colorado, U.S.A.

                           D.H. Tarling

                           Department of Geological Sciences,
                           Plymouth Polytechnic, U.K.

                           plus Director and Co-director.

We are grateful to the Scientific Affairs Division of NATO for
sponsoring this Workshop.

LIST OF PARTICIPANTS

Lecturers

Beer, J., Institute for Medium Energy Physics, Zurich
Cini Castagnoli, G., Universita di Torino
Creer, K.M., University of Edinburgh
Damon, P.E., University of Arizona, Tucson
DuBois, R.L., University of Oklahoma, Norman
Eddy, J.A., University Corporation for Atmospheric Research, Boulder
Feynman, J., Jet Propulsion Laboratory, Pasadena
Galli, M., Universita di Bologna
Isaak, G.R., University of Birmingham
McCrea, Sir William, University of Sussex
Mörner, N.-A., Stockholm University
Raisbeck, G.M., Laboratoire René Bernas, Orsay
Reid, G.C., National Oceanic and Atmospheric Administration, Boulder
Runcorn, S.K., University of Newcastle upon Tyne
Siegenthaler, U., University of Bern
Soderblom, D.R., Space Telescope Science Institute, Baltimore
Stephenson, F.R., University of Durham
Stuiver, M., University of Washington, Seattle
Tarling, D.H., Plymouth Polytechnic
Verosub, K.L., University of California, Davis
Weiss, N.O., University of Cambridge
Wigley, T.M.L., University of East Anglia
Wittmann, A.D., University Observatory, Goettingen
Wolfendale, A.W., University of Durham

Participants

Bailiff, I.K., University of Durham
Blinov, A., Leningrad Polytechnic
Bonino, G., Universita di Torino
Cecchini, S., Universita di Bologna
Doidge, C.M., Rutherford Appleton Laboratory, Chilton
Hanna, R.L., University of California, Davis
Harding, A.F., University of Durham
Jones, C.A., University of Newcastle upon Tyne
Jones, M.K., University of Durham
Kenworthy, J.M., University of Durham
Lowes, F.J., University of Newcastle upon Tyne
McCartan, D.G., University of Newcastle upon Tyne
Osborne, J.L., University of Durham
Papamarinopoulos, S.P., University of Patras
Parkinson, J.H., University College, London
Soward, A.M., University of Newcastle upon Tyne
Tulunay, Y.K., Istanbul Technical University

Wilkinson, D.A., University of Durham
Willis, D.M., Rutherford Appleton Laboratory, Chilton
Wright, D.A., University of Durham
Yau, K.K.C., University of Durham

VARIABILITY OF THE PRESENT AND ANCIENT SUN: A TEST OF SOLAR
UNIFORMITARIANISM

John A. Eddy
University Corporation for Atmospheric Research
Box 3000
Boulder, Colorado 80307
U.S.A.

ABSTRACT. Modern observations of the Sun demonstrate that surface
features vary on time scales of minutes (solar flares) to weeks and
months (the lifetimes of sunspots and active regions) to years (the
11 and 22 yr cycles of sunspot and magnetic activity). Less certain
in the available record of direct observations (now 378 years long)
are solar changes of longer period, including a possible "Gleissberg"
cycle of 80 to 90 yrs that is suggested in the envelope of annual
mean sunspot numbers. We examine what is known of solar variations
before the time of the telescope to test the thesis of solar uniformitarianism: whether the Sun behaved any differently in the past,
and whether the full spectrum of modern variations are found in ancient
and proxy records. For this purpose we consider naked-eye observations
of the Sun, records of the aurora, the tree-ring record of ^{14}C, initial
investigations of ice-borne ^{10}Be, and the possible records of solar
influence that are found in sedimentary deposits from the Precambrian.

1.   INTRODUCTION

Two hundred years ago James Hutton, who was born in Edinburgh, not
far north of Durham, put forth a _principle of uniformitarianism_ to
explain the natural evolution of the Earth. Later, Sir Charles Lyell,
another eminent Scot, would advance it even further. It held that
the geological features that appear today can be explained by natural
processes - all of which we now see at work. Nothing different or
divine was needed in the long reach of the past to explain the form
of the Earth. In time, this far-reaching notion became the foundation
of modern geology.
     For a long time we applied the same principle, without thinking
much about it, to the behaviour of the Sun and of the Earth's magnetic
field: namely, that what there is to know of either of them can be
described and explained by what we have observed in the last 100 years
of close examination. Solar variability was the regular pulsing of
the 11-year sunspot cycle, found by Schwabe in 1842. Geomagnetic
variability was fully described by the nature of the excursions noted

1

F. R. Stephenson and A. W. Wolfendale (eds.),
Secular Solar and Geomagnetic Variations in the Last 10,000 Years, 1–23.
© _1988 by Kluwer Academic Publishers._

on the magnetometers that Gauss and others brought into use at about the same time.

To me these long eras of solar and geomagnetic uniformitarianism were, like geology before the revelations of plate tectonics, a necessary but intrinsically limiting phase in the development of our science. We know now that the last 100 years of solar behaviour are not adequate to sample all that the Sun can do; nor will the era of magnetometers tell us of magnetic reversals or of the slow and possibly cyclic changes in the strength of the geomagnetic moment that fills the time between them. What uniformitarianism misses is what brought a NATO workshop to Durham: the secular changes of time scales of decades and centuries and millennia that are inadequately sampled in the brief period of modern, real-time examination.

In this review I shall compare, for the case of the Sun, the variations sensed in modern measurements with those that appear in the longer span - as far back as we can see before the time of the telescope. The purpose is to test the principle of solar uniform-itarianism. Did the Sun behave in any way differently in the past, compared to what we know of modern solar variability? And, as the question is more customarily asked, does the full spectrum of modern variations appear in ancient historical and proxy records? Answering these questions will also serve as an introductory review of the Sun and solar variability.

2.   OBSERVED NATURE OF SOLAR VARIABILITY

The last 100 years of close observation has demonstrated that the Sun varies in a number of ways. These can be divided, for our purposes, into two classes: the first are those that are seen on the surface of the Sun, almost always aided by the resolving power of a telescope; the second are the changes felt at the Earth, from the unresolved Sun, in the form of fluxes of energy, particles, or fields.

A.   SURFACE FEATURES

(i)   Sunspots.

The best known manifestations of solar variability are the dark and transitory spots that are seen on the white surface of the Sun when it is viewed in broad-band, visible light. Sunspots are so common and simply found that they were among Galileo's first discoveries with his little telescope, in 1609; they are of such high contrast compared to the bright light of the disk that the largest of them can be seen with the unaided eye by an observer with good acuity and a knowledge of what to look for.

Sunspots appear as small, dark grains in the normal pattern of photospheric granulation. They grow slowly, and last, on average, a week or two. The largest, which exceed the size of the Earth, live for several months, time enough for several rotations of the Sun.

The smallest, a few hundred kilometers in diameter, last but a day or so.

Sunspots are the surface manifestation of concentrated, sub-surface magnetic fields. Field strengths of several thousand gauss are commonly measured, spectroscopically, in the darker, center (or "umbra") of the spot; the less dark "penumbrae" that bound larger spots can be seen, on close examination, to be made up of filaments of matter that arch outward from the umbral core to the surrounding photosphere. The concentrated magnetic fields that produce sunspots are the bases for almost all other known forms of solar variability, including the most dramatic, which are solar "flares". The principal effect, however, of these ordered, radially-directed fields is to inhibit, locally, the outward flow of convective energy that pours from the center of the Sun. This local diversion of the flow of energy makes sunspots cooler and hence darker than the surrounding surface: the difference in temperature is, for umbrae, about 1700K, a drop of more than 25% from the 600K white-light surface; penumbrae are about 500K cooler than their surroundings. The contrast in brightness is for umbrae about 75% and for penumbrae about 20% (Allen, 1976).

The convective energy blocked by sunspots is not immediately re-radiated. As a result, the flow of energy from the surface of the Sun is not constant, but continually modulated. The largest sunspot groups can cover about 0.1% of the apparent disk of the Sun; since several of these can be present at one time we can expect the total radiative flux from the Sun, as measured at the Earth, to fluctuate at about this order of magnitude. Such is indeed the case (Willson, 1984). This daily-to-weekly modulation of several tenths of one percent in the flux of energy received at the Earth constitutes by far the greatest known modulation of the Sun's output of energy. The energy carried to the Earth in the form of solar-wind particles is, in contrast, about six orders of magnitude smaller than the radiative flux. Variations in the ultraviolet flux are also smaller, by almost two orders of magnitude, than the effect of sunspot blocking. The energy released in the largest solar flares is some 6 orders of magnitude less than that blocked by a single large sunspot group.

(ii) Faculae

Dark sunspots are almost always accompanied by associated bright regions that cover a much larger area than the spots themselves; they are also more difficult to see, because their contrast, relative to the undisturbed background, is less. These subtle brightenings appear before sunspots and persist after the associated sunspots have disappeared. They were found with the earliest telescopes of Galileo and his contemporaries and called "faculae", from the Latin for "torches". Like sunspots, faculae are also associated with concentrated magnetic fields and represent a net increase in the brightness of the active Sun relative to its undisturbed state. Faculae are typically 15% brighter than the surrounding photosphere and about 1000K hotter; they cover an area up to four times that of the associated spots. What is not certain is whether or how the convective energy blocked by sunspots is released by facular brightening, and whether the energy is precisely balanced between them. Complicating the case are other, more subtle features of the solar surface that also play a role in modulating the Sun's output of energy.

(iii)   The Sunspot Cycle

The number of dark spots seen on the visible hemisphere of the Sun, like the number of bright faculae, varies significantly from day to day, ranging from none to several hundred. Monthly or annual averaging of the number of sunspots that are seen reveals an obvious cycle between maxima or minima in their number, lasting about 11 years. The sunspot cycle is far from regular: cycles range from 8 to 14 years. The long-term average is about 11.3 years although in the present century the period is closer to 10 years. Power spectrum analyses demonstrate a distinctly bimodal distribution over the period of real-time observation (since about 1850) with peaks at about 9.8 and 11.0 years (Okal and Anderson, 1975).

The conventional index of sunspot activity is the Wolf sunspot number, an arbitrarily-defined index that varies in daily value from zero to several hundred and in annual averages from 0 (in 1810 and 1913) to 190 (in 1957) (McKinnon, 1987).

Solar activity is currently at the (minimum) end of a cycle that began, by terrestrial calendars, in mid-1976 and peaked with a smoothed monthly mean of 164 in the autumn of 1979. The annual sunspot number for 1979 was 155, making the cycle that just ended the second highest in the 186 years for which sunspot numbers have been observed or estimated. Solar activity rises more rapidly than it falls, and the next maximum is expected in about 1990, about two years from now.

Theory is as yet unable to provide an accurate prediction of either the amplitude or the phase of future cycles. We can guess, however, that the maximum of the new cycle will be lower than the last (155), for several reasons. The first is that the average maximum of the last 21 cycles is only 106. Two further reasons are evident on examination of Figure 1. Apparent is a long-term trend, possibly cyclic, from lower maxima at the turn of the century through an apparent peak in the all-time high of the cycle that reached maximum in 1957. Since then the trend appears to go downward. The longer record of historical sunspot numbers suggests that this slow rise and fall may be part of a "Gleissberg cycle" of about 90 years in the "envelope" one would draw by connecting the peaks in the annual sunspot number curve. The Gleissberg cycle may not be a real feature of solar activity; Barnes et al. (1980) have demonstrated that one can replicate all the features of the sunspot number envelope mathematically by passing a 22-year signal through a random filter. If the cycle is real, the next Gleissberg minimum should be soon upon us, for the last two minima in the envelope of annual sunspot numbers fell in about 1810 and about 1900. The height of the next cycle and the one that follows it are thus of particular interest as a test of the reality of this apparent 90 yr feature.

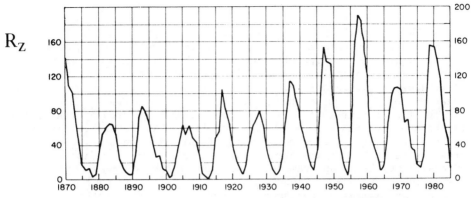

R_Z

Fig. 1.   Annual mean sunspot numbers (R_z), A.D. 1870–1986.

A final reason for expecting the next cycle to be lower than the
current cycle is a systematic alternation between higher and lower
cycles that is most apparent in the first five and the last three cycles
in Figure 1.   That trend is also borne out in the longer record.   The
alternation reflects, we think, the fundamental magnetic nature of
the solar cycle: the magnetic polarity of sunspot pairs reverses,
systematically, in alternate cycles.   Thus the magnetic cycle of the
Sun is not 11 but 22 years long.

Figure 2 depicts annual mean sunspot numbers for the full period
since the introduction of the telescope in 1609.   Before commenting on
it we must note that the record is of uneven quality (McKinnon, 1987;
Eddy, 1976).   Real-time records of the Wolf sunspot number exist only
since the index was defined, in 1848: all prior numbers were reconstructed from scattered data that become less continuous and reliable
as one looks farther back in time.   Beyond doubt, however, are the long-
term minima in solar activity at the turn of the last century (the
"Modern minimum"), in the early years of the 19th century (the "Dalton
minimum"), and between the period of about 1645 and 1715 (the "Maunder
minimum").   The Maunder minimum has been particularly well examined
(Eddy, 1976; 1983a), though we should note that the estimates of sunspot
numbers that are depicted for that period in Figure 2 are reliable to
at best a factor of two.

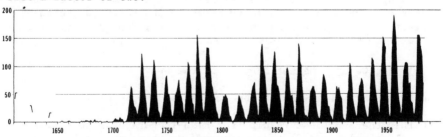

Fig. 2.   Annual mean sunspot numbers (R_z), A.D. 1609–1985.

(iv)   Solar Flares and Eruptive Prominences

The immense potential energies resident in the magnetic fields
associated with sunspots are at times catastrophically converted to
radiant and kinetic energy, with the sudden brightening of a localized
region and an accompanying release of high-energy, atomic particles.
Such events, called solar flares, last typically 5 to 10 minutes and
release up to 10^{30} ergs of energy (Sturrock, 1980). Even so they are
an insignificant perturbation in the total energy emitted by the visible
hemisphere of the Sun (2×10^{33} ergs sec^{-1}), and only very rarely can
they be seen in the broad-band, "white" light from the disk. Solar
flares are best seen in wavelengths where their thermal radiation peaks,
in ultraviolet, X-ray and gamma-ray regions of the solar spectrum,
and as emission in bright lines of the visible spectrum that originate
in the higher regions of the solar atmosphere. Atomic particles released at such times can reach the earth as "solar cosmic rays",
inducing a number of terrestrial effects, including displays of the
aurora and sudden disturbances in the Earth's magnetic field.
       Since solar flares are related to sunspot magnetic fields, their
intensity and occurrence follows the 11-year sunspot cycle, with
several large flares per day at high maxima of the cycle, and few
if any in years of sunspot minima.
       Solar "prominences" are clouds of denser material that are
suspended in the more rarified, upper atmosphere of the Sun by the
arching lines that connect regions of opposite magnetic polarity on
the surface of the Sun. Prominences typically persist for weeks or
months, appearing at the edge of "limb" of the Sun as semi-permanent
clouds.   Others are more active and can erupt in response to rapid
alterations in the topology of the sunspot magnetic fields in which
they are rooted. Since prominences are an effect of solar surface
fields their number also follows the sunspot cycle.
       Eruptive prominences can trigger ejections of matter from the
overlying solar corona, in the form of "coronal transients" that are
then expelled outward as large blobs of diffuse matter into the interplanetary medium. Though not discovered until the era of spaceborne
coronagraphs (ca. 1970) coronal transients are a very common feature
of the Sun: occurring, on average about once per week in years of
minimum solar activity and nearly every day at maxima of the cycle.
Under favourable geometry, the particles carried outward in coronal
transient disturbances can be intercepted by the Earth, where they
are detected as perturbations in the continuous, "solar wind" of
particles from the Sun.

(v)   The General Magnetic Field of the Sun

Sunspots, and many features of the 11-year sunspot cycle, can be
explained if we accept the presumptions of (1) unseen internal fields,
deep within the Sun (that result from high-temperature ionisation
of neutral atoms); and (2) the systematic ordering of these fields
in the radial direction by patterns of large-scale, sub-surface
convection. Dynamical coupling with the observed differential rotation

of the surface of the Sun will produce a systematic drift in the latitude at which spots are observed, and the observed alternation of the magnetic polarity of sunspot groups. The result is a solar surface magnetic cycle of 22 yrs, which is now taken to be the more fundamental cycle of solar variability.

The Sun has no apparent permanent global field. The composite surface field (made up of the aggregate of accumulated, local surface fields) (Howard, 1977) systematically drifts towards the rotational poles of the Sun in the course of the solar cycle, giving the appearance of a general field, chiefly bipolar, that is aligned with the Sun's axis of a rotation. This apparent field is visible in the structure of the solar corona that is seen at times of total solar eclipses, as radial plumes of trapped electrons that emanate, like lines of force from the ends of a bar magnet, from the rotational poles of the Sun.

The strength of the composite solar and the terrestrial magnetic field is in each case about 1 gauss. They differ fundamentally in that (1) the Sun's apparent general field is superficial as opposed to the more permanently rooted field of the Earth; the (2) as such, the solar field reverses regularly and much more frequently, at alternative maxima of the 11-year sunspot cycle. This regular reversal of the solar field, coupled with a "fixed" terrestrial field, offers a potential magnetic "switch" to theorists who would explain varied solar-terrestrial effects by postulated Sun-Earth magnetic coupling. As yet, the potential coupling seems to be more theoretical than practical.

B.   SOLAR VARIATIONS SENSED AT THE EARTH

Whether any of these telescopically observed features of the Sun is of more than academic interest depends on the effects, if any, that are produced at the orbit of the Earth. Terrestrial effects of solar changes involving particles and fields were identified more than a century ago, as displays of the aurora and as disturbances in the geomagnetic field. In the case of variations in the radiative flux, there was little more than conjecture until rockets and spacecraft came upon the scene. About 30 years ago, measurements of the ultra-violet and x-radiation made above the atmosphere demonstrated a strong, positive correlation between the flux from the Sun in these wavelengths and solar activity. The identification of more energetic, activity-related changes in the "solar constant", or total radiative flux, was not made until 1980.

(i)   Solar Particles and Fields

Changes in the flux of atomic particles from the Sun are influenced by conventional solar activity (as measured in the sunspot number) but are far from a simple function of it. The gross nature of the chief flow of particles from the Sun - the solar wind - shows no clear variation in either particle velocity, particle energy, or particle composition with the 11 yr sunspot cycle (Hundhausen, 1978). This is in part because

we receive, in addition to flare particles, a highly variable flux
of solar particles from "holes" in the solar corona   - extended
regions of radial or "open" magnetic field - whose occurrence and
placement are not simply related to sunspot number.

More energetic bursts in the flow are related to flares and
coronal transients, and hence to the sunspot cycle, but these are
added to an intrinsically variable flow that is to first order
independent of the 11 yr cycle.

Either of these two distinctly different classes of solar features
(sunspot-related solar flares and sunspot-unrelated coronal holes)
can provoke geomagnetic disturbances and produce the aurora borealis
and australis, by triggering the release of charged particles that
are resident in the Van Allen belts of the Earth. Records of the
occurrence of auroral displays have been traditionally used as a proxy
of sunspot activity (Schove, 1955), although an imperfect one, due to
reasons just noted. Aurorae are seen, at high latitudes, at all
phases of the solar cycle.

Aurorae occur in a roughly circular ring, or "auroral oval"
centred on the geomagnetic pole and at a range of geomagnetic latitudes:
sometimes nearer the pole, occasionally farther from it. Due to this
difference it is possible, in principle, to separate the two classes
of aurorae and hence to improve the utility of historical auroral
data as a proxy of sunspot number. Typical energies of particles
emitted by flare disturbances exceed those that accompany high-speed
disturbances in the solar wind. Activity-related aurorae result in
an auroral oval of larger diameter and are seen at lower geomagnetic
latitudes than those produced by high-speed wind streams from coronal
holes. The occasional aurorae noted at the low geomagnetic latitude
of Rome, or in Mexico City, for example, are clearly the result of
flares. Those noted in northern Scandinavia can be due to either.

This distinction, known only since coronal holes were first
identified in the early 1970's, explains why aurorae are seen at times
when there are no sunspots on the disk, and why a few aurorae were
reported in northern Europe during the span of the Maunder minimum.

(ii) Modulation of galactic Cosmic Rays

Another terrestrial effect of solar activity, and hence a potential
proxy for sunspot activity, is the modulation of the flux of high-
energy cosmic rays that are received at the Earth. Fluctuations in
the extended magnetic field of the Sun, carried outward toward the
Earth by the solar wind, deflect some of the charged atomic particles
that reach the Earth from all directions in the galaxy. The principal
perturbers of the cosmic ray flux are impulsive flare disturbances,
and hence are a direct measure of solar activity. Their effect is
to decrease the flux of high-energy galactic cosmic rays at the Earth
at times of maximum solar activity, resulting in an anticorrelation
of galactic cosmic ray flux with sunspot number. Today, at the
minimum of the sunspot cycle, we are receiving more of them.

The direct flux of flare-related particles or "solar cosmic rays"
will produce an opposite effect: more particles at times of high solar

activity. These can be distinguished, however, on the basis of
particle energy: solar cosmic rays are considerably less energetic
than those from the galaxy. The distinction proves important when
cosmogenic nuclides such as ^{14}C or ^{10}Be are used as proxies of solar
activity.

(iii)   Short-Wave Radiation

The ultra-violet and X-ray flux from the Sun varies considerably with
solar activity through a range that is a sharp function of wavelength
(White, 1977). These changes – systematically greater at shorter
wavelengths – are readily noted as day-to-day changes in spaceborne
measurements (Donnelly et al., 1982). As active regions grow, erupt,
or are carried across the solar disk by solar rotation, the ultraviolet
and X-ray flux sensed at the Earth varies through limits that range
from about 0.5% at 300 nm wavelength to several hundred per cent in
the shortest and hardest X-rays (White, 1977; Donelly, 1987). This
dependence on wavelength can be explained by the systematic difference
in the height in the solar atmosphere at which the continuum radiation
originates. The temperature of the solar atmosphere increases systematically above the photosphere, which is the source of the white-light
continuum. Shorter wavelengths originate at higher and hotter layers in
the more variable chromosphere and lower corona. Hence they vary more.

   More difficult to secure are continuous measurements of short-wave
radiation from the full disk of the Sun with a calibration sufficient
to establish year-to-year variations. Available data demonstrate,
as one might expect, that the flux of short-wave radiation from the
Sun rises and falls in phase with the 11-year sunspot cycle, through
limits that are roughly the same as those noted in rotational modulation.

   What is not known – though we can anticipate the result – are
the changes in short-wave flux that result from the longer term, or
secular changes on the Sun, such as the slow increase in the envelope
of sunspot activity that characterized the first half of the present
century. We can expect, however, that the ultraviolet and X-ray
fluxes from the Sun will follow the general envelope of the record
of annual sunspot numbers: low at the turn of the present century,
even at maxima of the 11-year cycle, and rising by a factor of two
to three at maxima of the high cycles of the last 40 years. By the
same reasoning we can presume that the ultraviolet flux from the Sun
was consistently low during the decades of the Dalton and Maunder
minima (A.D. 1795–1830 and 1645–1715).

(iv)   The Solar Constant

Far and away the most important parameter in the Sun-Earth connection
is the so-called "solar constant" – the total light and heat received
in all wavelengths above the atmosphere at mean Sun-Earth distance.
This fundamental quantity is only roughly obtained from the surface
of the Earth since the atmosphere, even above the highest mountains,
is a strong and highly variable absorber. A canonical value of about

2.00 cal cm^{-2}sec^{-1} was established about 30 years ago by combining data
from the ground with early rocket measurements (Johnson, 1954). The more
precise value, and the more interesting question of possible variations
in the solar constant could be addressed only recently, with the advent
of precision radiometers carried on long-lived spacecraft: specifically
the ERB experiment on Nimbus-7, launched in late 1978 (Hickey et al.,
1980), and the more precise ACRIM instrument carried on the Solar
Maximum Mission, launched in February 1980 (Willson et al., 1981). Both
are still in operation.

    Continuous records from both of these spaceborne radiometers show
systematic, day-to-day depletions that are clearly related to the
passage of large sunspot groups across the disk of the Sun, resulting
in regular dips lasting days to weeks that reflect the change in the
projected area of large sunspot groups with solar rotation (Willson,
1984). The deepest depletions are about 0.3%. These changes, though
interesting to solar physicists, are of negligible consequence in terms
of climatic response because of their short duration. Moreover, the
depletions are in time balanced to some extent by the more slowly
modulated, enhanced radiation from associated faculae.

    Recent analyses by Foukal and Lean (1987) have demonstrated the
existence of a third contributor to fluctuations in the solar constant,
which are extended magnetic regions unassociated with sunspots, where
a decreased gas pressure allows radiation to flow outward more readily
than from the undisturbed solar surface. Their work indicates that
on time scales of weeks, the sunspot and facular contributions nearly
balance. The third effect, however, contributes a net increase in total
solar flux at times of enhanced activity, resulting in a modulation of
about 0.05 to 0.07% that is in phase with the solar cycle.

    More interesting in terms of terrestrial effects is the systematic
downward drift in the solar constant that is found in the continuous
records from both spaceborne instruments. Since 1978 the solar constant
has been declining at a rate of nearly 0.02% yr (Willson et al., 1986).
It now seems to be flattening out, suggesting an 11 or 22 yr cycle
(Gurman, 1987; Frohlich and Eddy, 1984). Were it to continue downward
at the present rate, such a drift would soon be climatically significant
in terms of altering the mean surface temperature of the Earth. Foukal
and Lean (1987) interpret the drift as evidence of the extended magnetic
region effect, and predict a reversal of the trend in the next few years
with the growth of the new cycle of solar activity. Until the trend
reverses, however, it could also be explained as a modulation of longer
period.

    Reid (1987) and Reid and Gage (1988) have more recently interpreted
the trend as a reflection of the long-term envelope of the solar
activity cycle, as once suggested by Eddy (1976, 1977), based on the
coincidence of the Maunder minimum in solar activity with the coldest
extreme of the Little Ice Age. A sustained reduction of about 1% in
the solar constant (50 yrs at the present rate of decrease) is adequate
to induce a drop in the globally-averaged surface temperature of the
Earth of about 1°C, which is consistent with the climate of the Little
Ice Age. Reid notes that the sea surface temperature record of the
last 120 years bears out the correlation, suggesting, as do Foukal and

Lean, that the downward drop in the measured solar constant is a direct reflection of solar activity.

3. RECONCILING THE PAST: EVIDENCE FROM HISTORICAL AND PROXY RECORDS OF SOLAR CHANGES

A. Solar Variations in Historical Records

(i) Extent of historical records of the Sun

We must recognize that our observations of almost anything related to the Sun are either very short or extremely limited (Eddy, 1980). What we would most like to have is a long record of the solar constant, and today the available record is less than a decade long. The longest direct record is that of sunspots, which tells only of solar surface conditions. Our longest records – those from isotopic abundances in tree-rings and ice – are a further step removed, and at best partial indicators of conditions in the solar wind and of the extended magnetic field of the Sun. The nuclide data are also confused by the presence of other and more dominant modulations: due to geomagnetic shielding and, in the case of ^{10}Be, to climatic effects.

(ii) Limits of the Telescopic Sunspot Record

Given the uneven reliability of the telescopic record of sunspots (Figure 2), how certain are the major, secular changes in the level of solar activity, such as the Maunder minimum, the Dalton minimum, or the apparent rise in the present century?

We can be certain that the increase of a factor of three in maxima of the 11-year cycle in the first 60 years of the present century is real, coming as it does from controlled, real-time data. Further verification comes from the geomagnetic record, as demonstrated by Feynman and Crooker (1978): between 1900 and 1960 the base level of the geomagnetic aa index rose uniformly and even more steeply than did the sunspot number.

The Dalton minimum in sunspot numbers between about 1795 and 1830 has also been confirmed through the examination of auroral data, by Siscoe (1980) and by Feynman and Silverman (1980). The latter authors noted a dramatic drop in the number of middle to low latitude aurorae during the period, consistent with the hypothesis that aurorae during the time were dominantly weaker events induced by high-speed streams in the solar wind.

The reality of the Maunder minimum (1645-1715) is based on many lines of evidence that have been extensively tested and ultimately verified (Eddy 1976; 1983a). The primary evidence is found in con-temporary historical telescopic records, such as those in Figure 3. One of the weaker lines of evidence came from the Oriental dynastic histories that were written during the time. The largest sunspots can be seen with the unaided eye, when the Sun is sufficiently dimmed. In dynastic histories from China, Korea, and Japan are accounts of dark spots on the Sun that have been compiled to aid the recovery of

12

solar history.

APPENDIX

Animadverſiones ſuper I. Macula-

rum Solarium periodo.

1. **A**Nno à nato Chriſto, 1642. die 26. Octob.
in Sole nihil Macularum apparuit.

2. Die 27. Octobris cœlum non favit.

Inſignis Ma-
cula numbe
cernatta Oc-
tovat.
3. Die 28. Octob. magnam, oblongam, denſam, pulcherrimamq;
Maculam a, halone haud vulgari undique cinctam, ani-
madverti; præter hanc autem unicam nihil prorſus in Sole
deprehenſum.

4. Die 29. & 30. Oct. ob aëris turbulentiam, Maculam iſtam no-
tabilem, die præcedente conſpectam obſervare non licuit.

5. Die 31. Octobris, alterâ vice, Maculam ſuprà dictam, ſed in
formâ ampliori, duobusque nucleis denſiſſimis præditam,
nobis videre obtigit.

6. Die 1. Nov. illa ipſa Macula, tum quâ figuram, tum ſplendorem
magnitudinemque nihil planè ſe immutaverat.

7. Die 2. Novembris, nullus Sol affulſit.

8. Die 3. Nov. non ſolùm ulteriùs occaſum verſùs in Solis diſco
progreſſa erat; ſed & magis magisq; ad Aquilon. deflectebat.

9. Die 4. Novembris, nihil quicquam adhuc mutata viſa fuit.

10. Die 5. Nov. Cœli inclementia obſervationi fuit impedimento.

11. Die 6. Nov. paululùm oblongior videbatur; tum, & circa or-
tum tres novæ minores ſpectabantur, quæ in ſequens ſche-
ma reſeruantur.

12. Die 7. Novembris, aër fuit turbidus.

13. Die 8. Novemb. exitum hujus inſignis Maculæ perquàm
libenter obſervaſſem, ſed fruſtrà fuit, quia jam exiverat,
ſic ut nec veſtigium ampliùs de eâ apparuerit in Periphe-
riâ. Curſum quod attinet Maculæ, quantum colligere li-
Macula 13
confecit in
Sole dies.
cet, fuit 12. tantummodo dierum. Præterea, ex hoc motu
ſatis ſuperque liquet, viam hujus Maculæ fuiſſe concavam,
Qui vià in-
ceſſerit!
Aquilonem Borealem, & convexam Auſtrum occidenta-
lem verſùs.

Ani-

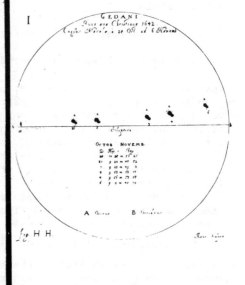

Fig. 3. Facing pages from Johannes Hevelius' Selenographia, published in Danzig in 1647, showing the passage of a single
sunspot across the disk of the Sun between October 26 and
November 8, 1642. The left-hand page is a daily log,
including those days when the sky was overcast. The
superposed table on the right lists the time of day to the
nearest minute, and the altitude of the Sun above the
horizon in degrees and minutes. This large spot, shown
in good detail, was sketched by the astronomer as projected
by a contemporary telescope. Though more than large
enough to be seen with the unaided-eye, it was one of the
countless others that were not recorded in Oriental
histories.

Although spots were noted as early as 28 B.C., none were recorded in
these official histories during the time of the Maunder minimum (Eddy,
1976). This initial assertion was later challenged by Xu and Jiang (1982)
who studied Chinese provincial histories written during the 17th century
and found 21 possible reports of sunspots. Not all of them were new information, however: all but six fell in the period before the Maunder
minimum began. Of these, three were duplicates of sunspots reported
telescopically in Europe at the time. This left three "new" possible

sunspots to add to the 600 that were known to have been seen with the telescope during the period of the Maunder minimum. One of these, reported in 1665, is particularly in doubt since it was seen at noon on the day of a total solar eclipse. A total eclipse does not make it easier to see sunspots with the naked eye. Moreover, noon seems the wrong time to look. Given the difficulty of looking directly at the Sun, most would agree that the bright disk could be examined with the naked-eye only when it is dimmed, as at sunrise and sunset, when reports of sunspots were presumably made.

At the same time, given the paucity of naked-eye sunspot reports in any period (the average is about one per decade over a period of 1700 years) one could argue that even one "new" naked-eye report is significant, in that it suggests the presence of many more that were too small to be seen without a telescope. Two facts argue against this: the first is that the number of spots seen with naked-eye during the 17th century drops during the Maunder minimum by the same ratio as did the telescopic reports (Eddy, 1983a). A second point is that by the time of the Maunder minimum the telescope had been introduced in China, by Jesuit missionaries in 1612 (D'Elia, 1960). The Chinese knew of Galileo's discoveries soon after he made them, including his "discovery" of spots on the Sun. How well this information had become disseminated in the subsequent years of the century we do not know, but it gives reason to question the value of any naked-eye sunspot reports made during or after the middle of the 17th century.

As we shall see, the strongest arguments for the reality of the Maunder and Dalton minima are now the confirmation of a dramatic increase in the production of ^{14}C and ^{10}Be during those periods.

(iii) Limits of the Pre-Telescopic Sunspot Data

An enigma of the Oriental pre-telescopic sunspot record is yet to be answered: namely, why were so few reported? There are now about 150 known reports, covering 17 centuries of dynastic and provincial accounts – or roughly one large spot per decade (Stephenson & Clark, 1978). The utility of naked-eye reports rests on the assumption that a continuous or nearly continuous watch was kept, as has shown to be valid for night-sky phenomena such as supernovae (Clark & Stephenson, 1977), or the returns of Halley's comet (Stephenson et al., 1985).

If we take the canonical figure of about one arc minute (1/30 of the angular diameter of the Sun) as the minimum size of a detectable sunspot or sunspot group we should expect not 150 but 170,000 naked-eye sunspot reports during the 17 centuries, if all were seen (Royal Greenwich Obs., 1955). The large, telescopic sunspot shown in Figure 3, for example, is one of many that apparently went undetected. The difference of a factor of 10^3 is not a trivial one. Could the error be in the 1 arc-minute criterion? Relaxing this to 3 arc minutes would rectify what were reported with what would have been seen, since only about one of these immense spot groups is seen per decade today.

Recently Richard Stephenson and I have been able to test that, through the work of an avid English amateur who watched the Sun set with unaided eye daily over the sea at Liverpool for a period of 13

months near the maximum of the last sunspot cycle. He took care not
to refer to a telescope until after he had looked with unaided eye.
The number that he saw, which were many, more than redeems the one arc
minute criterion.

For this reason I think it more likely that the Chinese catch was
one of random sampling: a continuous watch was not kept, and spots were
looked for only occasionally, prompted, perhaps, by augury. In that
case, however, one wonders why the record is as valuable as it is. In
fact, the thin Oriental naked-eye sunspot record shows all the major
secular features that are now found in the auroral record and in the
records of ^{14}C and ^{10}Be. In its heaviest parts, the Oriental naked-eye
sunspot record gives evidence also of the presence of the 11-year cycle,
which is one of the questions of uniformity that we sought to answer.
The cycle can be detected, statistically, in Oriental sunspot sightings
during the 12th and 13th centuries (Stephenson and Clark, 1978; Yunnan
Obs., 1977) - a time of apparent high activity, when the number of
sunspot reports is twice the long-term average.

(iv) Nature of the Historical Data

Reports of the aurora go back many hundreds of years in Europe, and
even further in Oriental accounts. For this reason they have been
extensively used as a proxy of solar activity, and particularly in the
period before the advent of the telescope (Schove, 1955). As noted
earlier, they are an imperfect proxy, unless one takes only those
reports made at low geomagnetic latitudes, where reports are few, or
sorts them judiciously by latitude. Several other factors complicate
their interpretation. Auroral reports are often hard to confirm, since
other natural phenomena of the sky, such as rainbows or haloes or
comets may be similarly described in sketchy records. Moreover, unlike
long-lived sunspots, the aurora is a short-lived and geographically
localized phenomenon, and the number that are reported at any place is
subject to meteorological conditions.

A modern auroral record has never been assiduously kept (for
longer than one person's interest) and, since the work of Fritz (1873)
and others, the full record has yet to be assembled on a systematic,
world-wide basis. Historical records, chiefly from Scandinavia, are
scattered. In spite of this, as with Oriental sunspots, the auroral
record is remarkably successful in demonstrating both the 11-year sunspot cycle and the existence of secular minima like the Dalton, Maunder
and Spoerer minima. Siscoe (1980) has demonstrated that the solar
90-year cycle can be found in the historical auroral record.

What has yet to be convincingly demonstrated is the effect on
auroral incidence of secular changes in the geomagnetic field. The
width and placement of the auroral oval is a function of the orientation
of the geomagnetic axis and the strength of the geomagnetic moment, as
well as a function of solar activity. Drifts of the magnetic pole
and changes in the strength of the magnetic moment should both appear
in a complete record of worldwide auroral incidence.

Changes in the incidence of aurora in the Orient (where reporting sites are at low geomagnetic latitudes) were invoked by Keimatsu et al. (1968) to postulate a westward shift of 15° to 20° in the geomagnetic pole in the 11th and 12th centuries. The case was further studied by Siscoe and Verosub (1983), who found it likely that a major change in pole position had occurred between A.D. 1400 and 1650. Keimatsu et al. endeavoured to reconstruct the auroral oval, and from this to estimate the orientation of the geomagnetic axis, by selecting those auroral reports that were made on the same night in the Orient and in Europe. Although the number of cases were few, they concluded that the coincidences could be best explained by a westward shift of the pole toward the Orient. They noted further that the number of aurorae that were reported in the Orient in this era rose dramatically, as they should were the pole directed more toward the Eastern hemisphere and the geomagnetic latitude of sites in China and Japan and Korea thereby increased.

One could also attribute the apparent increase in Oriental auroral reports to an increase in the level of solar activity at the time, as is suggested in the tree-ring record of ^{14}C. Such an increase would drive the aurorae toward lower latitudes, with no need for a shift of the geomagnetic pole. A decrease in the strength of the geomagnetic moment would have a similar effect. A further explanation, discussed by Stephenson and Clark (1978), is a sociological one: reports of the aurora, like reports of sunspots, are subject to changes in emphasis in the society in which they were made, and in the Orient coloured by the needs of astrology.

B. PROXY DATA FROM COSMOGENIC NUCLIDES

Our most valuable records of the past history of the Sun, and the most critical tests of solar uniformitarianism come from the proxy data from cosmogenic nuclides that now extend the record thousand – and even tens of thousands – of years into the past.

Whether the ^{14}C and ^{10}Be data refute or confirm the principle depends on the span over which it is tested. The ^{14}C data is the strongest evidence we have that the 100-year, modern record of solar behaviour is not adequate to describe the full range of solar activity. The 1000 yr precision radiocarbon record of Stuiver and Quay (1980) and Stuiver and Grootes (1980) is now the best available record of solar activity in the time before the telescope. Yet it is an inadequate sample to test for longer-term changes. Does the last 1000 yrs do any better job of catching significant changes than did the last 100 yrs of the modern historical record? Might there be classes of significant solar change on time scales of 1000 yrs or more?

(i) Tree-Ring Radiocarbon

^{14}C is formed in the upper atmosphere of the Earth by the incidence of high energy galactic cosmic rays. It finds its way into the new growth of tree-rings as a component of CO_2, through photosynthesis. The amount produced, about 2.5 atoms $cm^{-2}sec^{-1}$, varies temporally,

as a function of three factors: the flux of galactic cosmic rays
(presumed constant), the conditions of the solar wind in the interplanetary medium (that scatter cosmic rays away from the Earth at
times of high solar activity), and the strength of the geomagnetic
moment (that shields the upper atmosphere from the incidence of charged
particles). The amount of ^{14}C that reaches the trees will further
depend on a number of factors involved in the terrestrial carbon cycle,
including the effect of the oceans in taking up carbon. CO_2 reaches
the lower atmosphere and the trees by diffusion, on a time scale of
tens of years. The atmosphere acts in this sense as a low-pass filter,
attenuating the 11-year signal by about a factor of 100, from a
modulation of about 10% of 0.1% (Stuiver and Quay, 1980). For this
reason the 11-year signal has yet to be convincingly demonstrated in
tree-ring data. In terms of production, the solar and geomagnetic
effects will be always convolved, with the latter potentially stronger.

The long-term record (Figure 4), derived from combined analyses
of radiocarbon in the growth rings of the bristlecone pine, now extends
about 7500 years into the past (B.P.) (Damon et al., 1978). It is

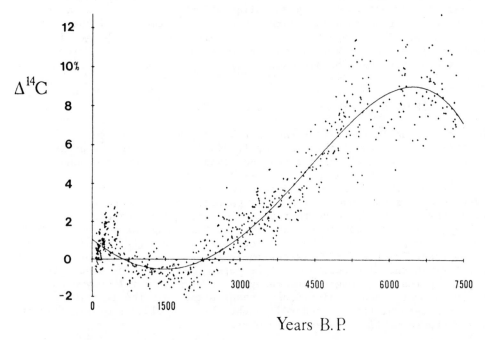

Fig. 4. History of deviations in the relative atmospheric ^{14}C concentration (^{14}C: ^{12}C ratio) from tree-ring analyses of the
bristle-cone pine, from Damon et al. The positive excursions
at the far left of the data set are the Maunder, Spoerer and
Wolf minima. The solid curve is from Bucha, depicting his
reconstruction of the strength of the geomagnetic moment.

dominated by a smooth and monotonic drop of about a factor of about 10% in the ^{14}C: ^{12}C ratio between about 6500 and 1500 B.P., and a gradual increase thereafter, suggesting, originally, a sinusoidal modulation with a full period of about 10,000 yrs. This major, long-term modulation has been conventionally attributed to the changing strength of the geomagnetic moment, that reached its maximum strength in about A.D.500 after a drop of about a factor of two (Bucha, 1969). Superposed on this are excursions of about 1-2%, typically 100 yrs long, that are attributed to solar modulation (Damon et al., 1978; Stuiver and Quay, 1980).

Stuiver and Quay's more precise analysis of the last 1000 yrs of this record, using wide-ring data from the U.S. Northwest, convincingly demonstrates the nature of the 100-yr wiggles. The Maunder minimum appears as expected as an increase of about 2% in the ^{14}C: ^{12}C ratio, corresponding to an increase of about 40% at the level of production. This implies that the solar modulation of galactic cosmic rays by secular (long-term) changes in solar activity exceeds that of the 11-year effect, suggesting that conditions in the solar wind are less disturbed during these epochs than at minima of the 11-year cycle. This at once refutes the notion of solar uniformitarianism, as applied to the last 100 yrs: the overall level of solar activity was significantly lower during the years of the Maunder and Spoerer minima than in the low activity years of the present century. The unspotted Sun (at minima of the 11-yr cycle) does not describe a zero level of solar activity, as far as conditions in the interplanetary medium are concerned. Stuiver finds as well the Spoerer minimum (A.D. 1420-1530), a Wolf minimum (A.D. 1280-1340) and an Oort minimum (A.D. 1010-1050) (Stuiver and Quay, 1980; Stuiver and Grootes, 1980). These features, including the "modern minimum" of 1880-1915, are neither evenly spaced in time nor uniformly long, suggesting that they are not a simple, periodic feature of solar behaviour. There is no obvious "Gleissberg cycle" of 90 yrs in the radiocarbon record, as is suggested in the last 230 years of telescopic sunspot data (Figure 2), and found by Siscoe (1980) in the auroral record. Using a model of radiocarbon production and of the carbon cycle, Stuiver and Quay (1980) have reconstructed a history of secular changes in solar activity that includes the Wolf, Spoerer, Maunder, and Dalton minima and that fits the envelope of the modern record of observed sunspot numbers (Figure 5).

The absence of the 11 yr cycle in the tree-ring record of ^{14}C says nothing about the persistence of that basic feature of modern solar behaviour, nor is there any reason to suspect that it was absent at any time in the last 10,000 years, including the time of the Maunder minimum. Uniformitarianism seems to apply to the 11 yr cycle.

Could the Wolf, Spoerer, Maunder, and Dalton minima found in the tree-ring record of ^{14}C be of geomagnetic origin, due to 100-yr decreases in the strength of the geomagnetic moment? The larger, longer-term modulation of ^{14}C in the tree-ring record is presumed to be due to a change of about a factor of two in the strength of the geomagnetic moment. Thus the increased radiocarbon production during events of the Maunder minimum type could be explained by 100-yr fluctuations of about 1/10 of this amount, which are below the

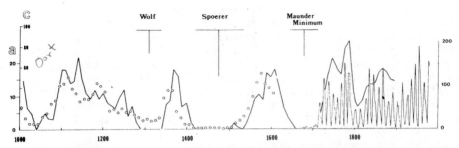

Fig. 5.   Three indices of solar activity compared as a function
of time for the period since A.D. 1000.   The observed
annual mean sunspot number (r, scale at right) is shown by
the thin line beginning in 1650.   The heavier curve, (c,
scale at left) extending from A.D. 1000 to A.D. 1900 is
a sunspot number index derived by Stuiver and Quay from
their precision measurements of tree-ring ^{14}C.   Open
circles (a, scale at left) are an index of the occurrence
of northern hemisphere aurorae, another measure of solar
activity.   The three independent indices confirm the
existence of significant, long-period changes in the level
of solar activity, including three prolonged periods of
depressed behaviour labelled as the Wolf, Spoerer, and
Maunder minima.

threshold of detection in archaeo- or paleomagnetic studies.   One
argument against this possibility is the direct historical observations
of the Sun made during the Maunder and Dalton minima (Eddy, 1976).
These leave little doubt that the Sun was unusually quiet during these
times.   Another is found in historical reports of the aurora.   During
the Maunder and Dalton minima aurorae were unquestionably much less
frequent (Siscoe, 1980).   Were these periods characterized by the
weaker magnetic moment needed to explain the accompanying increase
in tree-ring radiocarbon, aurorae would have been globally <u>more</u>
frequent, due to the decreased shielding of incoming solar particles
by the Earth's field.   Since the opposite is noted - i.e. far fewer
aurorae during the Maunder and Dalton minima - we can probably rule out
this possibility.

A final question still remains in the interpretation of the radiocarbon record: what is the norm from which long-term solar activity
is to be measured?   Radiocarbon analyses are made relative to the year
1890 as an arbitrary reference point (Damon et al. 1978), in part
because it describes the last usable era before the precipitous drop
in ^{14}C that followed increased burning of fossil fuel.   That year also

This example speaks to the need to utilize multiple data sets in
testing questions of past solar or geomagnetic behaviour.   The Keimatsu
shift, noted earlier (Keimatsu et al., 1968), could be tested more
critically, for example, by combining historical auroral data,
historical sunspot data, ^{14}C and ^{10}Be data, and available archaeomagnetic and paleomagnetic measurements.

falls in an era of moderately low solar activity. With this reference, Stuiver's interpretation of the last 1000 yrs shows only secular minima in the long-term level of solar activity. By shifting the point of reference, however, one could as well identify secular maxima, of which the modern mid-century is an example. What was the level of solar activity in the 12th century, compared to the levels of today? During that time the number of naked-eye sunspot reports was double the long-term average (Stephenson and Clark, 1978), and as noted earlier, many more reports of the aurora were made in China. Was it a time, as has been suggested, of a level of solar activity that exceeded the secular high of the middle of the present century?

(iii) Beryllium-10

Cosmogenic ^{10}Be found in dated ice cores from Greenland and Antarctica is today the "great white hope" of solar historians. This isotope, like ^{14}C, is formed in the upper atmosphere by the impact of galactic cosmic rays, and like ^{14}C its rate of production depends in part on the prevalent level of solar activity. ^{10}Be has the advantage of a longer half-life (1.5 million vs. 5730 yrs) and hence a potential of recovering a much longer history of solar behaviour. As important, it has a much shorter time of residence in the atmosphere, since it reaches the surface intact, by precipitation, in typically 2 yrs (Raisbeck and Yiou, 1980), as opposed to about 30 yrs in the case of ^{14}C. The first advantage makes it feasible to extend solar history, utilizing dated ice cores, as much as 100,000 yrs or more into the past. The second advantage makes it possible, in principle, to search for phenomena of much shorter time scale than can be hoped for in any ^{14}C analysis - including the 11 yr cycle.

A disadvantage is the dependence of ^{10}Be deposition on the rate of precipitation, since it ultimately reaches the surface in rain or snow (Raisbeck and Yiou, 1980). The rate of deposition of ^{10}Be is for this reason spatially nonuniform, unlike ^{14}C which in the form of CO_2 is globally well mixed. The interpretation of ^{10}Be deposits requires a knowledge of the local rate of rainfall or snowfall at the time it was laid down.

Moreover, if we are to read solar history to the limit of the iceborne record we shall have to have another record that is currently not available: a continuous record of the strength of the Earth's magnetic moment for 100,000 yrs or more into the past. The initial reading of the tree-ring record of ^{14}C in terms of solar modulation has been made easy by the existence of Bucha's estimates (1969) of the strength of the geomagnetic moment for the last 10,000 yrs. As far as I know, a continuous record of the accuracy required does not presently exist much further into the past.

It may be possible to separate effects of geomagnetic field strength, precipitation rate, and solar modulation by virtue of the dependence in deposition rate of ^{10}Be on geomagnetic latitude. Only the first of these is a function of that parameter. Comparison of ^{10}Be taken from Greenland or Antarctica and from low latitude Andean glaciers could provide a way to identify geomagnetic effects. Dated cores as old as 1500 years have now been taken for other purposes from the Quelccaya ice cap in Peru, only 14° south of the equator (Thompson et al., 1985; 1986). Similarly, the comparison of [10]Be data from sites as separate as Greenland and Antarctica should provide an independent check on estimates of precipitation rates at either site.

Initial measurements by Beer et al., (1983) of Greenland cores and by Raisbeck and his colleagues (1981) of Antarctic ice find clear evidence of the Maunder and Spoerer minima. In samples of recent Greenland ice, Beer et al. (1983) have found as well an indication of the 11 yr sunspot cycle, although it is apparently convolved with local precipitation rates.

C. THE AUSTRALIAN VARVES: A GLIMPSE OF THE PRECAMBRIAN SUN

Although this workshop focuses on the last 10,000 years it is pertinent to stretch that scale of time to consider another source of solar history that gives us, possibly, a glimpse of a much more ancient Sun, in the precambrian era of terrestrial history, when the Sun was a much younger star and the Earth a quite different planet. These intriguing data, taken by George Williams, a geologist, from sedimentary rock laid down in southern Australia 670 million years ago, may offer a powerful test of solar uniformitarianism.

The sedimentary rock that Williams examined exhibits a remarkably regular pattern of horizontal bands, arranged in repeated sequences of from 11 to 12 layers (Williams, 1981; 1983; 1986; 1987; Williams & Sonnett, 1985). Williams argues that the bands are annual varves, laid down by melting of a periglacial lake: by this interpretation the widths of the bands tell of the precambrian climate on the continent, which was then much nearer the equator. Detailed examination of the exposed surface and later, from a 30m core, reveals a striking similarity between band width and all features of the sunspot number in modern records. In each 11–12 band sequence, bandwidths begin narrow (low sunspot numbers), quickly widen (corresponding to the rapid rise of the 11-year sunspot cycle), and then slowly narrow again. At places in the sediment where bands are systematically thinner there appears an obvious alternation of relatively wide and relatively narrow sets of 11–12 bands, mimicking the 22 yr alternation of the modern sunspot cycle. The envelope of the sequence of band widths, like that of the envelope of sunspot numbers, slowly rises and falls. Power spectrum analysis of a sequence of 19,000 bands (50 times longer, if the sediments are annual varves, than the historical sunspot record) demonstrates every feature of the modern sunspot cycle, including a bimodal distribution about the mean number of about 12 bands.

Williams concludes, on the basis of strong circumstantial evidence, that the melting of the precambrian Australian glacier was dominated by solar influence, demonstrating an overwhelming Sun-weather connection. If so, the Australian sediments give us a surprisingly clear look at the behaviour of the precambrian Sun, and over a period of sampling that exceeds the longest records now in hand of the modern, holocene Sun. The period is long enough to allow him to clearly identify cycles of 175 and 315 bands (presumably years), and to search for evidence of Maunder minima. Williams finds epochs when the overall envelope of layer widths is systematically depressed by as much as a factor of three, corresponding, perhaps, to the modern Dalton minimum and 20th century minimum of solar behaviour, but never as much reduced as are sunspot numbers that have been used to describe the Maunder or Spoerer minima. Nor can he identify the "Gleissberg cycle" of roughly 90 bands (years).

Solar historians are loathe to use past climate as a proxy for solar behaviour, since any connection, particularly with the 11 yr cycle, has yet to be convincingly demonstrated (Eddy, 1983b). Williams' case seems to demonstrate one very clearly. Moreover, the tantalizing glimpse at the behaviour of the Sun in the Precambrian rests wholly on that presumption.

Why should Sun-weather connections, which are so elusive and con-troversial today, be so dominant a feature in the Precambrian? What conditions applied, 670 million years ago, that were so vastly different from today? Williams has suggested two possible answers. One of these (Williams, 1981) is that the difference might be due to a much weaker geomagnetic moment at the time, that would allow a strongly-modulated flow of solar particles to impinge upon the atmosphere, which through some unspecified connection might then indirectly affect the climate. His other possible explanation (Williams, 1986) rests on the known variability of solar ultraviolet emission and the fact that during the Precambrian the composition of the terrestrial atmosphere was much different from today. There were no plants then and the atmosphere was anoxic. Without oxygen there was no ozone layer to attenuate the strongly-varying flux of incoming radiation in the near and far ultraviolet. Thus highly modulated short wave flux might, through some unspecified mechanism, directly affect the weather.

If we accept the bands as annual varves, and the implied connection with solar activity, then the 30 meters of Australian rock give us a remarkably clear look at the precambrian sun, leaving little doubt about solar uniformitarianism. In that case we must conclude that the Sun of 670 million years ago behaved exactly as it does today, replicating throughout 19,000 bands (or years) every feature, but more so, of the 11 and 22 yr cyclicity of the Sun today.

Two apparent features of the modern sunspot record are absent in the proxy record of the varves: the 90 yr (Gleissberg) cycle, and the deep depressions like the Maunder minimum. The first of these, as we have already noted, was already in doubt, based on the questionable quality of the early telescopic sunspot record and the absence of any significant period of this length in the record of tree-ring radiocarbon. Perhaps the longer record of the varves (2 1/2 times the reach of the longest available tree-ring sequence) provides an even more conclusive test. But does the enigmatic record of the rocks refute the Maunder minimum and the sequence of other similar drops in solar activity identified in the histories of [14]C

and [10]Be, and establish the case for absolute solar uniformitarianism?
Should we correct the modern records to fit the implied standards of
670 million years ago? Or was the younger sun of the Precambrian era
less prone to deep depression? Might it operate, like the ocean-
climate system of today, in different modes, one of which we may have
sampled briefly in the Precambrian, another even more briefly in the
Holocene?

The answers to these questions should await, I suppose, a future
NATO Advanced Workshop, on Secular Solar and Geomagnetic Variations
in the last 1,000,000,000 Years.

REFERENCES

Allen, C.W., 1976, 'Astrophysical Quantities', The Athlone Press,
London, 310.
Barnes,, J.A., Sargent, H.H.III and Tryon, P.V., 1980, 'The Ancient
Sun', Pergamon press, 159.
Beer, J., Andree, M., Oeschger, H., Stauffer, B., Balzer, R., Bonani,
G., Stoller, C., Suter, M., Woefli, W., and Finkel, R.C., 1983,
Radiocarbon, 25, 269.
Bucha, V., 1969, Nature, 224, 681.
Clark, D.H., and Stephenson, F.R., 1977, 'The Historical Supernovae',
Pergamon Press, 233.
Damon, P.E., Lerman, J.C., and Long, A., 1978, Ann. Rev. Earth Planet
Sci., 6, 457.
D'Elia, P.M., 1960, 'Galileo in China', Harvard Univ. Press, 115.
Donelly, R.F., 1987, Solar Phys., 109, 37.
Donelly, R.F., Heath, D.F., and Lean, J.L., 1982, Journ. Geophys.
Res., 87, 10, 318.
Eddy, J.A., 1976, Science, 192, 1189.
Eddy, J.A., 1977, Climatic Change, 1, 173.
Eddy, J.A., 1980, 'The Ancient Sun', Pergamon Press, 119.
Eddy, J.A., 1983a, Solar Physics, 89, 195.
Eddy, J.A., 1983b, 'Weather and Climate Responses to Solar Variations',
Colo. Assoc. Univ. Press, 1.
Feynman, J., and Crooker, N.U., 1978, Nature, 275, 626.
Feynman, J., and Silverman, S.K., 1980, J. Geophys. res., 85, 2991.
Foukal, P., and Lean, J., 1987, Astrophys. J. (submitted).
Fritz, H., 1873, 'Verzeichniss Beobachter Polarlichter', C. Gerold's
Sohn, 255.
Frohlich, C., and Eddy, J.A., 1984, Adv. Space Res., 4, 121.
Gurman, J.B., 1987, 'NASA's Solar Maximum Mission; A Look at a New
Sun', NASA, Greenbelt, 34.
Hickey, J., Stowe, L., Jacobwitz, H., Pellegrino, P., Maschhoff, R.,
House, F., and Vonder Haar, T.H., 1980, Science, 208, 281.
Howard, R., 1977, Ann. Rev. Astron. Astrophys., 15, 153.
Hundhausen, A.J., 1978, 'The New Solar Physics', Westview Press, 59.
Johnson, F.S., 1954, J. Meteor., 11, 431.
Keimatsu, M., Fukushima, N., and Nagata, T., 1968, Jour. Geomag.
Geoelectricity, 20, 45.

McKinnon, J.A., 1987,'Sunspot Numbers: 1610-1985', W.Data Centre A, 112.
Okal, E. and Anderson, D.L., 1975, Nature, 253, 511.
Reid, G.C., 1987, Nature, 329, 142.
Reid, G.C., and Gage, K.S., 1988. (This volume).
Raisbeck, G.M., and Yiou, F., 1980, 'The Ancient Sun', Pergamon press,
    185.
Raisbeck, G.M., Yiou, F., Fruneau, M., Loiseu, J.M., Lieuvin, M.,
    Ravel, J.C., and Lorius, C., 1981, Nature, 292, 825.
Royal Greenwich Observatory, 1955: 'Sunspot and Geomagnetic Storm
    Data', HMSO, 41.
Schove, D.J., 1955, Jour. Geophys. Res., 60, 127, see also Schove,
    D.J., 1983, 'Sunspot Cycles', Hutchinson Ross Publ. Co., 393.
Siscoe, G.L., 1980, Rev. Geophys. Space Phys., 18, 647.
Siscoe, G.L., and Verosub, K.L., 1983, Geophys. Res. Letters, 10,
    345.
Stephenson, F.R., and Clark, D.H., 1978, 'Applications of Early
    Astronomical Records', Adam Hilger, 114.
Stephenson, F.R., Yau, K.K.C., and Hunger, H., 1985, Nature, 314,
    587.
Stuiver, M., and Grootes, P.M., 1980, 'The Ancient Sun', Pergamon,
    Press, 165.
Stuiver, M., and Quay, P.D., 1980, Science, 207, 11.
Sturrock, P.A., 1980, 'Solar Flares', Colo. Assoc. Univ. Press, 513.
Thompson, L.G., Mosley-Thompson, E., Bolzan, J.F., and Koci, B.R.,
    1985, Science, 229, 971.
Thompson, L.G., Mosley-Thompson, E., Dansgaard, W., and Grootes, P.M.,
    1986, Science, 234, 361.
White, O.R., 1977,'The Solar Output and Its Variation', Colo. Assoc.
    Univ. Press, 526.
Williams, G.E., 1981, Nature, 291, 624.
Williams, G.E., 1983, 'Weather and Climate Responses to solar
    variations', Colo. Assoc. Univ. Press, 517.
Williams, G.E., 1986, Sci. Amer., 255, (2), 88.
Williams, G.E., 1987, New Scientist, 25, 63.
Williams, G.E., and Sonnett, C.P., 1985, Nature, 318, 523.
Willson, R.C., 1984, Space Sci. Rev., 38, 203.
Willson, R.C., Gulkis, S., Janssen, M., Hudson, H.S., and Chapman,
    G.A., 1981, Science, 211, 700.
Willson, R.C., Hudson, H.S., Frohlich, C., and Brusa, R.W., 1986,
    Science, 234.
Xu, Zhen-tao and Jiang, Yao-tiao, 1982, Chin. Astron. Astrophys.,
    2, 84.
Yunnan Obs. Ancient Sunspots Records Research Group, 1977, Chin.
    Astron., 1, 347.
Zirker, J.B., 1977, 'Coronal Holes and High Speed Wind Streams', Colo.
    Assoc. Univ. Press, 454.

THE SUN AMONG THE STARS: WHAT STARS INDICATE ABOUT SOLAR VARIABILITY

David R. Soderblom,
Space Telescope Science Institute, 3700 San Martin Drive,
Baltimore, Maryland 21218, U.S.A.
Sallie L. Baliunas,
Harvard-Smithsonian Center for Astrophysics,
60 Garden Street, Cambridge, Massachusetts 02138, U.S.A.

ABSTRACT. We briefly review the kinds of solar-like phenomena seen
on other solar-type stars, including chromospheric and coronal activity,
spots, and magnetic fields. The stages of evolution of a one solar
mass star are described, particularly with respect to the levels of
magnetic activity that characterize those phases. Finally, we examine
evidence for long-term variability in solar-like stars. In hyper-
active cases, variability with time scales near 50 years are seen,
but magnetic field generation is highly non-linear in this regime:
this may reflect the decay time of active regions, not a true cycle.
    Among normal, single, main sequence stars, detecting cycles is
fairly straightforward, and nearly twenty years of data now exist.
Cycles of various periods are seen among stars of all masses and ages.
The only clear trends are that the most massive solar-type stars have
low-amplitude cycles; young stars are noisy, i.e., show chaotic
behaviour; and stars less massive than the Sun almost always show
a clear cycle. Also, the morphology of the cycles is consistent with
the solar cycle in that there appear to be no cases of a slow rise
followed by a rapid decline.
    Little direct information is available on Maunder-Minimum-like
phenomena in stars, but there appear to be no young stars that undergo
a complete loss of observable activity. There may be older stars
undergoing a Maunder Minimum phase, but, if so, the changes in their
overall activity levels are subtle.

1. THE SOLAR-STELLAR CONNECTION

1.1. The Sun as a Paradigm for Stellar Phenomena

A basic tenet in the study of stellar evolution, the Vogt-Russell
theorem, states that a star's mass, composition, and age determine
its gross properties, particularly temperature and luminosity. Short-
time-scale effects (e.g. flaring, convective motions) may well have
strong stochastic components, but this theorem predicts that other
aspects of a star change on long time scales. Recent evidence suggests

25

F. R. Stephenson and A. W. Wolfendale (eds.),
Secular Solar and Geomagnetic Variations in the Last 10,000 Years, 25–48.
© 1988 by Kluwer Academic Publishers.

that the Sun's 11 and 22 year sunspot and magnetic cycles have been operating for about the past 3/4 of a billion years (we refer to the Australian sediments), so that such behaviour is expected of an old star with mass and composition similar to the Sun's. Observations of solar-like cycles in other solar-type stars can test this hypothesis.

By solar-type stars we mean main sequence stars with masses between about 0.8 and 1.2 solar masses, or about K2V or F7V in spectral type. The solar-like magnetic phenomena that are observed in such stars can be attributed primarily to one property: the presence of a convective envelope. Although the details are poorly understood, we believe we understand the basic mechanisms. The interaction of convection and rotation leads to differential internal rotation, which then interacts with sub-surface convection to generate and maintain a magnetic field through the dynamo mechanism. These magnetic fields can force corotation of the star's wind well beyond the surface, leading to angular momentum loss (this process is witnessed for the Sun). The relation between rotation and magnetic field strength leads to a feedback mechanism, so that the relation between rotation rate and age is predictable. The magnetic fields manifest themselves as stellar activity, such as chromospheric and coronal emissions, flares, and sunspots.

The connection between the theoretically predicted presence of a convective zone and the loss of angular momentum is shown graphically in Figure 1. In this "bubble diagram", taken from Kraft (1967), the sizes of the circles are related to a star's rotation rate. The dramatic drop in rotation at about 1.25 solar masses is quite evident; this point coincides with the predicted onset of a sub-surface convective zone.

Figure 2 shows the relationships between some of the properties of solar-type stars and their age (Skumanich 1972). More recent analyses have shown that such simple relations probably do not apply, but nonetheless these quantities are clearly age-related. As noted, angular momentum loss appears to be an inevitable feature in a star like the Sun. Since the Ca II emission is generated (indirectly) by the rotation, it too declines with age, and, to first order, the Ca II emission strength is proportional to rotation. The decline of lithium with age is more complex, but it also appears to be related to convection, because relatively high temperatures (found at the base of the convective zone) are needed to destroy that element (Soderblom 1984).

On the one hand, the Sun is our role model for similar phenomena seen in other stars, but, on the other hand, one reason for studying those other stars is to examine those phenomena under circumstances different from the Sun's, and thus arrive at a better understanding of underlying causes of stellar activity. One fundamental difficulty presents itself; we cannot resolve the surfaces of other stars in the way we can for the Sun. For example, most of what's known about the solar cycle is determined from the sunspot number. Clearly some other indicator is needed to study cycles in stars.

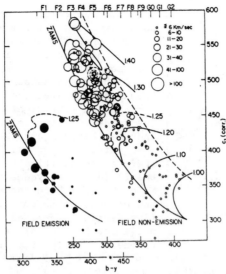

Fig. 1.   Rotation among F and early-G dwarfs (from Kraft 1967).
The sizes of the circles are related to a star's v sin i
(equatorial rotational velocity projected into the line of
sight) in the manner indicated. The curved lines with
numbers are post-main sequence evolutionary tracks for stars
of the masses (in solar units) shown. The abscissa, b-y,
is an observational color index. Bluer, or hotter, stars
are to the left. The ordinate c_1, is an observational
measure of a star's gravity, which indicates its distention
as it evolves away from the Zero Age Main Sequence (ZAMS).
The field emission stars have strong Ca II H and K emission,
and have larger rotation rates than non-emission stars of
the same mass. The key feature of this diagram is the
dramatic drop in rotation near 1.25 solar masses, which is
the point where convective zones start to become an important
feature of a star.

In this review, we will discuss solar-like phenomena as seen
on solar-type stars at the different phases of the evolution of such
a star, and then examine evidence for long-term changes and
variability, akin to the solar cycle. We will also discuss evidence
for Maunder-minimum-like phenomena in other stars. Our hope is to
see how activity cycles depend on mass, composition, age, or whatever,
in order to get clues to the underlying physics of the cycles, and
to understand better the solar cycle in particular.

1.2. What Stars can tell us about Solar Activity

Most solar phenomena are fairly subtle on a global spatial scale
(the Sun viewed as a star), so that high quality observations are
needed for detection. But the concept of extrapolating solar
phenomena to stars is not new. Many extreme examples of solar-like

28

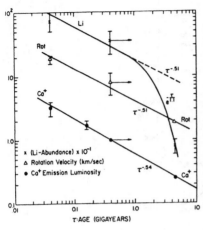

Fig. 2. Age-related quantities in solar-type stars (from Skumanich 1972). Three properties of solar-type stars are known to decline with age: their rotation rates, chromospheric emission, and lithium abundances. In this study, Skumanich compared the available data for the Sun (the rightmost points), the Hyades cluster (near the center), and the Pleiades cluster (the leftmost points). For the Ca II emission, there is an additional point between the Pleides and Hyades for the Ursa Major Group. Both the rotation and chromospheric emission are best fitted with power laws of slope $-\frac{1}{2}$, while the Li data are fitted with an exponential. The true relations are probably more complex than indicated here, and may reveal interesting phases in the evolution of these quantities.

magnetic activity have been studied for the "pathological" cases, i.e. hyper-active stars where the phenomenon is much more pronounced than on the Sun.

Accurate ground-based spectroscopic and photometric data, obtained over long time intervals, and ultraviolet and X-ray observations from space, have combined to open wide the field of solar-stellar comparisons. The uniqueness of the space-based data makes them especially useful for revealing interesting physical processes, but the comprehensiveness of the ground-based data - particularly their long time span - make them best suited for studying stellar activity cycles.

1.2.1 Chromospheres. More data are available for stellar chromospheres than for any other magnetic activity indicator. Chromospheric emission arises in a star's outer atmosphere, above the temperature minimum, at temperatures of about 8,000 to 100,000 K. It is necessary to observe in the ultraviolet to study the highest temperature portions of these plasmas, so that only limited information is available from rocket and satellite experiments (Avrett 1981; Dupree 1983; Linsky 1980, 1982; Vernazza, Avrett, and Loeser 1983). However, much more data are available from the visible part of the spectrum, where chromospheric features appear in the bottoms of deep absorption lines. The standard feature observed is the Ca II K line at 3933 Å, in part because the central 1Å or so is entirely chromospheric (Avrett 1981), and because this feature shows marked changes on a variety of time scales, both periodic and chaotic (White and Livingston 1981).

For example, the K line is very different at solar minimum than at solar maximum (Figure 3). This suggests that detecting solar-like cycles in other stars should be pretty straightforward, and indeed it is. As we will see, detecting cycles isn't that hard, but inter-

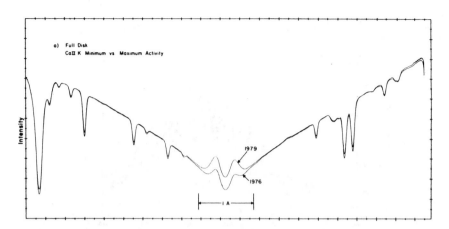

Fig. 3.   The Ca II K line at solar minimum (lower curve) and maximum
(upper curve), taken from White and Livingston (1981).   The
central 1Å of the K line is indicated; this is the portion
observed for stars.   Although there is an obvious change
in the spectrum when it is well resolved, as in this case,
for the central 1Å as a whole, the change is much less.

preting them and finding the significant trends is.

The data available for stars aren't nearly as good-looking as
that shown in Figure 3; one must be willing to settle for less, both
in terms of spectroscopic resolution and signal-to-noise ratio.   Instead
of fully resolved spectra, for stars the central 1Å is observed,
relative to the nearby continuum.   If the Sun is observed in an
analogous manner, the solar cycle is evident (Figure 4).

1.2.2.   <u>Coronae</u>.   Stellar coronae are very hot plasmas ($\sim$ 1 MK),
extending well above the chromosphere.   Their main diagnostic is X-ray
emission, and such observations have been made for many nearby solar-
type stars (Pallavicini et al. 1981; Rosner, Golub, and Vaiana 1985).
However, usually only one observation per star is available, and X-ray
satellites are short-lived, so that cycle variation data are not
available even for the Sun.   What coronal information exists provides
interesting clues for the processes in outer atmospheres, but in broad
terms it largely confirms the trends seen in the chromospheric data.

1.2.3.   <u>Spots</u>.   As mentioned, the surfaces of stars cannot be imaged
directly, as for the Sun, but many types of observations strongly
indicate the presence of spots.   Star spots were first invoked to
account for periodic photometric variations in very cool stars: an
uneven distribution of very large dark spots lowers the total output
of a star's light at visible wavelengths, and the variation is caused
by rotation changing the fraction of light blocked from the visible
hemisphere that is viewed.

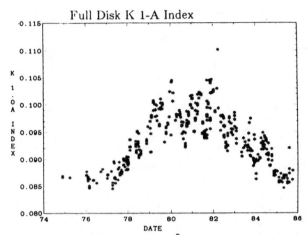

Fig. 4. The flux in the central 1Å of the Ca II K line, for the Sun observed as a star for ten years (from White, Livingston and Wallace 1981). One full solar cycle is shown here. There is substantial "noise" due in part to rotation, but the change due to the solar cycle is easily seen.

More direct evidence for star spots has recently come from "Doppler imaging" (Vogt 1981). This technique takes advantage of the fact that a line profile for a rapidly rotating star can be mapped back to longitude bands on the star. As a spot moves across the visible hemisphere from one limb to the other, the light removed by the spot produces a "bump" in the bottom of the line. The extent of that bump, as well as when it first enters and leaves the line, indicates the size and latitude of the spot (Figure 5).

The results from photometric observations and Doppler imaging are consistent:
Star spots can cover 10% or more of a star's surface, an order of magnitude greater than the coverage seen on the Sun.
Star spots are always cooler than the surrounding photosphere.
The spots appear to drift in latitude with time. In some cases, the spots drift toward the poles, which is the opposite of the behaviour seen on the Sun, but is in accord with some dynamo theories.
The missing light doesn't instantaneously appear elsewhere on the star - it must remain bottled up for later release (Hartmann and Rosner 1979), which again is the behaviour seen on the Sun.

Currently photometric spot data are available only for heavily spotted, extremely active stars, and are insufficient to say anything about periodicities in long-term variations.

1.2.4. _Magnetic Fields._ Magnetic fields broaden spectrum lines with large Landé g factors. This effect can be detected in the disk-integrated light of solar-type stars with very strong fields (Marcy 1984). The interpretation of these data in terms of field strength and spatial coverage is controversial, although the presence of fields

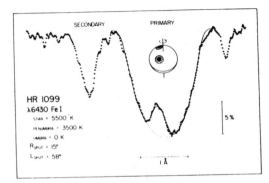

Fig. 5. An illustration of the technique of Doppler imaging (from Vogt 1984). The observations (dots) are for the RS CVn system HR 1099, which is a close binary. The Fe I absorption line in the primary (brighter) star shows a "bump" in the bottom, due to a spot removing light from that longitude. The solid line shows a fit to the observations from a simple model with circular spots. By following the spot over several rotations, one can infer spot distributions, sizes, and latitudes.

has been confirmed. These stars with strong fields are not good solar analogs, and, in any case, not enough measurements are available to investigate cyclic variability.

2. BACKGROUND: EVOLUTION AT ONE SOLAR MASS

It is useful to briefly review the stages of evolution of a solar mass star, to discuss the observational characteristics of these phases. In particular, stellar activity levels undergo interesting changes over a star's lifetime. The pre-main sequence evolutionary track of a one solar mass star is shown in Figure 6.

2.1. Formation: The Angular Momentum Problem

Little is known observationally about very young stars: We know that stars form, even if we're not sure just how. A key problem is explaining how protostars overcome the huge angular momentum of their natal cloud while condensing. Also to be explained is the observed random distribution of stellar rotation axes. Clusters do not show aligned rotation axes, nor do stars as a whole rotate in the same sense as the Galaxy. The easiest way out of these problems is to suppose that star formation occurs not as grand condensations in molecular clouds, but rather in turbulent eddies within or between clouds where the net angular momentum is very low and randomly oriented (Wolff, Edwards, and Preston 1982).

2.2. The T Tauri Phase: Turbulent Youth

A young solar-mass star first appears as a visible, luminous object at an age of about 1 My, and is called a T Tauri star. T Tauri stars are completely convective, and are characterized by strong mass outflows, and strong, chaotic activity. This activity is coherent enough that periodic signals can be extracted from rotational modulation of the light, but the dependence of the activity on other stellar parameters is not clear. In particular, T Tauri stars do not obey the same rotation-activity relation that main sequence stars do. The activity

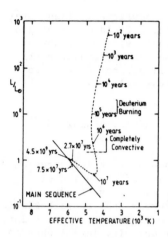

Fig. 6. Pre-main sequence evolutionary track of a one solar mass star (from Ezer and Cameron 1966). Stars first appear as luminous objects at ages of 10^5 to 10^6 years, as T Tauri stars. At this time they are completely convective. During evolution through the "hook" region, a radiative core develops. The Zero Age Main Sequence (ZAMS) is reached at an age of about 75 My. Note that the "main sequence" shown is not a ZAMS but rather an average of stars of the solar neighbourhood. These stars are, on average, fairly old, and so brighter than a ZAMS. The post-ZAMS evolution is also indicated, but difficult to see.

of T Tauris is qualitatively like that of main sequence stars, but greatly exaggerated. One interesting aspect of T Tauri activity is that regardless of the overall strength of chromospheric emission in a T Tauri, that emission is always saturated (Herbig and Soderblom 1980). Thus, the differing levels of activity arise from different fractional coverage by active regions, rather than different strength regions of the same areal coverage.

The descent down the Hayashi track covers roughly a decade in age for equal decrements of luminosity (Figure 6). Thus one expects many more low-luminosity pre-main sequence stars (the so-called "post-T Tauris") than T Tauris. Determined searches have been made for such older stars (age $\sim$ 10 My), largely without success. The observed dearth of post-T Tauris remains a fundamental problem.

Several T Tauri stars show signatures of proto-planetary disks, which is expected for young "solar systems" at such an age. These various phenomena make T Tauris a fascinating subject for study, but they are so different from main sequence stars that they are probably not useful for elucidating the problems encountered in understanding stellar activity in stars like the Sun.

2.3. Later Stages of Pre-Main Sequence Evolution: The Approach to the Main Sequence

Near the bottom of the Hayashi track, our young solar-mass star starts to develop a radiative core. This core grows in mass and size as the exterior of the star shrinks, leading to a major drop in the star's moment of inertia. After about 75 My, the center of the star is dense and hot enough to sustain the fusion of hydrogen to helium, and the star is said to have arrived on the Zero-Age Main Sequence (ZAMS).

This sharp drop in the moment of inertia during the last phases of pre-main sequence contraction is believed responsible for the observed high rotation rates of very young ZAMS stars, which can be as much as 100 times the solar rate. Other stars only slightly older rotate slowly. It is unclear how the star allows a rapid spin up as

the core condenses, yet provides efficient angular momentum loss once
the maximum velocity has been reached. It has been speculated that
after spin up, the star's convective zone decouples from the core,
so that relatively little angular momentum need be lost to slow the
outer layers (Stauffer et al. 1985; Hartmann et al. 1986). This phenomenon of ultrafast rotation, followed by immediate spin-down, is
illustrated in Figure 7.

The development of the radiative core appears to make the activity
of these stars better behaved than for the T Tauris. Although these
almost-main-sequence stars are extremely active, they fit the extrapolated relations of activity-age-rotation for main sequence stars.
Unlike the T Tauri stars, where magnetic activity seems independent
of rotation, later pre-main sequence stars seem to obey the magnetic
activity-rotation rules of main sequence stars (Hartmann and Noyes
1987). This is a clue to the nature of the physics of magnetic activity.

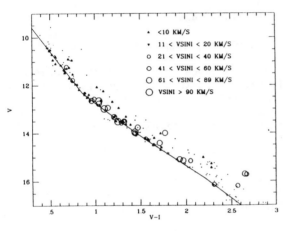

Fig. 7. Rotation on
the lower main sequence
of the Pleiades cluster
(from Stauffer et al.
1984). The solid line
is a theoretical ZAMS,
and the points show
rotation rates for indi-
vidual stars. Note the
sudden increase in rota-
tion near $(V - I) \cong 1$,
$V \cong 12.5$, which corresponds
to a mass of about 0.8
solar masses. Some of
those stars rotate as
much as 100 times the
solar rate. Note also
that there are many stars
of similar mass that rotate slowly.

2.4. Young Main Sequence Stars

After passing through this phase of rapid rotation, stars settle onto
the ZAMS with a rotation rate about 10 times solar. They are also
about an order of magnitude stronger in chromospheric activity than
the Sun, while their coronae can be as much as 100 times more active
than the Sun's. In addition to strong magnetic activity, these young
stars clearly have exceptionally large spots, for detecting the
rotational modulation of the light cause by those spots is easy-
amplitudes and spot coverage areas are a few percent.

If chromospheric activity is used as an indicator, a solar-type
star is "young" until it is about 1/3 the age of the Sun, i.e., for
the first 1 or 2 Gy of its main sequence lifetime. These young stars
also have abundant lithium. Despite their deep convective zones,
the interior temperatures of T Tauri stars must not get high enough
to destroy Li; that must only occur gradually over a star's main sequence

lifetime.

2.5 Older Main Sequence Stars

The gradual, inexorable, and lessening loss of angular momentum leads
to slower rotation as a solar-type star ages, with a concomitant
reduction in chromospheric and coronal activity. Lithium is also
depleted in old stars, except that low metallicity stars formed early
in the Galaxy's history deplete their Li at a very slow rate (Spite
and Spite 1982). As we shall see, detecting activity cycles in these
old stars is fairly easy, despite their low levels of activity. These
old stars are generally ignored in studying the physics of magnetic
activity because they are so sedate.

2.6 The Solar Condition : Classic Middle Age

This brings us to the present Sun, which presents the classic
symptoms of main sequence middle age:
  The Sun's reduced level of activity comes and goes, with occasional
  loud outbursts (wilder activity is seen in the younger generation).
  It's getting bigger around the middle.
  It thinks the world revolves around it.
One goal of the next section is to see how typical the Sun is for an
older star.

3. STELLAR ACTIVITY AND LONG-TERM VARIABILITY

3.1 Rotation-Activity-Age Relations

The overall scheme relating rotation, convection, and the dynamo
mechanism was laid out above. A key to understanding stellar activity
has been the demonstration of a tight rotation-activity relation: fast
rotators always have high levels of activity, while slow rotators are
always inactive. Rotation is also related to age. The only serious
exceptions to the rotation-age relation are binaries that are close
enough together to force tidal synchronization of the rotation with
the orbit. In other words, close binaries rotate faster than they
would if they were left alone as single stars.
     Thus the rotation-activity-age relations are deterministic, not
just a tendency. Nevertheless, the physical mechanisms responsible
for stellar activity are not well understood in detail, and are being
studied further.

3.2 Surveys of Chromospheric Emission

Because the Ca II H and K lines can be observed from the ground and
are such useful diagnostics of stellar chromospheres, they have been
extensively studied in many stars. In particular, Vaughan and Preston

(1980) have surveyed H and K emission among late-type dwarfs of the
solar neighbourhood. Such stars are intrinsically variable, due to
cycles, rotation, and random activity changes (such as flares).
However, the variability is generally less than about 10%. Vaughan
and Preston typically observed the survey stars one or two times each,
yet such a snapshot is likely to catch the star at a representative
level of activity, and it is the overall properties of the survey that
are most of interest.

This survey, with attention restricted to the solar-type stars,
is shown in Figure 8. R'_{HK} is an index of the HK emission that ratios
the H and K flux to the star's bolometric flux, and corrects for photospheric light in the instrumental bandpasses (Noyes et al. 1984a).
The most obvious feature of the survey is the "gap", a lack of stars
at intermediate levels of activity. This was noticed by Vaughan and
Preston, but it is still not clear whether it is caused by an uneven
rate of star formation, an uneven relation between the HK emission
and age, a selection effect in the sample, an upper and lower limit
to chromospheric emission levels, or a combination of these factors
(Hartmann et al. 1984).

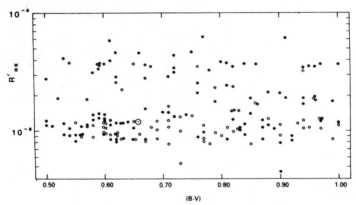

Fig . 8.   The Vaughan-Preston survey of a Ca II H and K emission
          for solar-type stars (from Soderblom 1985). The R'_{HK} index
          is the observed H and K emission, corrected for photo-
          spheric light, and ratioed to the star's bolometric flux
          (see Noyes et al. 1984a). Open circles are high-velocity
          stars of the old disk and halo populations. They are
          systematically older than young disk stars (full circles),
          and have appropriately lower HK emission. The Sun is
          a relatively inactive star. There is a lack of stars with
          intermediate levels of HK emission; this is the so-called
          "Vaughan-Preston gap".

3.3 Rotational Modulation and Evidence for Diffferential Rotation

The presence of spots on stars produces periodic variations in the
light. These are particularly obvious in H and K emission, and
enable one to derive stellar rotation periods, free of aspect dependence,

even for stars rotating at half the speed of the Sun. A fundamental
result of these studies has been the parameterization of activity
as a function of the Rossby number (the ratio of rotation period to
convective turnover time), as shown in Figure 9. This relation makes
it possible to use HK emission strengths to predict rotation periods
with good accuracy (Soderblom 1985). Using the relation in Figure
9, one can convert R'_{HK} to the rotation period, or, equivalently, the
angular velocity (Figure 10).

Another important result from rotational modulation studies has
been the detection (Baliunas et al. 1985) of differential rotation
in a few stars (Figure 11). Patient monitoring of more stars over
several seasons should reveal more such cases, and lead to a better
understanding of a fundamental ingredient of the magnetic dynamo.

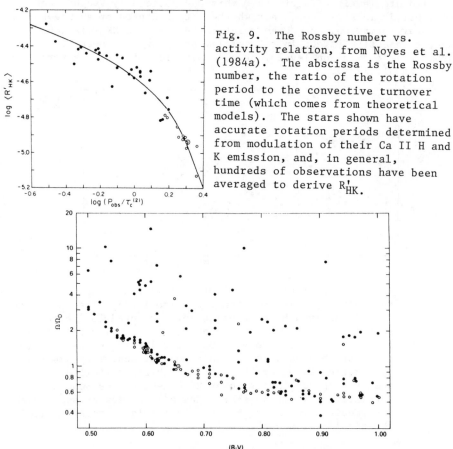

Fig. 9. The Rossby number vs.
activity relation, from Noyes et al.
(1984a). The abscissa is the Rossby
number, the ratio of the rotation
period to the convective turnover
time (which comes from theoretical
models). The stars shown have
accurate rotation periods determined
from modulation of their Ca II H and
K emission, and, in general,
hundreds of observations have been
averaged to derive R'_{HK}.

Fig. 10. Rotation among solar-type stars of the solar neighbourhood (from Soderblom 1985). These are the same stars
shown in Fig. 8, transformed using the relation of Fig. 9.
Comparison of predicted to true rotation periods shows
that the predicted periods are good to ± 20%

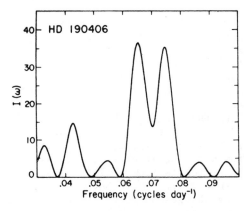

Fig. 11. Evidence for differential rotation in HD 190406 (from Baliunas et al. 1985). The power spectrum of the H and K data (upper panel) clearly shows two strong peaks. A double sine wave is a good fit to the observations (lower panel). The two periods presumably reflect different rotation rates for different regions that give rise to the H and K emission.

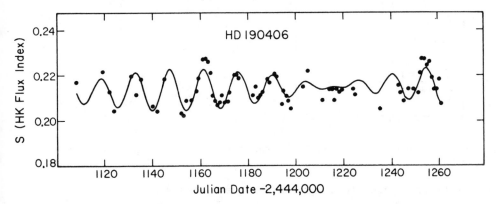

3.4 Variability in the Hyper-Active Stars

By "hyper-active" stars we mean those that exhibit solar-like activity, but at a grossly exaggerated level. Such stars are either close binaries, such as the RS Canum Venaticorum variables, where tidal forces cause the rotation to synchronize with the short-period orbit, or they are very young stars (which consequently rotate rapidly), like many of the BY Draconis stars (some BY Dra's are also close binaries). Although not strictly representative of an old, inactive star like the Sun, these hyper-active stars exhibit interesting magnetic activity.

These hyper-active stars are the ones for which Doppler imaging is possible, because their spots are enormous compared to sunspots. Their spots can decay on time scales of a few years (Vogt 1981). Larger amplitude variations also occur on very long time scales. For example, BY Dra itself and BD + 26°730 have shown changes in their light of 40 to 60% on time scales of 50 to 60 years (Figure 12). It is not yet clear whether these long time scales represent cycles in these stars, or whether they are, say, the time needed for the decay of a single, large active region. The magnetic dynamos of these stars

38

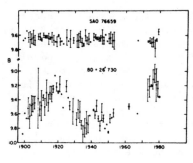

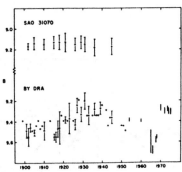

Fig. 12. Long-term variability in two hyper-active stars. The data shown are estimates of the star's brightness, determined from photographic plates. Similar measures for a comparison star on the same plates is shown along the top of each panel to indicate the typical errors of measurement. The top panel is from Hartmann et al. (1981), while the bottom is from Phillips and Hartmann (1978). In both cases, major changes on 40 to 60 year time scales are seen.

lie well beyond any reasonable extrapolation of our experience with solar behaviour.

3.5 Activity Cycles in Single Main-Sequence Stars

3.5.1 Historical Background. We now discuss solar-like cycles in single, main sequence stars. Seventy years ago it was recognized that detecting such cycles would be interesting and possibly revealing (Hale 1915), but the subsequent attempts made were inadequate because of the limited dynamic range of photographic plates. Olin Wilson was the pioneer in the study of stellar activity, and his construction of photoelectric spectrophotometers began the study of cycles in the mid 1960s.

When Wilson retired, he reported on the progress made in twelve years in his Russell Prize lecture (Wilson 1978). His major point was simply that cycles were indeed readily observable (Figure 13), particularly in stars cooler than the Sun. Old stars somewhat hotter (and more massive) than the Sun tended to show little or no variability in their Ca II K emission. Many cycle periods could be seen, typically in the range of about 6 to 10 years. Young stars were noisy, and in general obvious periodicities were not seen for them.

The importance of continuing such a long-term project has been apparent, but it is the work of Baliunas, Noyes, and their co-workers at the Harvard-Smithsonian Center for Astrophysics and at the Mount

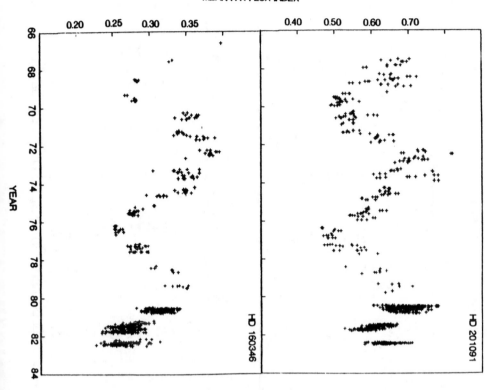

MEAN H-K FLUX INDEX

Fig. 13.   Observations of stellar activity cycles for two cool stars
          (from Vaughan 1983).   In these cases, detection of the
          cycles is straightforward.

Wilson and Las Campanas Observatories that has led to the success
of this effort in spite of many difficulties. Other groups are now
beginning projects for similar long-term monitoring of stellar
activity, but the efforts started by Wilson at Mount Wilson have now
produced twenty years of unique data to analyze.

3.5.2 Results After Twenty Years. The Mount Wilson project has now
followed 99 stars, most for the full 20 years. Nearly all are main
sequence stars, although a few evolved stars have also been observed.
The results reported here have been distilled from a forthcoming paper
by Baliunas et al. (1987).
     Figures 14, 15, and 16 show representative examples of the
observations. The overall trends reported by Wilson remain, although
the greater span has meant that many more cycles have been detected.
They range in period from about 2 years, up to 18 years or more.
Some of the stars just exhibit long-term trends, with no reversal
yet seen that would indicate a cycle. Such cycles if they exist must

40

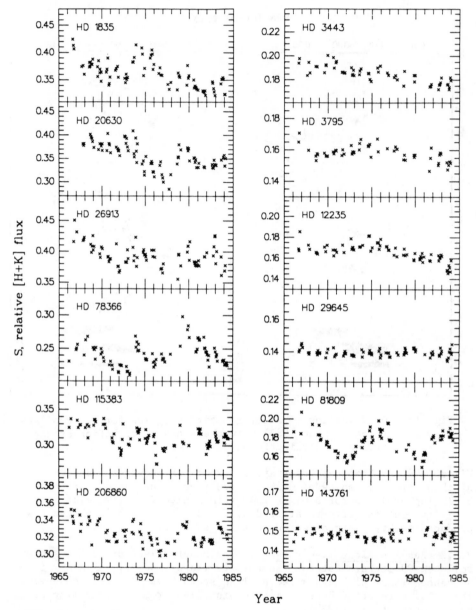

Fig. 14.   Stellar activity cycles in stars near one solar mass
(from Baliunas et al. 1987). The left panel shows six
young stars, the right panel six old stars. The old star
HD 81809 has a cycle very much like the Sun's. Note that
both old and young stars show cycles, but the young stars
are also noisy. The solar analog HD 81809 has a more
assertive cycle than most old solar-type stars.

41

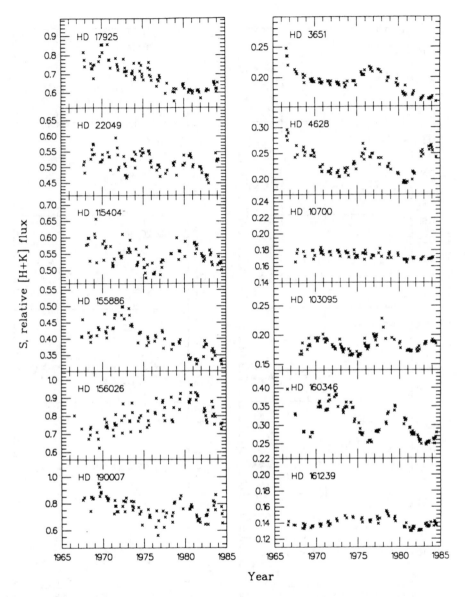

Fig. 15. Stellar activity cycles in stars somewhat less massive
than the Sun (from Baliunas et al. 1987). Again the left
and right panels show young and old stars, respectively.
Among these less massive stars, cycles are almost always
seen. It is interesting to compare HD 10700 and HD 103095,
which are both very old, metal-poor stars formed early
in the Galaxy's history; one has a strong cycle, the
other has little or none.

42

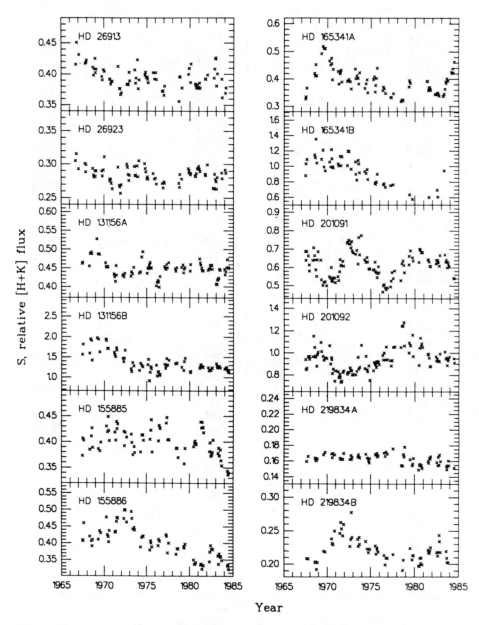

Fig. 16. Stellar activity cycles in the components of visual binaries (from Baliunas et al. 1987). Six pairs of two are shown. There appears to be little or no relation between the presence, period, or amplitude of a cycle in one star and its companion.

be decades long.

3.5.3 Where Are The Trends? A major goal of these observations has

been to determine how cycle properties depend on gross stellar characteristics. Is cycle period or amplitude related to age or mass? Do
very old, metal-poor stars behave much differently than the Sun? What
are the trends, and what do they mean? One problem with any such
analysis is that the current twenty-year baseline may be too short
to properly assess long periodicities.

Examination of Figures 14 and 15 shows that young stars are
apparently noisy, and show lots of random fluctuations. Despite this,
they also reveal cycles, with periods like those of the older stars.
The old stars have quieter (smoother and better defined) cycle curves,
but not necessarily lower amplitude ones. The most obvious cycles
are seen in old stars less massive than the Sun (Figure 15).

The right-hand side of Figure 14 compares several stars whose
other characteristics (mass, age, and composition) are like the Sun's.
The amplitude and shape of the curve for HD 81809 is very much like
that for the Sun. One is immediately struck by how few old stars have
cycles that are as assertive as the Sun's. The solar cycle, as
measured by sunspot number, is stronger now than it has been in the
past; perhaps this is consistent.

Figure 16 compares cycles in the components of visual binaries.
Clearly even when two stars are of the same age and composition, and
similar mass, their activity cycles are not closely related.

Figure 17 shows the strength of H and K emission (which is S,
the observational quantity before conversion to R'_{HK}) versus stellar
color, an indirect indicator of mass. Both young (active, or high
S-value stars) and old (inactive, low-S) stars have comparable prop-
ortions of stars with cycles, but the young stars have most of the
"variable" stars (no trend of any kind discernible), and the old stars
have nearly all of the stars that just show long-term trends. For
the stars with cycle periods, there is no clear dependence of cycle
period on either color (Figure 18), or rotation period (Figure 19),
or Rossby number (Figure 20). Note that the earlier result of Noyes,
Weiss, and Vaughan (1984), relating cycle period to Rossby number,
disagrees with the present results because a much broader region of
the data plane is now sampled.

At least one trend exists in these data: The morphology of the
cycle curves is such that most show a rapid rise followed by a slow
decline-the same sense as the Sun. Some curves are nearly sinusoidal,
but none show the opposite sense. The nature and time scales of these
activity cycles clearly mean that many years of work are ahead of us
before a better understanding can be reached.

3.6 DO STARS SHOW MAUNDER-MINIMUM-LIKE PHENOMENA?

What evidence is there for solar-type stars showing significant changes
on very long time scales? Do any show the virtual absence of activity

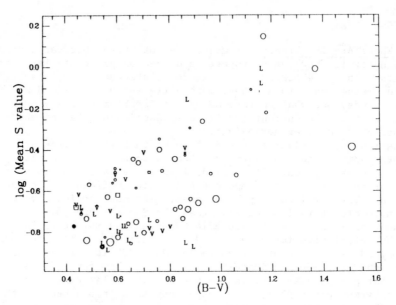

Fig. 17.   Activity cycles in solar-type stars (from Baliunas et
al. 1987).   The abscissa, (B-V), is a color index which
increases with decreasing mass on the main sequence.
The ordinate, S, is the observed H and K flux, relative
to the nearby continuum, before conversion to R'_{HK}.   The
broad upward slope of S with increasing (B-V) is due to
the decreasing ultraviolet continuum of cooler stars.
Circles and squares denote stars with cycles, while V
denotes a star that shows only random variations, and L
denotes a star with a long-term trend.

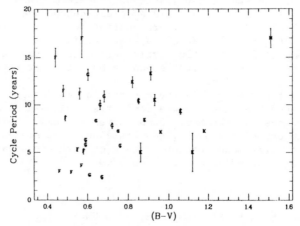

Fig. 18.   Observed cycle period versus stellar color (from Baliunas
et al. 1987).   No relation is apparent.

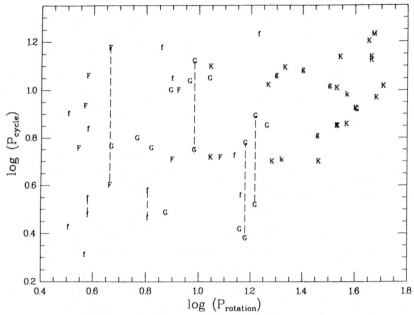

Fig. 19. Observed cycle period versus observed (capital letters)
or predicted (lower case letters) rotation periods (from
Baliunas et al. 1987). Again, there appears to be no
correlation.

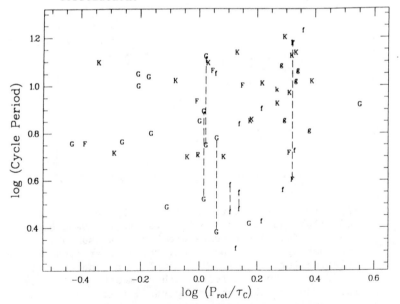

Fig. 20. Observed cycle period versus Rossby number (from Baliunas
et al. 1987).

that characterized the Maunder Minimum? Clearly the Mount Wilson data
cannot address such questions for individual stars because of the
limited time span covered, but it may be possible to say something
very rough in a statistical sense.

In Figure 15, two of the old stars (HD 10700 and HD 103095) are
very old, metal-poor stars, formed early in our galaxy's history. The
second of them shows an assertive cycle, while the first shows none.
Since cycles seem ubiquitous among these cool stars, it has been
speculated that perhaps HD 10700 is in a Maunder-Minimum-like state
(Vaughan 1984).

One can make a stronger statement about young stars. Specifically,
it is possible to rule out that young stars go through any significant
phase where their activity drops to very low levels. Figure 21 shows
R'_{HK} versus color for solar-type stars in the Hyades cluster, which is
about one-tenth the age of the Sun. All of them fit a nice, tight
relationship; there are no stars deviating significantly on either
the high or low side. The dispersion seen can be entirely attributed
to rotational modulation, random variations, and small amounts of
cyclic variability and observational error.

It is, in fact, difficult to quantify the Maunder Minimum for
the Sun. What were the levels of various magnetic activity indicators
during that time? We have only a rough idea of the sunspot number,
and even that is affected by historical problems. We cannot say just
how little activity the Sun had, except that it was probably lower
than what we now see at solar minimum. Other indirect solar activity
indicators support this conclusion.

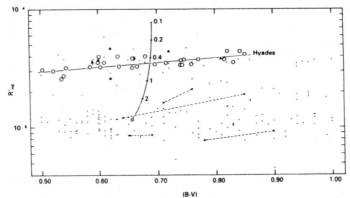

Fig. 21. R'_{HK} versus color   for stars of the Hyades cluster (from
Soderblom 1985). The small dots are the same stars as
Fig. 8. The open circles are Hyads, presumably stars of
the same age and composition. The scatter about the mean
relation (straight line) is small, and can be attributed
entirely to known effects such as measurement error,
rotational modulation, cyclic variability, and random
variability. For these young stars, there are no apparent
cases where there is a large drop in overall activity like
the Maunder Minimum.

ACKNOWLEDGEMENTS

This paper is based on work supported by the National Science Foundation, under grant AST-8616545, the Smithsonian Scholarly studies Program, the Langley-Abbot Fund of the Smithsonian Institution, and the National Geographic Society (grant 2548-82).

REFERENCES

Avrett, E.H., 1981, In:'Solar Phenomena in stars and Stellar Systems', ed. R.M. Bonnet and A.K. Dupree, (Boston, Reidel), 173.

Baliunsa, S.L., Donahue, R.A., Horne, J.H., Noyes, R.W., Porter, A., Gilliland, R.L., Duncan, D.K., Frazer, J., Lanning, H., Misch, A., Mueller, J., Soyumer, D., Vaughan, A.H., Wilson, O.C., and Woodard, L.A., 1987, Ap. J., submitted.

Baliunas, S.L., Horne, J.H., Porter, A., Duncan, D.K., Frazer, J., Lanning, H., Misch, A., Mueller, J., Noyes, R.W., Soyumer, D., and Vaughan, A.H., 1985, Ap. J., 294, 310.

Dupree, A.K., 1983, In: 'Activity in Red Dwarf Stars', ed. P.B. Byrne and M. Rodono, (Boston, Reidel) 447.

Ezer, D., and Cameron, A.G.W., 1966, In:'Stellar Evolution', ed. R.F. Stein and A.G.W. Cameron (New York, Plenum), 203.

Hale, G.E., 1915, 'Ten years' Work of a Mountaintop Observatory', (Washington, D.C.: Carnegie Institution).

Hartmann, L.W., Bopp, B.W., Dussault, M., Noah, P.V., and Klimke, A., 1981, Ap. J., 249, 662.

Hartmann, L.W., Hewett, R., Stahler, S., and Mathieu, R.D., 1986, Ap. J., 309, 275.

Hartmann, L.W., and Noyes, R.W., 1987, Ann. Rev. Astron. Astrophys., 25, 271.

Hartmann, L.W., and Rosner, R., 1979, Ap. J., 230, 802.

Hartmann, L.W., Soderblom, D.R., Noyes, R.W., and Burnham, J.N., 1984, Ap. J., 276, 254.

Herbig, G.H., and Soderblom, D.R., 1980, Ap. J., 242, 628.

Kraft, R.P., 1967, Ap. J., 150, 551.

Linsky, J.L., 1980, Ann. Rev. Astron. Astrophys., 18, 439.

Linsky, J.L., 1982, In: 'Advances in Ultraviolet Astronomy: Four Years of IUE Research', ed. Y. Kondo, J.M. Mead, and R.D. Chapman, NASA CP-2238, 17.

Marcy, G., 1984, Ap. J., 276, 286.

Noyes, R.W., Hartmann, L.W., Baliunas, S.L., Duncan, D.K., and Vaughan A.H., 1984a, Ap. J., 279, 763.

Noyes, R.W., Weiss, N.O., and Vaughan, A.H., 1984b, Ap. J., 287, 769.

Pallavicini, R.P., Golub, L., Rosner, R., Vaiana, G.S., Ayres, T.R., and Linsky, J.L., 1981, Ap. J., 248, 279.

Phillips, M.J., and Hartmann, L.W., 1978, Ap. J., 224, 82.

Rosner, R., Golub, L., and Vaiana, G.S., 1985, Ann. Rev. Astron. Astrophys., 23, 413.

Skumanich, A., 1972, Ap. J., 171, 565.

Soderblom, D.R., 1984, In: 'Third Cambridge Workshop on Cool Stars, Stellar Systems and the Sun', ed. S.L. Baliunas and L.W. Hartmann, (New York: Springer), 205.

Soderblom, D.R., 1985, Astron. J., 90, 2103.

Spite, M., and Spite, F., 1982, Nature, 297, 483.

Stauffer, J.R., Hartmann, L.W., Burnham, J.N., and Jones, B., 1985, Ap. J., 289, 247.

Stauffer, J.R., Hartmann, L.W., Soderblom, D.R., and Burnham, J.N., 1984, Ap. J., 280, 202.

Vaughan, A.H., 1983, In: Solar and Stellar Magnetic Fields, I.A.U. Symp. 102, ed. J.O. Stenflo (Dordrecht: Reidel), 113.

Vaughan, A.H., 1984, Science, 225, 793.

Vaughan, A.H., and Preston, G.W., 1980, Pub. Astron. Soc. Pac., 92, 385.

Vernazza, J.E., Avrett, E.H., and Loeser, R., 1983, Ap. J. Suppl., 45, 635.

Vogt, S.S., 1981, Ap. J., 250, 327.

Vogt, S.S., 1984, In: 'The Physics of Sunspots', ed. L.E. Cram and J.H. Thomas (Sunspot, N.M.: Sacramento Peak Observatory), 455.

White, O.R., and Livingston, W.C., 1981, Ap. J., 249, 798.

White, O.R., Livingston, W.C., and Wallace, L., 1987, J. Geophys. Res., 92, 823.

Wilson, O.C., 1978, Ap. J., 226, 379.

Wolff, S.C., Edwards, S., and Preston, G.W., 1982, Ap. J., 252, 322.

LONG AND SHORT CYCLES IN SOLAR ACTIVITY DURING THE LAST MILLENNIA

M.R. Attolini
Istituto TESRE-C.N.R.
Via Castagnoli, 1, Bologna, Italy.

M. Galli
Dipartimento di Fisica, Universita di Bologna
via Irnerio, 46, Bologna, Italy.

T. Nanni
Istituto FISBAT-C.N.R.
Via Castagnoli, 1, Bologna, Italy.

1. INTRODUCTION

It is agreed that during the last billion years the Sun, as the analysis of the terrestrial climate has demonstrated, behaved as a stationary system (Mitchell, 1977) and that stars like the Sun evolve on a timescale of a billion years (Soderblom and Baliunas, 1988). Thus one should expect that its activity is subjected to more or less regular oscillations. The study of the likely laws ruling these oscillations (periodicities, power spectra etc.) constitutes one of the essential pieces of information needed to understand solar physics.

Solar activity may be observed directly from the Earth through undulatory and corpuscular emission, or indirectly detected through proxy data of the solar wind, as geomagnetic activity, polar aurorae, or cosmogenic isotope concentration in datable matter, as ^{10}Be in polar ice and ^{14}C in tree rings.

The most evident cycle variation in solar activity is the eleven year (or Schwabe cycle) of the sunspot number. Its existence is well confirmed since about 250 years ago, as the observation of sunspots since that time has been adequately systematic. Between 1645 and 1715, when the instrumental observation of the Sun seems to have been active enough, historical research (Eddy, 1977) has shown a sunspot minimum, currently named the "Maunder Minimum".

We have been able to show that Schwabe cycles existed in solar wind activity before that epoch. Longer cycles have also been discovered. Here we will review recent findings on already described cycles and will also present evidence of new cycles. The work in the field is still in progress. No attempt will be made in order to fit the various types of cycles into a physical model of the solar activity.

F. R. Stephenson and A. W. Wolfendale (eds.),
Secular Solar and Geomagnetic Variations in the Last 10,000 Years, 49–68.
© 1988 by Kluwer Academic Publishers.

50

2. THE SCHWABE CYCLE

2.1 The Schwabe Cycle since 1700

The daily number of visible spots on the solar disc counted according
to the Wolf method clearly shows, since 1700, cycles ranging from
8 to 15 years (Figure 1). Their amplitude and the total number of
sunspots per cycle (Figure 2) seem to vary with a period of roughly
80-90 years.

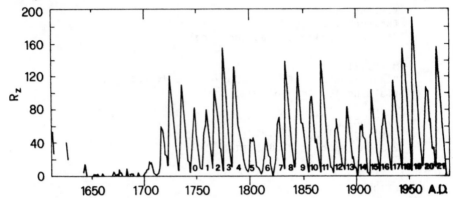

Fig. 1.   Annual sunspot number R_z according to Waldmeier and Eddy
(adapted from the article of G.E. Williams, Sci.Amer.
Oct. 1986 p.80); from 1750 the cycle number is shown.

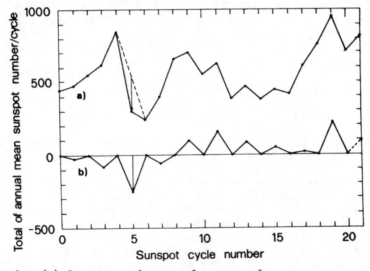

Fig. 2.   (a) Sunspot number totals per cycle.
(b) Difference of an odd cycle total, from the mean of the
two adjacent even cycles, as indicated by the arrow.
Note the change of phase between cycle 7 and 9.

Moreover the Schwabe cycle appears also, in a less evident way, by eye inspection, in the series of the annual number of aurorae observed in about the same epoch (Legrand and Simons, 1987) or in the annual series of the (aa) geomagnetic indexes (Mayaud, 1973) (Figure 3). This may be attributed to the fact that aurorae and geomagnetic activity are both due to geomagnetic disturbances caused by solar wind, and that only a part of the solar wind is related to sunspot activity (Legrand and Simons, 1985). The remaining part may be attributed to solar wind emitted mainly by the solar surface related to coronal holes.

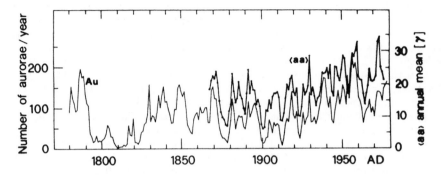

Fig. 3.   Annual frequency of auroral (Au) occurrence, up to 62 degrees of geomagnetic latitude, from 1780 to 1979, according to Legrand and Simon (1987); and the annual mean levels of geomagnetic activity (<aa>) from 1868 to 1979 (Mayaud, 1973).

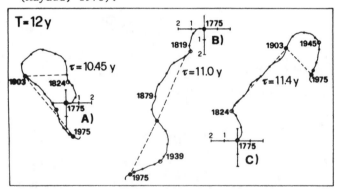

Fig. 4.   The phase cyclogram of annual R_z from 1775 to 1980. The best extension is obtained for the interval 1824–1903 and the period 11.4y (C) and for the interval 1903–1980 and the period 10.45y (A). The 11.0y cyclogram (B) indicates a frequency modulation around 1c/11y. Note the turning point at about 1824.

In order to study the time variation of the Schwabe cycle in
the annual Wolf numbers, we computed their phase cyclogram (Attolini
et al., 1985) for subseries of 12y and test periods of 10.45y (Figure
4A), 11y (Figure 4B) and 11.4y (Figure 4C). We specify here that a
cyclogram (or amplitude cyclogram) of a series of N equispaced terms,
as explained elsewhere (Attolini et al., 1984), is the geometric sum of
the Fourier vector components of a test period τ of any subseries of
length T that one can obtain moving its starting point, term by term,
from the beginning of the series itself. The cyclogram will be a
straight line if the series represents a sine wave of period τ and will
be a line turning respectively to the right or to the left according
to whether the period is longer or shorter than τ. The square of the
extension of a cyclogram is equal to the corresponding value of the
power spectrum. We call phase cyclogram the sum of all vectors with
the original phase but with the same amplitudes. Marks in the cyclogram indicate vectors computed from completely different subseries.
Two subsequent marks correspond to a time interval of T terms.
It appears that not only the amplitude but also the phase of the
11y wave varies rather gradually, in such a way that the period appears
to be 11.4y between 1825 and 1903 and about 10.45y between 1903 and
1975; the corresponding first harmonics of 5.7y, 5.5y and 5.2y show
phase discontinuities of 90 — 180 degrees around 1820 and around
1905 (Figure 5). This means that the Schwabe cycle undergoes a sudden
change of shape at cycle 6 and cycle 14; the same dates correspond
also to a change of the size of the cycles and of their relative
minima.

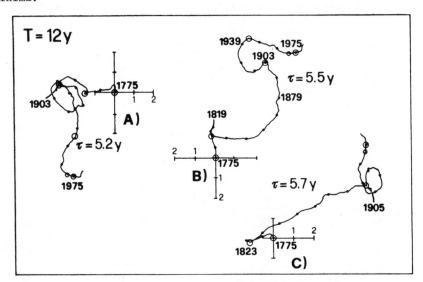

Fig. 5. The phase cyclograms for the first harmonics of the
frequencies of Figure 4. Notice the sharp changes of
phase at about 1823 and at about 1905.

As the magnetic polarity structure of sunspots and the Sun's general magnetic field reverses every cycle, one should more correctly talk of a 22y cycle often referred to as magnetic or Hale cycle. However, this cycle can be physically real only if we can find some type of symmetry between odd and even cycles. Indeed one can observe that the total number of sunspots per cycle alternates every cycle with a possible change of phase at cycle 8 (Figure 2). An additional double Schwabe cycle effect of geophysical origin ought to be expected in the auroral frequency of occurrence, due to the change in polarity of the solar wind interacting with the Earth's magnetosphere.

2.2   The Schwabe Cycle before the Maunder Minimum

As mentioned above, historical studies strongly suggest that, between 1645 and 1700 sunspot activity was unusually low. This is confirmed by the study of the cosmogenic isotopes ^{14}C (Stuiver and Quay, 1980) and ^{10}Be (Beer et al., 1985).

No direct evidence is found for the Schwabe cycle during the period before the Maunder Minimum (Eddy, 1977). It is no wonder that the Schwabe cycle had not been discovered in the records of historical naked eye observations of sunspots that are available for 2000 years (Wittman and Xu, 1988), as any correlation between cycles rapidly disappears in a few decades, as we will see below.

On the contrary the 11y cycle might appear in cosmogenic isotopes especially in ^{10}Be, whose residence time in the atmosphere is about 1 or 2 years. In fact, from analysis of ^{10}Be concentration measurements done by the Bern Group in a dated ice core from Milcent in Greenland (Beer et al., 1985), the existence of a cycle of 11.4 years in the interval 1180-1500AD was clearly demonstrated.

The original series of ^{10}Be measurements, corresponding to intervals of 3 - 7 years, in the time interval 1180-1800 (Figure 6a), first had the trend removed, thus obtaining a new series of non equispaced data (Figure 6b). In order to find the average wave of period τ to be used for Fourier transform, cyclograms, etc. we first transferred, by folding, all measurement points into a single interval of length τ and then averaged the data into bins of not less than 6-8 equal intervals.

In such a way we have found the phase cyclogram depicted in Figure 7, which shows coherent 11.4y cycles in the data, from 1180 to 1505. This is confirmed by the average cycle of period 11.4, 11.0 and 11.8 years of Figure 8 whose significance level with respect to a random distribution corresponds to 9.7, 0.2, 0.9 sigma respectively.

54

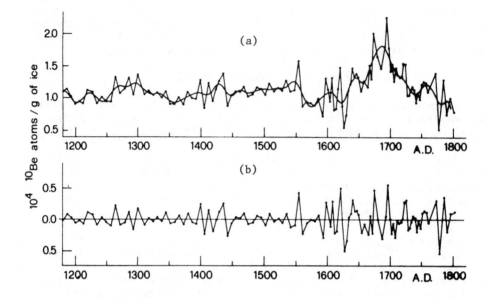

Fig. 6. (a) ^{10}Be concentration in Milcent (Greenland) ice core
plotted for the period 1180–1800AD ($\smile$). Superposed
is the long term trend obtained by a 2y-step moving
least-square-fit over 50y with a 3rd order polynomial
(——).
(b) Residual oscillation after subtracting the long term
trend.

    Further positive indications of the existence of the 11y cycle
in the past solar activity have been obtained using the series of
the annual records of aurorae (Figure 9) whose description has been
found by de Mairan (1733), Link (1962, 1964), Stothers (1979), Dall'Olmo
(1979), Schroder (1979), Keimatsu (1970-1976, 1968) and Matsushita
(1961). The series may be considered as a sample of a realization of
a random process, whose time expectation values are given by the product
of the actual frequency of aurora occurrence with the probability that
an aurora could be effectively visible, and then recorded in a

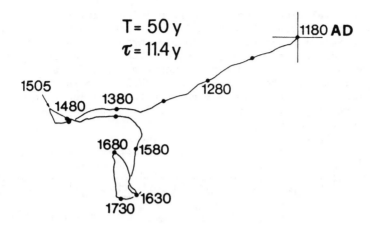

Fig. 7. Phase cyclogram of ^{10}Be for the interval 1180–1730AD obtained by best fitting a sinusoid of period τ = 11.4y to the residuals of the experimental data contained in a 50y interval and moving the interval 5 years at a time.

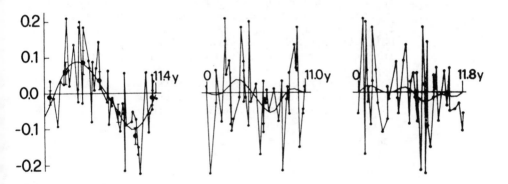

Fig. 8. Epoch point folding of the residuals of Figure 6b for 11.4, 11.0, 11.8y respectively. Cubic polynomial fits to the data points are also shown (●). Data point averages computed over $\tau/6$ intervals.

document that could be eventually found. In order to partially avoid these epochs having more abundant observations biasing the estimates of auto-correlation and power spectrum, the following normalization has been made; each annual or decadal total has been replaced by the corresponding fraction of the mean value computed over a certain interval centered in the year considered.

The autocorrelation function of the original series, so normalized for intervals of 150 years, is shown in Figure 10. The 11 year signal in the autocorrelation is clearly visible through a wave that initially reaches a significant level of about 0.2%. The wave vanishes after a 25–30 year lag, as one would expect from the already mentioned variability of the Schwabe cycle visible in Wolf numbers. In our case, from the autocorrelogram, one can see that the length of the cycles can range approximately from 9.5 to 11.5 years.

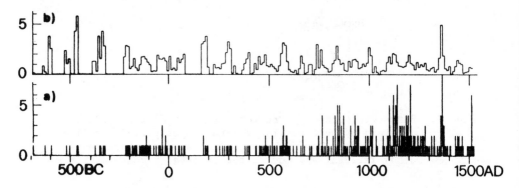

Fig. 9. (a) The annual series of 2207 historical aurorae recorded from 687BC to 1500AD, obtained combining the lists of western reports of De Mairan, Link, Stothers, Dall'Olmo and Schroeder with the list of all events in Far East reports of Keimatsu, with reliability degree 1,2,3, and the list of Matsushita.

(b) The normalized series over 150 years, i.e. the series of decennial totals of historical aurorae in units of the centered running mean of 15 totals.

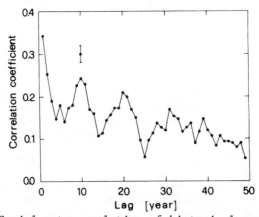

Fig. 10. Serial autocorrelation of historical aurorae. The significance of the 11-year oscillation can be seen by comparing the coefficient with the error bar.

In Figure 11 we show the power spectra of the same series for the intervals 687BC - 1000AD and 1000AD - 1721AD and for comparison that of modern aurorae. Here by power spectrum of a time series $\{x(t); t=0, \Delta t..., (N-1)\Delta t; \Delta t=1\}$ for the frequency f, we mean the square of the Fourier amplitude per unit time $P(f) = |x(f)|^2/N$. The comparison of the spectra shows a general enhancement in the band 9.5 - 11.5 years, but a large variability in position and height of the main peaks, which may indicate a modulation of the 11y cycle, which is likely to occur, both in amplitude and frequency. The phase cyclogram of Figure 12 shows some details of the pattern of the frequency variability of the 11y cycle. A clear frequency modulation of the 11 year cycle having an oscillation timescale of about 200 years can be seen throughout in the cyclogram.

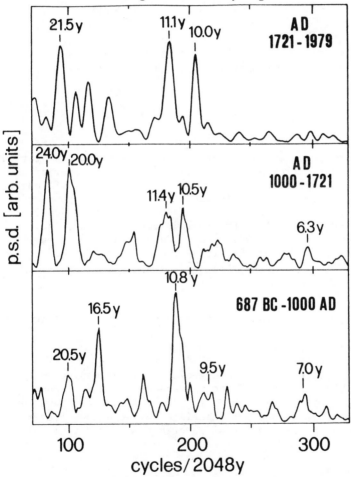

Fig. 11. The comparison of the power spectra of historical aurorae before (a) and after (b) 1000AD to be compared with the spectrum of modern aurorae (c).

58

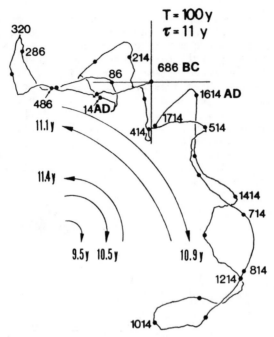

Fig. 12. The phase cyclogram of the annual series of historical
aurorae from 687BC to 1715AD for the test period of
11y; observing the bending to right or to left one can
notice a large scale bending radius corresponding to
the periodicity of 10.9y from ∿ 300BC to ∿ 900AD
and   of 11.1y from ∿ 1000AD to ∿ 1700AD; the inversion
of phase at ∿ 300BC and ∿ 1000AD and finally an
oscillation of the 11y period between 9.5y and 11.4
years in   cycles of ∿ 200 years.

3.   THE GLEISSBERG CYCLE AND THE 130 YEAR CYCLE IN HISTORICAL AURORAE

3.1   The Gleissberg cycle

Another cycle, called the Gleissberg cycle, corresponding to the
modulation of the amplitude of the Schwabe cycle in Wolf numbers
has been studied by Gleissberg (Eddy, 1977); it is about 88y long
and has also been found and well confirmed by Feynman & Fougere (1984)
through the analysis of historical aurorae from 450AD to 1450AD.

3.2   The 130 year cycle

A cycle of about 130 years has also been found by our group through
the analysis of historical aurorae (Attolini et al. 1988).   The series

we have used is that of the 10-year mean to the annual frequency
of historical aurorae normalized with respect to an interval of 150
years. The autocorrelation function of this series computed up to
a lag of 1500 years is shown in Figure 13. A clear wave of about
130 years is present in the correlation, showing among other things
that a wave of 130 years was, on the average, in phase with a wave
of the same period at a distance of about 1500 years.

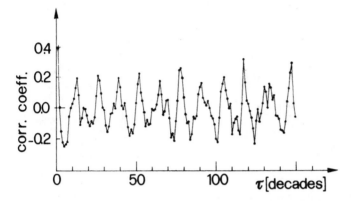

Fig. 13. The autocorrelogram of historical auroral series summed
per decades and normalized over 150 years. Notice the
persistence of the 130y wave and the phase consistency
for lags greater than 1000 years. Similar results have
been also obtained normalizing to 130, 250y and to 350y.

The Fourier Transform of the autocorrelation function which
gives an estimate of the power spectrum of the series shows (Figure
14) significant peaks corresponding to the period of 130.4y and to

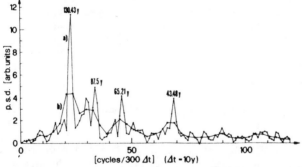

Fig. 14. The power spectrum of historical aurorae obtained by
Fourier transforming the autocorrelation function of
Fig. 13. The smoother curve has been obtained by Fourier
transforming the series corresponding to the first 50 lags.
Notice the 87.5 peak exactly half way between the 130y
and 65y peaks.

its first and second harmonics. Another significant peak corresponds
to a period of 87.5 years that clearly coincides with the Gleissberg
cycle. The cyclograms for the periods of 125y, 131y and 135y (Figure
15) show that the 130 year cycle continues through the entire series
although a disturbance due to the 88y wave becomes more important
towards the end of the series.

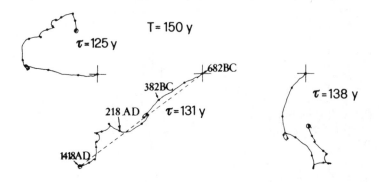

Fig. 15. The amplitude cyclogram of historical aurorae for the
          test periods 125y, 131y, 138y showing details about
          the persistence of the 130y wave. The bendings to the
          right after about 100AD indicate the coexistence presence
          of the 130y cycle with that of 87y.

The average shape of the 130 year wave which depends on the
relative phase and amplitude of the first and the second harmonic,
has been found by the forementioned method of point epoch superposition. The results (Figure 16) confirm the existence of the cycle,
the values of which range between 0.5 and 1.5 times the average number
of aurorae in an interval of 150 to 350 years. Although the 130y
cycle of aurorae might also originate from a cyclic variation either
of the geomagnetic latitudes of the observing stations or of the geomagnetic dipole moment, since paleomagnetic studies have failed to
prove the existence of such periods (Dubois, 1988), we conclude that
most probably the 130 cycle of the historical aurorae, has to be
attributed to solar activity.

3.3 The Gleissberg and 130y cycles in radiocarbon series since
     5000 BC

In order to answer the question of whether the above mentioned long
cycles of 88 and 130 years were also present in epochs preceding
the one just considered, we have searched for them in the series
of ^{14}C concentration in tree rings. Since ^{14}C has different residence
and exchange time constants in different reservoirs as atmosphere,
biosphere and oceans, the variation with time of its concentration
in dated organic matter, can be thought of as derived from the
production time series by means of a filter whose response has been

computed by Siegenthaler et al. (1980).

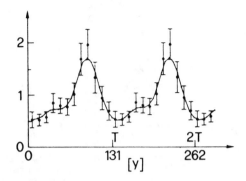

Fig. 16.   The full 10-yearly data series of historical aurorae
           folded into a 131 year interval, binned in (131/12)
           years, expressed in units of 150-year-running-mean value.
           Also shown is the best fitted 3rd degree polynomial;
           the wave has been repeated in order to show its shape
           more clearly. The increase of errors with amplitude
           is due to the years of missing reports, which may be
           present at any time and produce a larger dispersion
           when the frequency of auroral observations is higher.
           The average 130y variation ranges from about 0.5 to
           2 times the mean value of the series over 150 years.

If we ignore the possible effects due both to the large scale
atmospheric and ocean circulation and to the changes of ^{14}C reservoirs,
we can hope to discover in the ^{14}C series the signals of the Gleissberg
and 130y cycles, as they should be reduced from their original
amplitudes, by not more than a factor of 20.

The 130 year cyclogram of a 20-yearly series of tree ring radiocarbon from 5210BC to 1830AD (Pearson, et al., 1986) shows a large
stretched part for the interval 1550BC - 1830AD (Figure 17a) which
certainly corresponds to a significant peak in the power spectrum.

The point epoch superposition for various periods around 130y
over the same interval gives average waves whose significance is
shown in Figure 17c. The maximum significance is 5.2 sigma obtained
for the period of 129 years; the corresponding average wave is shown
in Figure 18.

The 86y cyclogram of the same series shows the largest stretched
part during the time interval 4410BC to 2530BC (Figure 17b); the
corresponding average waves having period around 86y reach a maximum
significance of 6.2 sigma (Figure 17c) for the period 86.3y. The
average wave of 86.3y is shown in Figure 19.

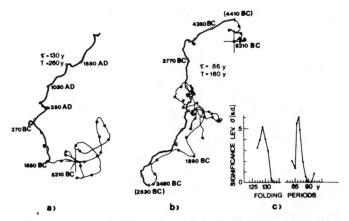

Fig. 17. The 130 year and Gleissberg cycles in the 20-yearly radiocarbon concentration series found in Irish oak tree rings.
(a) The 130 year cyclogram indicating a sequence of 130
year oscillations from about 1550BC to 1700AD.
(b) The 86 year cyclogram indicating a sequence of 86
year oscillations from about 4400BC to 2530BC.
(c) Showing the significance level against a random distribution of the data of the mentioned intervals, folded
into periods around 130 and 87 years respectively.

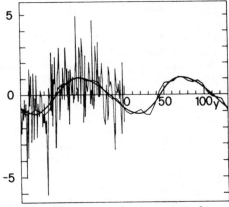

Fig. 18. The mentioned radiocarbon data series from 1550BC to
1700AD folded into a 129 year interval. The best fitted
3rd degree polynomial over a running interval of 80 years
(heavy line) and 40 years (thin line) is also shown;
the slight oscillation of the thin line over the heavy
line might indicate the Hale cycle. The significance
of the mean 129 year oscillation is about 5.2 sigma.

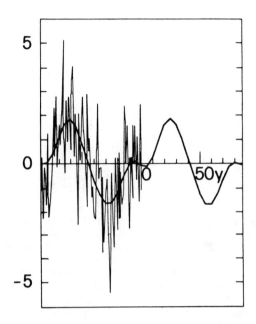

Fig. 19. The radiocarbon data series from 4400BC to 2530BC folded into a 86.3 year interval; the significance level of the mean oscillation is about 6.2 sigma.

4. FREQUENCY STRUCTURE OF SOLAR ACTIVITY SPECTRA

4.1 The modern records of sunspots and aurorae

The series of solar activity data or their proxy data utilizable for our purposes are the forementioned series of sunspot numbers, aurorae and aa data of Figures 1 and 3. They show an amplitude modulation and large scale variation that might be a part of a systematic variation due to longer cycles.

A striking similarity between portions of auroral series repeating themselves at a distance of about 90 years, suggests that many wiggles visible in its graph may not be random. In other words, the peaks 1-1', 2-2', etc. (Figure 20) of the corresponding section a-a', b-b', etc. of the series, seem rather to belong to a deterministic structure of the series.

In order to study in these series the presence of long periods, we have to prevent the interference of the lobes that generally appear in the power spectrum, when the length of the series is not an exact multiple of the considered period, so that some peaks of the spectrum may depend on the length of the series. We have done this by computing the Fourier complex amplitudes with the forementioned method of the point epoch superposition.

64

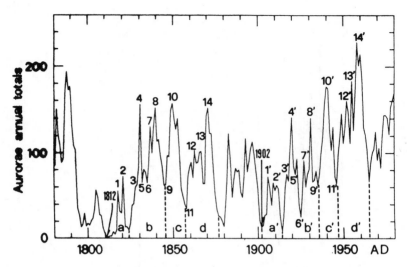

Fig. 20.   The modern auroral series of Figure 3 where the sections
           a, b, c,.. seem to be repeated into a', b', c'... after
           an interval of 90 years.   Also shown are supposedly
           corresponding wiggles 1, 2, 3... and 1', 2', 3',.. The
           marked dates of 1812 and 1902 indicate the beginning
           of the two smallest sections.

     In Figure 21 we show the spectra of the modern auroral and
sunspot series so computed.   Both show evident peaks at 21.3, 11.2
and 10.0 years, with minor peaks, most of which having frequency
distance from each other of 1c/88y or an integral multiple of it.
In the 11 year band of the auroral spectrum we see two more prominent
peaks at 11.5 and 10.7 years.   One may notice also that the 11.2y
peak in sunspots and aurorae corresponds to the mean value of the
mentioned peaks in aurorae, at 1c/11.5y and 1c/10.7y.   One can roughly
attribute the two sunspot peaks at 10.0 and 11.2 years to the
Gleissberg modulating effect on Schwabe cycles; and the two peaks
at 11.5y and 10.7y to a further modulating cycle on aurorae.   One
should remember here that the spectrum of a sine-wave of frequency
f_c modulated by a sine-wave of frequency f_m, shows lines at the
frequencies f_c+f_m, f_c-f_m and generally also at f_c.
     Of course, due to the paucity of the data and the finite numerical
computing approximations, one can expect in the above frequency
figures, an error of one or two percent.   The following deviations
in percentages of some conspicuous lines of the auroral spectrum
are found from combinations of the Hale frequency $f=1c/21.85y$ as
determined by Bracewell in sunspots (Bracewell, 1986), and $F=1c/87.5y$
as determined by us in the historical aurorae: (f) 0.7; (2f F/2)
-1.2; (2f - 7/4) 1.3; (2f + 3 F/4) 0.2; (4f) 1.0; (5f) 0.5; (6f) -0.9;
(7f) -0.7; (8f) -0.7.

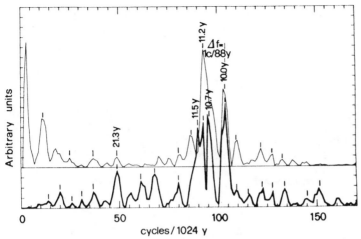

Fig. 21. High frequency part of the spectra of R_z from 1755 to 1977
(thin line) and of modern aurorae from 1780 to 1979 (heavy
line) obtained with the mentioned point folding method.
Marks over peaks indicate frequency values spaced from
some other peaks of frequency intervals integer multiples
of 1c/88 years.

4.2 The Spectra of Historical Aurorae

We have then analysed in a more detailed way the spectrum of
historical aurorae, in order to find the quantized frequency spacing
mentioned above. The spectrum actually shows a large number of
prominent frequency lines corresponding to integral multiples of the
frequencies 1c/88y and 1c/132y or of lines spaced from each other
by integral multiples of such two frequencies (Figures 22 and 23).
One may also notice the frequent occurrence of the spacing 1c/44y,
clearly visible also in Figure 14, which is double that of 1c/88y and
triple that of 1c/132y.

5. DOES THE SUN BEHAVE AS A CHANGEABLE QUASIPERIODIC SYSTEM?

The above spectral frequency structure of auroral and sunspot data
series immediately suggests the possibility that the solar activity
system contains at least two or more fundamental oscillators, sometimes locked together, one of which might have the frequency 1c/132y
or 1c/88y and the other one the Hale frequency of about 1c/22y.
From the data we have analysed we find that the 132y and 88y
cycles are not uniformly present. Sometimes they seem to have coexisted
as during the interval 0–1700AD. At other times it seems that only
one prevails, as seen in aurorae and sunspots from 1700 to the present
time – where only the 88y cycle seems to exist – or as seen from

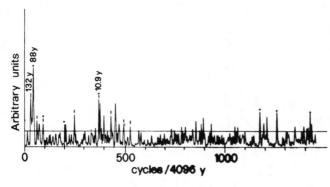

Fig. 22. The power spectrum of historical aurorae from 1AD to 1000AD. Dot(.) and dash (') marks over peaks indicate some frequency values spaced, from some other peaks, of frequency intervals integral multiples of 1c/88 year or 1c/131 years.

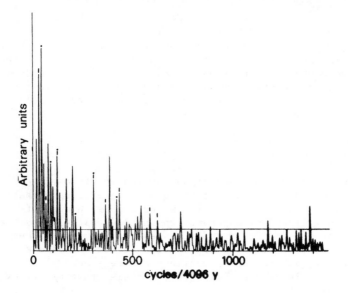

Fig. 23. Same as Fig. 22 for the historical aurorae from 1001 to 1731AD.

radiocarbon between 4400BC to 2500BC for the Gleissberg cycle and between 1500BC to 1700AD for the 130y cycle. Furthermore we find many remarkable rapid changes in phase, as for the Schwabe cycle at 300BC, 1000AD, 1645AD, 1825AD or for its first harmonics at 1825AD and 1905AD. In the longer cycles rapid changes of phase are also found at 2500BC for the Gleissberg cycle and at 1500BC for the 130y cycle.

The changes of phase suggest the possibility that important changes may occur in some individual convective cells of the sun or in its global convective system.

The sun might therefore behave as a quasiperiodic non linear dynamic system, with changes of state and frequency lockings as described in the recent literature (e.g. Eckman et al., 1985).

6. CONCLUSIONS

From the analysis of the series of data of solar activity such as sunspots or of proxy data of it such as aurorae and cosmogenic isotopes, the following conclusions can be drawn.

1 - The Schwabe cycle has been found to be present also before the Maunder Minimum. This has been proven through the analysis of the ^{10}Be variation from 1200 to 1500AD. Historical aurorae have also proven the existence of a frequency variable 11y cycle, on the whole with a coherence no longer than a few cycles. The Schwabe cycle appears to be variable in length from 9.5 to 11.5 years with supercycles of 100-200 years.

2 - The Hale cycle with the odd-even oscillation in the total number of sunspots per Schwabe cycle, appears evident throughout cycles 0-21 with a likely change in the alternation at cycle no. 7-8 which favours the similarity between the solar cycles and the varve cycles of Williams, (1985).

3 - A cycle of about 130y has been discovered in historical aurorae, that can be attributed to solar activity. Its amplitude ranges from 0.5 to 1.5 times its average value. During the interval 0-1700AD it coexisted with the cycle of 88 years.

4 - Coherent 86.3y cycles have been found in tree ring radiocarbon series from 4400 to 2500BC. Cycles of 129y have been found in the same series during the interval 1500BC-1800AD.

5 - Recurrent wiggles at time distances of about 90 years can be observed in the graphical representation of modern auroral series. Its spectrum shows frequency lines at about 1c/88y, 1c/22y and 1c/11y and a number of prominent peaks spaced in frequency from each other of the frequency interval 1c/88y of integer multiples of it.

6 - Frequency spacings of about 1c/88y and 1c/132y or of integral multiples of them appear also in line spacings of the spectra of historical aurorae, suggesting that the Sun might behave as a quasi-periodic system containing two or more basic oscillators, one of which might be the Gleissberg frequency or the 1c/130y frequency, and the other might be the Hale frequency.

7 - Finally the cycles do not appear to be perennial; the above mentioned longer cycles sometimes appear to coexist, at other times only one of the two is clearly evident. Rapid changes of phase on all cycles considered, can be observed; some of them appear to be related to some change in the Earth's climate.

No investigation has been made so far to find in our series indication of longer cycles as, for example, the 200y cycle found by Sonett (1984) in tree ring radiocarbon.

ACKNOWLEDGEMENTS

Many thanks are due to Dr. Stefano Cecchini for many stimulating discussions, frequent advice and assistance in computation.

REFERENCES

Attolini, M.R., Cecchini, S. and Galli, M., 1984, Nuovo Cimento, 7C, 245.
Attolini, M.R., Galli, M. and Cini Castagnoli, G., 1985, Solar Physics 95, 391.
Attolini, M.R., Cecchini, S., Galli, M., and Nanni, T., 1988, A 131-year Periodicity in Historical Auroral and Possibly in the Solar Activity (submitted to Nature for publication).
Beer, J., Oeschger, H., Finkel, R.C., Cini Castagnoli, G., Bonino, G., Attolini, M.R., and Galli, M., 1985, Nucl. Instr. Meth. Phys. Res. B10/11, 415.
Bracewell, R.N., 1986, Nature, 329, 516.
Dall'Olmo, U., 1979, J. Geophys. Res., 84, 1525.
de Mairan, J.D.D., 1733, 'Traite Physique et Historique de l'Aurore Boreale', Imprimerie Royale, Paris.
Dubois, R.L., 1988, Archeomagnetic Results from United States and Meso-America 10,000 BC to the Present, this volume.
Eckman, J.P., and Ruelle, D., 1985, Rev. Mod. Phys., 57, 617.
Eddy, J.A., 1977, 'Solar Output and its Variation', Colorado University Press.
Feynman, J., and Fougere, P.F., 1984, J. Geophys. Res., 89, 3023.
Keimatsu, M., Fukushima, N., and Nagata, T., 1968, J. Geomagn. Geoelec., 20, 45.
Keimatsu, M., 1970, Ann. Sci. Kanazawa Univ., 7, 1; 1971, 8, 1; 1972, 9, 1; 1973, 10, 1; 1974, 11, 1; 1975, 12, 1; 1976, 13, 1.
Legrand, J.P. and Simons, P.A., 1985, Astro. Astrophys., 152, 199, 1986, 155, 227.
Legrand, J.P. and Simons, P.A., 1987, Annales Geophys., 5A, 168.
Link, F., 1962, Geofysikalni Sbornik, 173, 297.
Link, F., 1964, Geofysikalni Sbornik, 212, 501.
Matsushita, S., 1961, J. Geophys. Res., 61, 297.
Mayaud, P.N., 1973, IAGA Bull., 33.
Mitchell, M., 1977, 'Solar Output and its Variation', White, O.R. ed., Colorado University Press, Boulder.
Pearson, G.W., Pilcher, J.R., Baillie, M.G.L., Corbett, D.M., and Qua, F., 1986, Radiocarbon, 28, 911.
Schroeder, W., 1979, J. Atmosph. and Terr. Phys., 41, 445.
Siegenthaler, U., Heiman, M., and Oeschger, H., 1980, Radiocarbon, 22, 177.
Soderblom, D.R. and Baliunas, S.L., 1988, 'The Sun Among The Stars', this volume.
Sonett, C.P., 1984, Review of Geophysics, 30, 239.
Stothers, R., 1979, Astron. and Astrophys., 77, 121.
Stuiver, M. and Quay, P.D., 1980, Science, 207, 11.
Williams, G.E., 1985, Aust. Phys., 38, 1027.
Wittman, A.D. and Xu, Z.T., 1988, 'Solar Activity as Inferred from Sunspot Observations 165BC to 1986', this volume.

IS THE SOLAR CYCLE AN EXAMPLE OF DETERMINISTIC CHAOS?

N.O. Weiss
Department of Applied Mathematics & Theoretical Physics
University of Cambridge
Cambridge CB3 9EW

ABSTRACT. Recent developments in nonlinear dynamics make it possible
to distinguish between random (stochastic) behaviour and deterministic
chaos. Trajectories on a chaotic attractor show sensitive dependence
on initial conditions. Solar activity exhibits apparently chaotic
behaviour on several different timescales, associated with the eruption
of active regions, the 22-year magnetic cycle and modulation at grand
minima. Although it has not yet been possible to construct a fully
convincing self-consistent model of the nonlinear solar dynamo,
qualitative features of the solar cycle are reproduced by simple toy
systems, such as that investigated by Weiss, Cattaneo & Jones (1984).
This sixth order system shows a sequence of three Hopf bifurcations,
so that trajectories in the six-dimensional phase space are attracted
to a three-torus. Frequency locking is followed by a period-doubling
cascade that leads to chaos. The "ghost attractor" still survives
and modulates the cycles in a manner reminiscent of the Maunder
minimum. This signature of chaos associated with quasiperiodicity
is consistent with modulation of the ^{14}C record but differs from
the envelope of the Elatina cycles, which seem to be quasiperiodic
rather than chaotic. The Elatina record does not appear to be consistent with the behaviour of the solar cycle over the last 10,000
years.

1. INTRODUCTION

The last decade has seen remarkable advances in our understanding
of nonlinear dynamic systems. In particular, we have become aware
of the prevalence of chaos in deterministic systems, with the
implication that aperiodic behaviour need not be ascribed to random
or stochastic forcing. There are many natural examples of deterministic
chaos, of which the most obvious is the weather. Among various
examples of aperiodic behaviour in astrophysics (Spiegel 1985) solar
activity shows irregular cycles that are apparently chaotic (Zel'dovich
et al. 1983).

F. R. Stephenson and A. W. Wolfendale (eds.),
Secular Solar and Geomagnetic Variations in the Last 10,000 Years, 69–78.
© 1988 by Kluwer Academic Publishers.

Hydrodynamic dynamo theory has provided several simple models that exhibit chaotic behaviour in the nonlinear regime. Disc dynamos governed by the Lorenz equations have been used to explain sporadic reversals of the earth's magnetic field (Allan 1962; Cook & Roberts 1970; Robbins 1977; Krause & Roberts 1981). Aperiodic dynamo waves appear as solutions of a number of simple mean field dynamo models designed to represent the solar cycle (Yoshimura 1978; Jones 1983; Zel'dovich et al. 1983; Weiss et al. 1984). In this survey I shall argue that the observed record of solar activity is indeed consistent with the behaviour of a chaotic oscillator.

First of all, I shall outline the basic features of deterministic chaos and contrast chaotic and stochastic systems. Then, in Section 3, I shall summarize the relevant properties of the solar cycle. In Section 4 I shall describe chaos in simple dynamo models, emphasizing the relationship between quasiperiodic behaviour (with trajectories that lie on tori in phase space) and chaotic behaviour with modulation associated with "ghost tori". This distinction is important in assessing the relevance of Precambrian varve sequences, which are discussed in Section 5.

2. DETERMINISTIC CHAOS

The traditional picture of the solar cycle is of a periodic oscillator disturbed by random impulses, leading to stochastic behaviour. Yule (1927) drew the analogy with a pendulum surrounded by small boys with peashooters: each impact would then affect the phase and amplitude of the oscillations. Others have thought in terms of a periodic signal distorted as it is transmitted through the outer layers of the Sun (e.g. Bracewell 1985). Barnes et al. (1980) showed that it was possible to construct a random time series with prescribed statistical properties that closely resembled the record of solar activity but the observed record is not long enough to establish whether the sunspot cycle is stochastic or chaotic.

Deterministic systems are governed by differential equations of the form

$$\dot{x}_i = f_i(\underline{x}) \qquad i = 1, 2, \ldots, n, \qquad (1)$$

where the f_i are (nonlinear) functions of the variables x_k. At any instant the state of a system is represented by a point in an n-dimensional phase space with co-ordinates $\{x_i\}$, which describes a trajectory as the system evolves. The set of trajectories represent a flow in phase space, with a velocity $\underline{v}$ such that $v_i = f_i$, and we can form the divergence of the flow,

$$\Delta = \sum_i \partial v_i / \partial x_i. \qquad (2)$$

We shall be concerned with dissipative flows, for which $\Delta < 0$ and trajectories are attracted to some set of measure zero in the phase space.

The simplest type of attractor is a fixed point, corresponding
to a steady solution of the differential equations (1). Next comes
a limit cycle, a closed curve in phase space that corresponds to a
periodic solution. For a two-dimensional system (n = 2) the phase
space is a plane and these are the only possible attractors. For
n ≥ 3 there may be quasiperiodic (i.e. multiply periodic) solutions,
with trajectories that lie on tori. The simplest case is a doubly-
periodic solution with two distinct frequencies; if the ratio of these
frequencies (the winding number) is irrational then each trajectory
covers the surface of a 2-torus embedded in the phase space (cf. Figure
1). It is easy to visualize a 2-torus embedded in a 3-space but harder
to envisage a 3-torus in a higher-dimensional phase space. Finally,
there may be chaotic solutions with trajectories that lie on strange
attractors. They have more complicated structure, corresponding to
the product of some manifold in the phase space with a Cantor set.

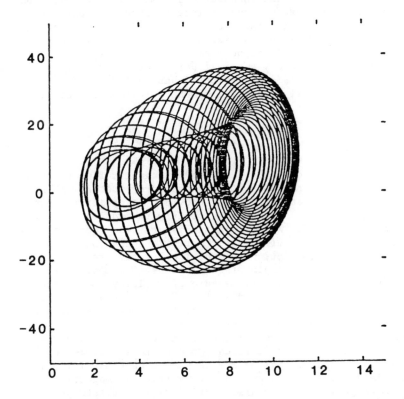

Fig. 1.   Projection of a 2-torus onto a plane for the reduced 5th
          order system derived from (6) with D = 3.6. The abscissa
          measures the amplitude of the poloidal field while the
          ordinate is proportional to the real part of the amplitude
          of the toroidal field.

Deterministic chaos is characterized by sensitive dependence on initial conditions. Neighbouring trajectories on the attractor diverge exponentially and so solutions started from almost identical initial conditions may behave very differently. Conversely, if the initial state of a system is prescribed with finite precision then its future development cannot be accurately predicted. The simplest way of recognizing chaos is from the power spectrum of a signal. Whereas periodic or quasiperiodic solutions have line spectra, chaotic solutions have noisy spectra. Near the onset of chaos broadened lines will still be visible but as the system becomes more chaotic the lines become submerged in noise. Other methods of demonstrating the presence of a chaotic attractor are by calculating Lyapunov exponents (which measure the rate at which trajectories diverge) or showing that the attractor has a low fractal dimension. In many nonlinear systems the attractors are indeed strongly attracting. As a result it has been possible to detect periodic, quasiperiodic and chaotic behaviour in laboratory experiments and in numerical simulations despite the imperfections of the systems actually investigated. In particular, a small amount of external noise has no qualitative effect on chaotic behaviour. Trajectories jiggle about in the neighbourhood of the attractor but do not escape from it. To be sure, any individual trajectory will be shifted by external perturbations and its detailed development will be significantly altered but so long as it stays near the attractor statistical properties of the solution will be virtually unaffected.

3. THE SOLAR CYCLE

Solar activity is caused by the eruption of magnetic flux to form active regions in the photosphere. Sunspots are located within these active regions and the sunspot number provides an arbitrary measure of activity which correlates satisfactorily with indices that are more rationally defined. We can distinguish between four different timescales for variations of magnetic activity on the Sun. The first is associated with the eruption of individual active regions, which might be regarded as an example of spatiotemporal chaos. The daily sunspot numbers provide a global measure of activity, modulated by rotation of the Sun. Spiegel and Wolf (1987) have analysed data accumulated over 100 years and find that it is consistent with the presence of a low-dimensional chaotic attractor.

Next comes the well-known 11-year activity cycle which recurs aperiodically with variable amplitudes and periods. Since the toroidal field in active regions reverses at sunspot minima, while the polar fields reverse around sunspot maxima, the magnetic cycle has an average period of 22 years. Active regions first appear at sunspot latitudes around ±30° and the zones of activity extend towards the equator at sunspot maximum. Activity at higher latitudes diminishes until the last spots of the old cycle occur near the equator just as the new cycle begins at higher latitudes. Active regions therefore appear as waves travelling towards the equator, though small scale

features display more complicated patterns. These activity cycles
are modulated on a longer timescale, leading to episodes of reduced
activity such as the Maunder minimum of the late seventeenth century.
The lack of spots was noted by contemporary observers, rediscovered
by Spörer and subsequently discussed by Maunder and by Eddy (1976).
Spörer (1889) pointed out that the systematic drift of sunspot zones
towards the equator was present in the earliest telescopic observations
(around 1620) but that between 1670 and 1710 sunspots were not only
rare but also confined to very low latitudes, as though the cycle
had been interrupted. Finally, the overall pattern of activity has
changed as the Sun has evolved over periods of order 10^9 yr and
gradually spun down; its magnetic history can be studied by observing
younger stars of the same spectral type, as discussed by Soderblom
in these Proceedings.

Modulation of the activity cycle can be traced back by measuring
anomalies in the production rates of ^{14}C and ^{10}Be, as described
elsewhere in these Proceedings. These proxy records are consistent
with behaviour of the envelope of the activity cycle since Galileo's
telescopic observations. Moreover the ^{14}C data show that there have
been several episodes of reduced activity in the last 1000 years
(Stuiver & Quay 1980; Stuiver & Grootes 1980). There is now a reliable
record of ^{14}C abundance variations over the last 9000 years (Stuiver
et al. 1986); as Stuiver explains in these Proceedings, grand minima
recur aperiodically with a characteristic timescale of around 200 yr.
This aperiodic modulation suggests that the solar cycle is chaotic.
It should therefore be possible to develop some nonlinear dynamical
system that describes its behaviour and has a chaotic attractor. The
short-term variations associated with the local appearance of active
regions can then be regarded as stochastic disturbances with no
particular qualitative effect on the behaviour of the system.

4. THEORY OF CHAOTIC DYNAMOS

Magnetic fields in late-type stars are maintained by dynamos that
rely on differential rotation and helicity associated with cyclonic
eddies (Parker 1979; Zel'dovich et al. 1983). Hydromagnetic dynamo
theory is reviewed by Soward in these Proceedings. Several types
of oscillatory dynamo model have been investigated. Gilman (1983)
and Glatzmaier (1985) have shown that dynamo waves occur in fully
nonlinear self-consistent models. There are many examples of mean
field dynamos whose behaviour depends on a single parameter, the
dynamo number

$$D = \frac{\alpha\Omega'D^4}{\eta^2} , \qquad (3)$$

where α is proportional to the helicity, Ω' is the shear in angular
velocity, η is the (turbulent) diffusivity and D is a characteristic
length scale. In these models there is a trivial solution ($\underline{B} = 0$)

of the dynamo equations that loses stability at an oscillatory (Hopf)
bifurcation when $D = D_{crit}$; solutions to the linear problem typically
give dynamo waves which propagate towards the equator if Ω increases
inwards. In nonlinear models the amplitude of the field increases
with D and for simple models $D \stackrel{\sim}{\sim} (\Omega\tau_c)^2$, where τ_c is an appropriate
convective timescale. The observed magnetic activity in lower main-
sequence stars apparently depends only on the inverse Rossby number
$\Omega\tau_c$, as required by simple theories (Baliunas & Vaughan 1985). In
nonlinear dynamos magnetic torques affect differential rotation. Since
the Lorentz force is quadratic in the magnetic field $\underline{B}$ we would expect
to find torsional waves with twice the frequency of the magnetic cycle
and such waves have been observed on the Sun (LaBonte & Howard 1982).

Another approach to nonlinear dynamo waves is to seek the simplest
model that captures the essential physics of a stellar dynamo. Plane
dynamo waves, with a magnetic field of the form

$$\underline{B} = e^{ikx}(B(t)\hat{\underline{y}} + ikA(t)\hat{\underline{z}}) \tag{4}$$

propagating along the x-axis of cartesian co-ordinates, satisfy the
scaled linear equations

$$\left.\begin{array}{l} \dot{A} = 2DB-A , \\[2ex] \dot{B} = iA-B \end{array}\right\} \tag{5}$$

(Parker 1979). Once again the trivial solution is unstable to
oscillatory modes (waves) for $D > D_{crit} = 1$. In order to saturate
growth of the field some non-linearity must be included. Torsional
waves are described by the sixth-order system

$$\left.\begin{array}{l} \dot{A} = 2DB - A, \\[2ex] \dot{B} = iA - \frac{1}{2}iA^*\omega-B, \\[2ex] \dot{\omega} = -iAB - \nu\omega, \end{array}\right\} \tag{6}$$

where ν is a viscous diffusivity, which is a complex generalization
of the Lorenz system (Weiss et al. 1984). This system has a rich
bifurcation structure which has been analysed in some detail. The
oscillatory bifurcation at $D = 1$ leads to periodic solutions, with
trajectories attracted to a limit cycle in the six-dimensional phase
space. When $\nu = \frac{1}{2}$ there is a second oscillatory bifurcation at
$D \stackrel{\sim}{\sim} 2.07$ leading to quasiperiodic motion with trajectories attracted
to a 2-torus. At $D \stackrel{\sim}{\sim} 3.47$ there is a third oscilatory bifurcation,
after which trajectories lie on a 3-torus. Further bifurcations lead
to frequency locking, followed by a period-doubling cascade that leads
to chaos for $D > 3.84$. The chaotic regime apparently persists as
$D \to \infty$ (Jones et al. 1985).

For this sytem it is possible to follow the variation of the
field strength after the basic oscillation has been taken out. Then
the 3-torus that appears after the third bifurcation is reduced to

a 2-torus, as shown in Figure 1 for D = 3.6. The two frequencies
that remain are a fast oscillation introduced at the second bifurcation
and the slow modulation introduced at the third bifurcation. The
former leads to rapid motion round the surface of the torus, accompanied
by a slow drift parallel to its axis. Trajectories move along the
outer surface of the torus (with large amplitude) towards the origin
and then decrease in amplitude before returning through the central
orifice. For this value of D the slow modulation is periodic and
leads to episodes of reduced magnetic activity (corresponding to grand
minima of the solar cycle) that recur at regular intervals. When
D = 4 the torus no longer exists but a "ghost attractor" survives
to produce fairly regular modulation. Figure 2(a) shows how the
amplitude of the basic cycle varies in this chaotic regime. For D
= 8 in Figure 2(b), the "ghost attractor" can still be recognized
but the modulation is aperiodic. Both the amplitude of the cycles
and the intervals between grand minima vary aperiodically, mimicking
the behaviour of the solar cycle.

Similar transitions to chaos can be found in even simpler systems
(Langford 1983). Recurrent grand minima are the signature of systems
where chaos is preceded by quasiperiodic behaviour with a slow
modulating frequency. The sunspot record, as extrapolated through
^{14}C data, suggests that the solar dynamo is an example of this process.
Moreover, the persistence of a "ghost attractor" can explain the
correspondence between different minima in the ^{14}C record, despite
the fact that they recur aperiodically (cf. Stuiver 1980). All these
arguments suggest that stellar magnetic cycles are intrinsically
chaotic but there is another important data set which points to a
different conclusion.

5. QUASIPERIODIC CLIMATIC OSCILLATIONS

The Elatina formation in South Australia was laid down 6.8×10^8 yr
ago. Williams (1981, 1985; Williams & Sonett 1985) recognized glacial
varves in this Precambrian deposit with about 20,000 "annual" layers.
The layer thicknesses vary cyclically and successive cycles are
separated by layers of darker, finer material. The number of layers
in each cycle varies between 10 and 14 with an average of 12, which
immediately suggests control of the deposition rate by climatic
variations modulated by solar activity. Moreover the total thicknesses
of successive cycles also vary, with periods of 2.1, 13.1 and 26.1
times the average cycle period. The last of these are astonishingly
regular, implying that the whole pattern is quasiperiodic, with a
period of about 310 years.

The regular modulation is quite different from the aperiodic
occurrence of Maunder-like minima in the ^{14}C record. If the Elatina
varves provide a fossil record of the solar cycle then the solar
dynamo must have behaved differently in the Precambrian era (Weiss,
1987). There are, however other difficulties. It seems more likely
that the period of the solar cycle was shorter, rather than longer,
7×10^8 yr ago (Noyes et al. 1984) and, more important, there is

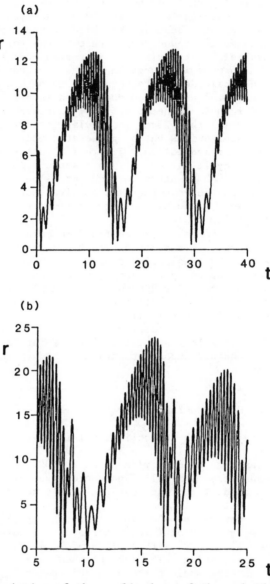

Fig. 2. Variation of the amplitude r of the poloidal field, such
that $A = re^{i\theta}$, with time for dynamos in the chaotic regime
with (a) D = 4 and (b) D = 8. Note the persistence of
a "ghost attractor" and the appearance of marked
aperiodicity as D is increased.

no clear physical mechanism by which solar activity could produce such significant climatic oscillations. In recent times there has been no convincing evidence of long-term correlations between ^{14}C anomalies and climatic variations. Unfortunately, it therefore seems unlikely that the Elatina varves contain information about the solar cycle, though they remain of great intrinsic interest and provide a record that should be analysed in detail. Further research may reveal other mechanisms that can lead to cyclic variations of this kind.

6. CONCLUSION

I have argued that the available evidence is consistent with the hypothesis that the aperiodic behaviour of the solar cycle is an example of deterministic chaos and that such behaviour is indeed to be expected from nonlinear dynamos. Moreover, the slow modulation that leads to grand minima resembles the pattern to be expected after a transition from quasiperiodic behavour (with trajectories that move slowly round a torus) to chaos. A specific prediction that follows from this picture is that the phase of the cycle should not be preserved and, in particular, that it should be lost during the passage through a grand minimum. Gough (1987) has considered the phase of the activity cycle since 1713 and finds that it is marginally more consistent with random drift than with a distorted periodic signal. Prior to the Maunder minimum the only sunspot minimum whose date can be reliably determined is that of 1619, which is out of phase with the minima of the eighteenth century. Wittmann, in these Proceedings, reports that the early Chinese sunspot observations are not clustered round the maxima of an extrapolated 11-year cycle. So the evidence available at present does not support the assumption that phase is preserved. Beer has reported that the 11-year activity cycle can be detected from variations of ^{10}Be abundance in the Milcent ice-core from Greenland, which also shows the Maunder minimum. It is important that this core should be measured carefully, for it can finally settle the issue of whether the phase of the solar cycle is preserved or not.
   Finally, I should like to emphasize the distinction between quasiperiodic and chaotic behaviour. Much experimental work has been dominated by an obsessive search for periodicity in limited data-sets. Complicated nonlinear systems like the terrestrial climate or the solar dynamo are, however, likely to be chaotic. Power spectra may still yield peaks of finite width but sensitive dependence on initial conditions ensures that events will not repeat indefinitely at regular intervals. Chaotic behaviour is richer and more interesting.

REFERENCES

Allan, D.W., 1962, Proc. Camb. Phil. Soc. 58, 671.
Baliunas, S.L. and Vaughan, A.H., 1985, Ann. Rev. Astron. Astrophys. 23, 3.
Barnes, J.L., Sargent, H.H. and Tryon, P.V., 1980, in 'The Ancient Sun', ed. R.O. Pepin, J.A. Eddy and R.B. Merrill, p. 159, Pergamon, New York.
Bracewell, R.J., 1985, Aust. J. Phys. 38, 1009.
Cook, A.E. and Roberts, P.H., 1970, Proc. Camb. Phil. Soc. 68. 547.
Eddy, J.A., 1976, Science 192, 1189.
Gilman, P.A., 1983, Astrophys. J. Suppl. Ser. 53, 243.
Glatzmaier, G.A., 1985, Astrophys. J. 291, 300.
Gough, D.O., 1987, Proc. Int. Sch. Phys. E. Fermi (Varenna), 95 (in press).
Jones, C.A., 1983, in 'Stellar and Planetary Magnetism', ed. A.M. Soward, p. 159, Gordon & Breach, London.
Jones, C.A., Weiss, N.O. and Cattaneo, F., 1985, Physica 14D, 161.
Krause, F. and Roberts, P.H., 1981, Adv. Space Res. 1, 231.
LaBonte, B.J. and Howard, R.F., 1982, Solar Phys. 75, 161.
Langford, W.F., 1983, in 'Nonlinear Dynamics and Turbulence', ed. G.I. Barenblatt, G. Looss and D.D. Joseph, p. 215, Pitman, London.
Noyes, R.W., Weiss, N.O. and Vaughan, A.H., 1984, Astrophys. J. 287, 769.
Parker, E.N., 1979, 'Cosmical Magnetic Fields', Clarendon Press, Oxford.
Spiegel, E.A., 1985, in 'Chaos in Astrophysics', ed. J.R. Buchler, J. Perdang and E.A. Spiegel, p. 91, Reidel, Dordrecht.
Spiegel, E.A. and Wolf, A., 1987, Ann.N.Y.Acad.Sci. 497 (in press).
Spörer, G., 1889, Nova Acta Ksl. Leop.-Carol. Deutschen Akad. Naturf. 53, 272.
Stuiver, M., 1980, Nature 286, 868.
Stuiver, M. and Grootes, P.M., 1980, in 'The Ancient Sun', ed. R.O. Pepin, J.A. Eddy and R.B. Merrill, p. 165, Pergamon, New York.
Stuiver, M., Pearson, G.W, and Braziunas, T., 1986, Radiocarbon 28, 980.
Stuiver, M. and Quary, P.D., 1980, Science 207, 11.
Weiss, N.O., 1987, in 'Physical Processes in Comets, Stars and Active Galaxies', ed. W. Hillebrandt, E. Meyer-Hofmeister and H.-C. Thomas, p.46, Springer, Berlin.
Weiss, N.O., Cattaneo, F. and Jones, C.A., 1984, Geophys. Astrophys. Fluid Dyn. 30, 305.
Williams, G.E., 1981, Nature 291, 624.
Williams, G.E., 1985, Aust. J. Phys. 38, 1027.
Williams, G.E. and Sonett, C.P., 1985, Nature 318, 523.
Yoshimura, H., 1978, Astrophys. J. 226, 706.
Yule, G.U., 1927, Phil. Trans. R. Soc. A 226, 267.
Zel'dovich, Ya.B., Ruzmaikin, A.A. and Sokoloff, D.D., 1983, 'Magnetic Fields in Astrophysics', Gordon & Breach, London.

FAST DYNAMOS WITH FLUX EXPULSION

A.M. Soward,
School of Mathematics,
The University,
Newcastle upon Tyne,
NE1 7RU, U.K.

ABSTRACT. Many of the important advances in dynamo theory have depended on results from meanfield magnetohydrodynamics. In particular, it is argued that small scale turbulence leads to an α-effect from which mean electromotive forces may be produced parallel to the large scale mean magnetic field. It is well known that, in conjunction with differential rotation, the ω-effect, oscillatory $\alpha\omega$-dynamos may occur which can possibly account for solar and stellar activity cycles. The theory upon which the magnitude of the α-effect is based is only robust when the microscale magnetic Reynolds number is small. In most stellar applications that approximation is inappropriate and alternative theories for the α-effect are required. Here recent developments will be described which consider the possibility that even on the small scale flux expulsion occurs and that the magnetic field is confined to ropes and sheets.

1. FAST AND SLOW DYNAMOS

Dynamo theory is concerned with the generation and maintenance of magnetic field, $\underset{\sim}{B}$, by the motion of electrically conducting fluid. It is central to our understanding of astrophysical and geophysical magnetohydro-dynamics. At the simplest conceptual level of kinematic theory we are interested in solutions of the magnetic induction equation,

$$\frac{\partial \underset{\sim}{B}}{\partial t} = \underset{\sim}{\nabla} \times (\underset{\sim}{u} \times \underset{\sim}{B}) + \eta \nabla^2 \underset{\sim}{B} \ , \quad (\underset{\sim}{\nabla} \cdot \underset{\sim}{B} = 0) \ , \qquad (1.1a,b)$$

in which $\underset{\sim}{u}$ (given) is the fluid velocity and η is the magnetic diffusivity. If U is a typical velocity and L is the length scale of the system, we can introduce the convective and diffusive time scales,

$$T_C = L/U \quad \text{and} \quad T_D = L^2/\eta, \qquad (1.2a,b)$$

F. R. Stephenson and A. W. Wolfendale (eds.),
Secular Solar and Geomagnetic Variations in the Last 10,000 Years, 79–96.
© *1988 by Kluwer Academic Publishers.*

respectively. The relative importance of convection and diffusion
of the magnetic field can then be measured by the magnetic Reynolds
number,

$$R = T_D/T_C = LU/\eta. \tag{1.3}$$

The system is said to operate as a dynamo if, in the absence of any
external electric current sources, the magnetic field can be maintained
over a time scale long compared to the magnetic diffusion time, T_D.

For sufficiently complicated flow we expect a dynamo to operate
when advection of the magnetic field overcomes the inevitable ohmic
losses. That should happen for moderate values of the magnetic
Reynolds number when the convection time, T_C, is shorter than the
diffusion time, T_D. In the case of steady motion,

$$\underset{\sim}{u} = \underset{\sim}{u}(\underset{\sim}{x}) \, , \tag{1.4}$$

which depends only on the position, $\underset{\sim}{x}$, and not on the time, t, the
idea can be made more precise by seeking solutions of the magnetic
induction equation (1.1) in the form,

$$\underset{\sim}{B}(\underset{\sim}{x},t) = \underset{\sim}{\hat{B}}(\underset{\sim}{x})e^{pt} \, , \tag{1.5}$$

where p (generally complex) is the growth rate $(\partial \underset{\sim}{B}/\partial t = p\underset{\sim}{B})$. When
there is no motion (R=0) the magnetic field decays and the eigenvalue,
p, of our linear problem has negative real part. As the magnitude
of the motion (or equivalently R) is increased, the decay rate
decreases until at some value of R we find that Re p vanishes. That
critical value of R yields a dynamo and any further increase leads
to a growing magnetic field (Re p > 0).

In astrophysical plasmas the magnetic Reynolds number is generally
very large,

$$R \gg 1 \, , \tag{1.6}$$

and the time scales separate, $T_D \gg T_C$. Since magnetic field is
magnified by convection, it cannot grow on a time scale shorter than
the convective time scale, T_C; but, can it grow that fast? In the case
of steady motion, to which we restrict attention throughout this
paper, the question can be put another way. Do "Fast" dynamos exist
with the property,

$$pT_C = 0(1) \quad \text{as} \quad R \to \infty, \tag{1.7a}$$

or are all dynamos "Slow", meaning that magnetic field amplification
occurs but not on the convective time scale i.e.

$$pT_C \downarrow 0 \quad \text{as} \quad R \to \infty ? \tag{1.7b}$$

The question whether or not fast dynamos exist focuses our
attention on the key issues of large magnetic Reynolds number dynamo

theory. In particular we must consider in detail the mechanisms by which such dynamos operate. The question has led to much controversy (see, for example, Zel'dovich, Ruzmaikin and Sokoloff, 1983) largely because it is extremely difficult to solve the magnetic induction equation (1.1), when R >> 1, for flows sufficiently complicated to produce dynamos. The matters raised here may seem esoteric but they do have important implications, particularly in meanfield magneto-hydrodynamics. There it is argued (see, for example, Krause and Radler, 1980) that small scale turbulence can produce an α-effect. The combination of the α-effect and differential rotation (ω-effect) leads to the idea of an αω-dynamo which is often used to account for solar and stellar activity cycles. As we explain in Section 3 below, the calculation of the magnitude of α is a delicate matter, when the magnetic Reynolds numbers based on the short turbulent scales is large, and is linked to the "Fast" dynamo question, which we have raised. In the following sections we isolate the key ideas by a number of illustrative examples. Since much of the material covered has been discussed recently by Soward and Childress (1986) and Fearn, Roberts and Soward (1987), both of which contain extensive lists of references, we will refer here mainly to standard textbooks and sparingly to original articles.

2. BASIC MECHANISMS

2.1 Perfectly conducting fluid, $\eta = 0$

When the fluid has perfect electrical conductivity, $\eta = 0$, magnetic field is frozen to the fluid and moves with it. Provided the trajectories of fluid particles are known, the evolution of an arbitrary initial magnetic field is readily obtained. Further insight can be obtained by expressing the magnetic induction equation in the form,

$$\frac{D\underset{\sim}{B}}{Dt} \equiv \frac{\partial \underset{\sim}{B}}{\partial t} + \underset{\sim}{u} \cdot \nabla \underset{\sim}{B} = \underset{\sim}{B} \cdot \nabla \underset{\sim}{u} \, , \tag{2.1}$$

appropriate for an incompressible fluid ($\nabla \cdot \underset{\sim}{u} = 0$). It says that the rate of change of $\underset{\sim}{B}$ following the motion of fluid particles, $D\underset{\sim}{B}/Dt$, is proportional to the strain, $\nabla \underset{\sim}{u}$, which measures the relative motion of neighbouring fluid particles,

$$D(\underset{\sim}{x}_2 - \underset{\sim}{x}_1)/Dt = \underset{\sim}{u}(\underset{\sim}{x}_2) - \underset{\sim}{u}(\underset{\sim}{x}_1) \sim (\underset{\sim}{x}_2 - \underset{\sim}{x}_1) \cdot \nabla \underset{\sim}{u} \, , \tag{2.2}$$

Indeed, the notion of material magnetic field lines follows by comparison of (2.1) and (2.2) with the neighbouring points, $\underset{\sim}{x}_2$ and $\underset{\sim}{x}_1$, lying on the same field line. To investigate the role of source term, $\underset{\sim}{B} \cdot \nabla \underset{\sim}{u}$, on the right of (2.1), it is sufficient to consider the case of constant $\nabla \underset{\sim}{u}$. In the following subsections that case is investigated together with the added influence of small but finite magnetic diffusivity.

2.2 Pure straining motion

We consider the steady two-dimensional stagnation point flow which relative to rectangular cartesian coordinates (x,y) is given by

$$\underset{\sim}{u} = (x,-y)/T_C , \qquad (2.3)$$

where the constant, T_C, is the convection time scale. The flow stretches magnetic field lines parallel to the x-axis and in particular the uniform magnetic field,

$$\underset{\sim}{B} = (B_C(t),0) , \qquad (2.4a)$$

is intensified by the flow (2.3) as illustrated in Figure 1. In terms of the particle paths,

$$\underset{\sim}{x}(t) = [x(0)e^{t/T_C}, y(0)e^{-t/T_C}] , \qquad (2.4b)$$

it is given by

$$B_C(t)/B_C(0) = x(t)/x(0) = e^{t/T_C} . \qquad (2.4c)$$

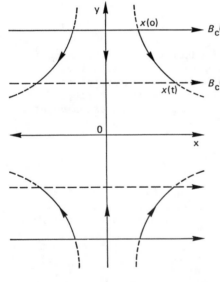

Figure 1. The stagnation point flow (2.3). The horizontal solid lines and long broken lines denote the uniform magnetic field initially and after time, t, respectively.

When the magnetic field is non-uniform, the frozen field picture though helpful must be modified to take into account the effect of finite magnetic diffusivity. So, for example, if the uniform magnetic field (2.4a) is replaced by

$$\underset{\sim}{B}(y,t) = [B(t)\sin\{y/\ell(t)\},0] , \qquad (2.5a)$$

its oscillatory profile evolves as though frozen to the fluid, where by (2.4b)

$$\ell(t) = \ell(0)e^{-t/T_C} , \qquad (2.5b)$$

while ohmic diffusion reduces its amplitude from its frozen field value (2.4c) and yields

$$B(t) = B_C(t)\exp\left[-\frac{1}{2}\left[\frac{T_C}{T_D(t)} - \frac{T_C}{T_D(0)}\right]\right] , \qquad (2.5c)$$

where

$$T_D(t) = \ell^2(t)/\eta \qquad (2.5d)$$

is the "local" magnetic diffusion time scale. Note that (2.5) provides an exact solution of (2.14) below. Clearly, considerable amplification of the magnetic field is possible when the magnetic Reynolds number,

$$R = T_D(0)/T_C , \qquad (2.6)$$

based on the initial conditions is large. The maximum amplitude,

$$B(t) \sim B(0)R^{\frac{1}{2}} , \qquad (2.7a)$$

is achieved when the local diffusion time equals the convection time $(T_D(t) = T_C)$ or equivalently,

$$t = \frac{1}{2} T_C \ell n\ R. \qquad (2.7b)$$

Subsequently, the ratio, $T_C/T_D(t)$, grows exponentially and the magnetic field is rapidly annihilated.

The results (2.5) are interesting because they show how structures frozen to the fluid can have their length scales reduced indefinitely, typically at an ever increasing rate. They say also that, though considerable field amplification may be expected in a highly conducting fluid, the length scales are broken down until eventually the diffusive scales are reached. Once the "local" magnetic Reynolds number, $T_D(t)/T_C$, becomes small, magnetic field is destroyed rapidly. The process is analogous to energy cascade in a turbulent viscous fluid.

2.3 Linear Shear

For comparison we mention briefly the special case of the simple shear flow defined by

$$\underset{\sim}{u} = (y,0)/T_C , \qquad (2.8)$$

for which fluid particles initially on the line, x = 0, move to

$$\underset{\sim}{x}(t) = [(t/T_C)y(0), y(0)], \quad (x(0) = 0) . \tag{2.9}$$

If the initial magnetic field is uniform and aligned with the y-axis, $\underset{\sim}{B}_C(0) = (0, B_C(0))$, it remains frozen to the fluid particles and increases at a linear rate,

$$\frac{\underset{\sim}{B}_C(t)}{B_C(0)} = \frac{\underset{\sim}{x}(t)}{y(0)} = (\frac{t}{T_C}, 1) , \tag{2.10}$$

as illustrated in Figure 2. The mechanism described is the basis of the ω–effect of dynamo theory mentioned in Section 1, whereby a star's poloidal magnetic field is stretched by differential rotation to induce a strong toroidal field.

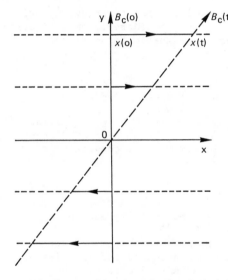

Figure 2. The shear flow (2.8). The solid Oy-axis is an initial magnetic field line. The long broken line is that magnetic field line after time, t.

If the amplitude of the initial magnetic field varies spatially and is given, for example, by

$$\underset{\sim}{B}(x, 0) = \underset{\sim}{B}_C(0) \sin kx, \tag{2.11}$$

then as in Section 2.2 the structure is frozen to the fluid,

$$\underset{\sim}{B}(x, t) = \underset{\sim}{B}(t) \sin k[x - (t/T_C)y] , \tag{2.12a}$$

but the amplitude is reduced from its frozen field value (2.10) and is given by

$$\underset{\sim}{B}(t) = \underset{\sim}{B}_C(t) \exp \{ - \int_0^t \frac{dt}{T_D(t)} \} \tag{2.12b}$$

instead, where $T_D(t)$ is the "local" magnetic diffusion time scale defined by

$$1/T_D(t) = \eta \, k^2 \, \{1 + (t/T_C)^2\} \, . \tag{2.12c}$$

When the magnetic Reynolds number, $R = T_D(0)/T_C$, is large, the magnetic field (2.12c) attains its maximum amplitude,

$$|\underset{\sim}{B}(x,t)| \sim R^{1/3} B_C(0) \, , \tag{2.13a}$$

when the time is the same as the local diffusion time $(t = T_D(t))$, or equivalently

$$t = T_C R^{1/3} \, . \tag{2.13b}$$

2.4 Flux ropes and sheets

The magnetic field annihilation outlined in Sections 2.2 and 2.3 above occurs because the mean value of the magnetic field vanishes. When that mean value is non-zero the total magnetic flux is conserved despite the diffusive processes. So, for example, in the case of the pure straining motion (2.3) the non-uniform but unidirectional magnetic field,

$$\underset{\sim}{B} = [B(y,t),0] \, , \tag{2.14a}$$

evolves according to the equation,

$$\frac{\partial B}{\partial t} = \frac{\partial}{\partial y} \, [\frac{y}{T_C} B] + \eta \frac{\partial^2 B}{\partial y^2} \, , \tag{2.14b}$$

with the total magnetic flux,

$$F = \int_{-\infty}^{\infty} B(y,t) dy \, , \tag{2.14c}$$

remaining constant. To illustrate the effect on a localised magnetic field, we note that Gaussian profiles preserve their shape,

$$B(y,t) = \{F/\pi^{\frac{1}{2}} \ell(t)\} \exp \{-[y/\ell(t)]^2\} \, , \tag{2.15a}$$

where the sheet width, $\ell(t)$, satisfies

$$\ell^2(t) = (\ell^2(0) - \ell_D^2) e^{-2t/T_C} + \ell_D^2 \tag{2.15b}$$

and

$$\ell_D = (2\eta T_C)^{\frac{1}{2}} \tag{2.15c}$$

is the diffusion length scale. Whatever the initial width, $\ell(0)$, the flux sheet finally achieves a steady state configuration of width, ℓ_D, in which inward convection balances outward diffusion.

The case of converging axisymmetric flow is very similar. If we take the stagnation point flow, which relative to cylindrical polar coordinates (s,ϕ,z) is given by

$$\underset{\sim}{u} = (-s,0,2z)/T_C \; , \tag{2.16}$$

then the unidirectional axisymmetric magnetic field,

$$\underset{\sim}{B} = (0,0,B(s,t)) \; , \tag{2.17a}$$

aligned with the symmetry axis satisfies

$$\frac{\partial B}{\partial t} = \frac{1}{s}\frac{\partial}{\partial s}[\frac{s^2}{T_C}B] + \frac{\eta}{s}\frac{\partial}{\partial s}[s\frac{\partial B}{\partial s}] \; . \tag{2.17b}$$

The total magnetic flux,

$$F = 2\pi \int_0^\infty s\, B(s,t) ds \; , \tag{2.17c}$$

is conserved and we find as before that the Gaussian profile of flux ropes is preserved,

$$B(s,t) = \{F/\pi \ell^2(t)\}\exp\{-[s/\ell(t)]^2\} \; , \tag{2.18}$$

where $\ell(t)$ is still given by (2.15b,c).

The analogy is often stressed between vorticity dynamics in a viscous fluid and the evolution of magnetic field satisfying the induction equation (1.1). In the vorticity context the final steady state solutions, (2.15) and (2.18) with $\ell(t) = \ell_D$, are well known and are discussed, for example, by Batchelor (1967, pp. 271-272).

3. THE ALPHA EFFECT

One of the main ideas, which emerges from meanfield magnetohydrodynamics, is the possibility that turbulence on a short length scale, ℓ, can regenerate a mean magnetic field, $\bar{B}$, on the long length scale, L, of the system. The analysis of the problem proceeds by decomposing the flow velocity, $\underset{\sim}{u}$, and the magnetic field, $\underset{\sim}{B}$, into their mean and fluctuating parts,

$$\underset{\sim}{u} = \bar{\underset{\sim}{u}} + \underset{\sim}{u}' \; , \qquad \underset{\sim}{B} = \bar{\underset{\sim}{B}} + \underset{\sim}{B}' \; . \tag{3.1}$$

The mean magnetic field which varies on the long length scale, L, then satisfies

$$\frac{\partial \bar{B}}{\partial t} = \nabla \times (\bar{u} \times \bar{B} + \bar{e}) + \eta \nabla^2 \bar{B} \, , \qquad (3.2a)$$

where

$$\bar{e} = \overline{u' \times B'} \, . \qquad (3.2b)$$

Upon subtracting (3.2a) from (1.1a) we obtain an equation for B'. The solution of that equation, which with (3.2b) in turn determines $\bar{e}$, provides the main obstacle of the theory.

On the short length scale, ℓ, of the turbulence, the mean field is almost uniform and so it is argued that B' and hence $\bar{e}$ are related linearly to $\bar{B}$. When higher order effects are considered, the spatial variation of $\bar{B}$ must be included and such considerations lead to the expansion,

$$\bar{e}_i = \alpha_{ij}\bar{B}_j + \beta_{ijk} \, \partial\bar{B}_j/\partial x_k + \dots \, , \qquad (3.3)$$

for the components of $\bar{e}$. The leading term, $\alpha_{ij}\bar{B}_j$, is called the α-effect. For isotropic turbulence, we have

$$\alpha_{ij} = \alpha\,\delta_{ij} \, , \quad \beta_{ijk} = \beta\varepsilon_{ijk} \, , \qquad (3.4)$$

in which case β can be identified with a turbulent diffusivity. The special feature of α predicted by theory is that it is generally proportional to the helicity,

$$\overline{u' \cdot \omega'} \, , \qquad (3.5)$$

of the fluctuating motions, where $\omega' = \nabla \times u'$ is the vorticity. Fortunately, helical turbulence is to be expected in rotating systems, and, moreover, it is often argued that the magnitude of α in a star is proportional to its rotation.

In general, the mathematical derivation of α_{ij} in (3.3) is robust only when the microscale magnetic Reynolds number,

$$R_M = u'\ell/\eta \, , \qquad (3.6)$$

is small. Then the fluctuating field, B', is small compared to the mean field, $\bar{B}$, and the perturbation analysis upon which the calculation is based is justified. When, on the other hand,

$$R_M \gg 1, \qquad (3.7)$$

the fluctuating field, B', is no longer small and the standard perturbation methods described, for example, by Krause and Radler (1980), Moffatt (1978) and Parker (1979) encounter servere difficulties. An alternative approach to the large R_M case has been initiated by Childress (1979). He recognised that the fluctuating fields would be dominant, $|B'| \gg |\bar{B}|$, and that for steady flows most of the magnetic

field is likely to be concentrated into flux ropes and sheets.
Nevertheless, the meanfield equations (3.2) remain valid and an α-effect
is calculated. These ideas are explained in the final Section 4 which
follows below.

4. FAST DYNAMOS

4.1 The Vainstein-Zel'dovich rope dynamo

The possibility of "Fast" dynamos is not a straightforward matter and
the cases of perfect electrical conductivity, $\eta = 0$, and large magnetic
Reynolds number, $R \gg 1$, must be carefully distinguished. In Section
2.1 we explained that in a perfectly conducting fluid the amplification
of magnetic field was a local property. The results of Section 2.2
showed that persistent straining of a fluid element, which is generic
for a fully three-dimensional motion, leads to exponential growth on
the convective time scale. According to (1.7a), therefore, it has
the "fast" dynamo property. Nevertheless, the results of that section
also illustrated that, if the length scales are broken down, fields
with zero mean are ultimately destroyed by diffusion.
      A simple model, which appears to satisfy our criterion for a "Fast"
dynamo in the case of small but finite electrical conductivity,
$R \gg 1$, is the rope dynamo of Vainstein and Zel'dovich (1972). It
is a refinement of an earlier model by Alfven (1950). They consider
a loop of magnetic flux, which by analogy is to be thought of as an
elastic band. Quite simply the band is stretched to twice its length,
one half is then twisted and folded to form a doubled up band of the
original length, as illustrated in Figure 3. The effect is to form
a flux rope with some fine structure but of double the original strength.
If the process is repeated at regular intervals it is magnified after
N operations by a factor 2^N, i.e. exponential growth. Evidently, as
a result of the twisting and folding, a considerable amount of fine

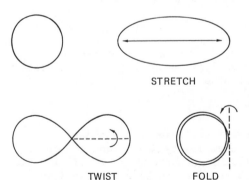

STRETCH

TWIST   FOLD

Figure 3. The sequence of
motions in the stretch-twist-
fold dynamo.

structure develops in the rope. Nevertheless, locally the mean magnetic field on a cross section is non-zero and, as the results of Section 2.4 indicate, such flux should be preserved. A critical discussion of the details of the model has been given by Moffatt and Proctor (1985).

4.2 The G.O.Roberts dynamo

The effect of fully three-dimensional flow on magnetic field is difficult to analyse in detail, especially as in the rope dynamo just mentioned the motion is also time dependent. A simple dynamo proposed by Roberts (1972), which avoids these complexities, invokes steady spatially periodic two-dimensional flow. Relative to rectangular cartesian coordinates (x,y,z), the motion is independent of z and lies on stream surfaces, $\psi(x,y) = $ constant. The flow velocity is

$$\underset{\sim}{u}(x,y) = (\partial\psi/\partial y, -\partial\psi/\partial x, U) \qquad (4.1)$$

For his particular model, the motion illustrated in Figure 4a has

$$\psi = (L^2/T_c)\sin(x/L)\sin(y/L) , \qquad (4.2a)$$

with U constant on stream surfaces and antisymmetric in ψ,

$$U(-\psi) = -U(\psi) , \quad (U(0) = 0) . \qquad (4.2b)$$

Unlike fully three-dimensional flow, neighbouring fluid particles (except on $\psi = 0$) separate on average at a linear rate just like the linear shear of Section 2.3. Consequently, when the fluid is perfectly conducting only the stagnation points have the "Fast" dynamo property!

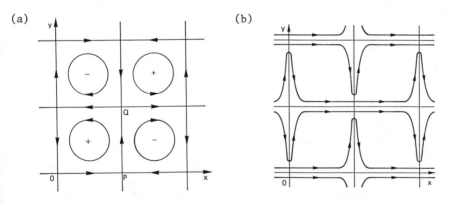

Figure 4(a) The streamlines for the G.O. Roberts motion (4.1). The sign of the vertical motion in each of the cells is indicated.
(b) The mean magnetic field, $\underset{\sim}{\bar{B}} = (\bar{B},0)$, expelled by the motion into sheets.

When the magnetic Reynolds number is large but finite, Childress (1979) addressed the problem in the following way. He considered the effect of the motion on a uniform mean magnetic field orientated in (say) the x-direction. The motion expels the flux from the centre of the eddies (Weiss, 1966) and confines it to sheets on the stream surfaces, $\psi = 0$, connecting the stagnation points, as illustrated in Figure 4b. The dynamo properties of the flow are determined by the local behaviour close to $\psi = 0$ between, for example, the origin 0:(0,0) and P:(Lπ,0). The motion defines a magnetic Reynolds number,

$$R = \eta^{-1} \int_0^P (\partial\psi/\partial y)\,dx = \eta^{-1} \int_{-\infty}^{\infty} |\underset{\sim}{u}|^2 dt , \qquad (4.3a)$$

and total helicity,

$$H = \int_0^P (\partial U/\partial y)\,dx = \int_{-\infty}^{\infty} \underset{\sim}{u}\cdot\underset{\sim}{\omega}\, dt , \qquad (4.3b)$$

where the integrands of the time-integrals are evaluated at points following the motion of fluid particles moving from 0 to P.

It is a relatively straightforward matter to calculate the boundary layer structure of the magnetic field. The results can be used to calculate the mean electromotive force,

$$\underset{\sim}{\bar{e}} = \overline{\underset{\sim}{u} \times \underset{\sim}{B}} , \qquad (4.4a)$$

averaged over a cell $[0,Lπ] \times [0,Lπ]$ (say). The non-vanishing contribution originates from the tongues of flux like PQ of Figure 4. It is

$$\underset{\sim}{\bar{e}} = \alpha\underset{\sim}{\bar{B}} , \qquad (4.4b)$$

where

$$\alpha = - \text{CONSTANT} \times R^{-\frac{1}{2}} H . \qquad (4.4c)$$

The total helicity, H, has the dimensions of velocity and the constant is independent of both η and the magnitude, L/T_C, of the velocity. By symmetry the result (4.4) also holds if $\underset{\sim}{\bar{B}}$ is directed in the y-direction and so it is valid for any mean magnetic field, $\underset{\sim}{\bar{B}}_H$, lying in the horizontal xy-plane. If, like Roberts (1972) we seek solutions to the mean field dynamo equation (3.2a) with $\underset{\sim}{\bar{u}} = 0$ of the form,

$$\underset{\sim}{\bar{B}}_H = B_0 e^{pt}(\cos \beta z, \sin \beta z, 0) , \qquad (4.5a)$$

then we find that the growth rate is

$$p = -\alpha\beta - \eta\beta^2 . \qquad (4.5b)$$

It is maximised when

$$\beta = \beta_{max} = -\alpha/2\eta \qquad (4.6a)$$

and magnetic field grows on the time scale,

$$T_{max} = 1/p = 4\eta/\alpha^2 , \qquad (4.6b)$$

which by (4.4c) is independent of the diffusivity.

The dynamo mechanisms have been discussed recently in detail by Soward (1987). Though the result (4.6b) has the "Fast" dynamo property, it has been applied outside the limits of validity of the Childress (1979) theory and so must be interpreted with caution. The central difficulty is the vertical length scale, $|1/\beta_{max}|$, of the fastest growing mode, which is comparable in magnitude with the boundary layer thickness, ℓ_D, (see 2.15c)) of the flux sheets. On such short vertical length scales the assumptions upon which the result (4.4c) depend begin to fail. Soward (1987) shows that (4.6b) underestimates T_{max} and that it is not strictly a "Fast" dynamo. Nevertheless, the distinction is very fine and in spirit the results (4.5) and (4.6) are correct. The realised magnetic field configuration resembles a stack of horizontal flux ropes whose direction varies with height, z. Like the Vainstein-Zel'dovich rope dynamo, magnetic field is stretched by the horizontal motion (see Figure 4b) and twisted by the vertical motion. Both the fold operation and the temporal sequence are, however, unnecessary because of the spatial periodicity of the flow.

4.3 The Childress dynamo mechanism

Childress (1979) extended the idea developed in the previous section to axisymmetric flows. By analogy with (4.1) he considered the motion, which relative to cylindrical polar coordinates (see Section 2.4), is given by

$$s\underset{\sim}{u} = (-\partial\psi/\partial z, h(\psi), \partial\psi/\partial s), \qquad (\psi = \psi(s,z)) \qquad (4.7a)$$

and represented the axisymmetric magnetic field in the form,

$$\underset{\sim}{B} = [-\frac{1}{s}\frac{\partial\chi}{\partial z}, sK, \frac{1}{s}\frac{\partial\chi}{\partial s}] . \qquad (4.7b)$$

The flow is assumed to be periodic in the axial z-direction with

$$\psi(s,2(n+1)L+z) = \psi(s,2nL+z) = -\psi(s,2nL-z), \quad (0\leq z\leq L), \quad (4.8a)$$

$$h(-\psi) = -h(\psi) , \qquad (4.8b)$$

for all integer n, while both χ and K satisfy

$$\chi(s,2(n+1)L+z) = \chi(s,2nL+z) = \chi(s,2nL-z) \qquad (4.8c)$$

The fluid is confined by a perfectly conducting boundary at s = d and a mean axial magnetic field, $\bar{B}_z$, permeates the fluid giving the constant total flux,

$$F = \pi d^2 \bar{B}_z .$$
(4.8d)

(a)   (b)   (c)

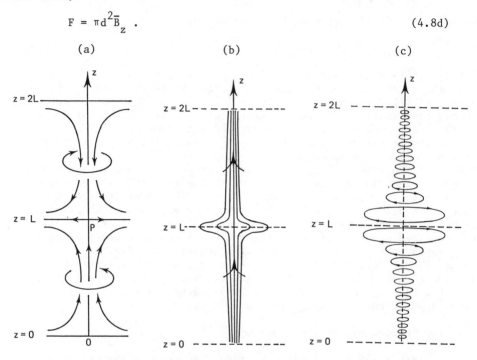

Figure 5(a)   The axisymmetric flow (4.7a). The stream surfaces, ψ = constant, and the direction of the swirl velocity, h/s, are indicated.
        (b)   The axial magnetic flux rope (4.11). Lines of χ = constant are shown.
        (c)   The azimuthal magnetic field, sK.

    As in the G.O. Roberts two-dimensional model the mean magnetic field, $\bar{B}_z$, is expelled from the middle of the eddies. It is confined to a rope close to the axis and sheets on the stream surface, ψ = 0, connected to the axis at the stagnation points: that includes the planes, z = nL (integer n), and the cylinder s = d. The key to the dynamo calculation is the nature of the axial rope which contains almost all the magnetic flux. For definiteness, we consider vertical motion on the axis, s = 0,

$$w(z) = s^{-1} \partial\psi/\partial s ,$$
(4.9)

which is upward (w > 0) between the stagnation points, 0 (z = 0) and P(z = L), as illustrated in Figure 5a. In terms of the coordinates (ψ,z), the magnetic flux in the axial rope is governed by the boundary layer equation,

$$\frac{\partial \chi}{\partial z} = 2\eta\psi \frac{\partial^2 \chi}{\partial \psi^2} \quad , \quad (\psi \sim \tfrac{1}{2}s^2 w) \quad . \tag{4.10}$$

The solution downstream of the stagnation point, 0, is

$$\chi = \tfrac{1}{2}d^2\bar{B}_z(1-e^{-\zeta}) \quad , \quad (0 \leq z \leq L), \tag{4.11a}$$

where ζ is the similarity variable,

$$\zeta = \psi/2\eta z \quad . \tag{4.11b}$$

Near 0, where

$$\psi \sim s^2 z/T_C \quad , \tag{4.12}$$

the solution merges with the steady state solution (2.18) with $\ell(t) = \ell_D$. Our new solution (4.11) indicates that the rope thickens downstream of the stagnation point as indicated in Figure 5b.

The azimuthal swirling motion, $s^{-1}h(\psi)$, which has no singularity on the axis ($h(0) = 0$), causes the flux rope to wind up and generate azimuthal magnetic field, sK, as illustrated in Figure 5c, where $K(s,z)$ satisfies the boundary layer equation,

$$\frac{\partial K}{\partial z} = \frac{2\eta}{\psi} \frac{\partial}{\partial \psi} [\psi^2 \frac{\partial K}{\partial \psi}] + \tfrac{1}{2}h'(0) \frac{dw}{dz} \frac{\partial \chi}{\partial \psi} \quad , \quad (h'(\psi) = \frac{dh}{d\psi}) \tag{4.13}$$

The azimuthal magnetic field generated in this way downstream of the stagnation point, 0, is

$$K = \frac{h'(0)}{2z} \left(\int_0^z z \frac{dw}{dz} dz \right) \frac{\partial \chi}{\partial \psi} \quad , \tag{4.14}$$

where χ is given by (4.11). When it emerges below the streamsurface, $z = L$, it initiates the radial flux,

$$\int_{-\infty}^0 K(2\pi su_s) dz = 2\pi \int_0^\infty K d\psi, \tag{4.15a}$$

in the horizontal boundary layer, $0 < s \leq d$, $z = L$, where near the stagnation point $(0,L)$

$$K = -\frac{1}{2L} H \frac{\partial \chi}{\partial \psi} \tag{4.15b}$$

and

$$H = \int_0^P h'(0)wdz \equiv \int_{-\infty}^{\infty} \underset{\sim}{u} \cdot \underset{\sim}{\omega} \, dt \qquad (4.16)$$

is the total helicity defined as in (4.3b). The boundary layer equations near the plane, $z = L$, and the symmetry condition, $\partial K/\partial\psi = 0$ on $z = L$, imply that the flux, $\int_0^{\infty} Kd\psi$, is conserved. Consequently, the mean electromotive force, averaged on the cylinder, $0 \leq z \leq L$, $0 \leq s \leq d$, is readily obtained. It is

$$\underset{\sim}{\bar{e}} = \overline{\underset{\sim}{u} \times \underset{\sim}{B}} = (1/L) \int_0^{\infty} Kd\psi = -(d/2L)^2 H \underset{\sim}{\bar{B}} , \qquad (4.17)$$

where $\underset{\sim}{\bar{B}} = (0,0,\bar{B}_z)$. There are additional contributions to $\underset{\sim}{\bar{e}}$ from the planes $z = 0$ and $z = L$, of equal but opposite signs. They cancel on averaging.

   As Childress (1979) explains, the complete boundary layer structure is complicated. The central rope is, however, intrinsically different from the surface boundary layers. All boundary layers, both line and surface, are of thickness, ℓ_D, but because of the geometry the axial boundary layer carries less fluid flux. The recirculating fluid from the surface layers is confined to a thick axial boundary layer whose width is of order $R^{\frac{1}{4}}\ell_D$, where $\ell_D = O(R^{-\frac{1}{2}}L)$. Diffusion on this relatively long length scale is negligible. It is this effective low diffusivity on the axis which causes flux to be concentrated there as a rope rather than as a sheet on the outer boundary, $s = d$.

4.4   Beltrami flows

The Childress dynamo mechanism is exciting because, unlike the G.O. Roberts dynamo, which produces an α-effect of order $R^{-\frac{1}{2}}(L/T_C)$ (see (4.4c)), the magnitude of the α-effect predicted by (4.17) is of order (L/T_C). In principle, mean magnetic field varying on the length scale, L, can be expected to grow on the convective time scale, T_C, and so lead to a "Fast" dynamo. There are, however, two difficulties. Firstly, the theory assumes that the mean magnetic field varies on a length scale long compared with L. Secondly, the unidirectional electromotive force (4.17) cannot, by itself, cause a dynamo to operate.

   To overcome the second difficulty, Childress (1979) suggested that the flow (4.7a) should be embedded in a fully three-dimensional flow. A natural example much studied in the dynamo context, is the class of Beltrami flows, which, relative to rectangular cartesian coordinates (x,y,z), have components,

$$u_x = (L/T_C)\{A \sin(z/L) + C \cos(y/L)\} , \qquad (4.18a)$$

$$u_y = (L/T_C)\{B \sin(x/L) + A \cos(z/L)\} , \qquad (4.18b)$$

$$u_z = (L/T_C)\{C \sin(y/L) + B \cos(x/L)\} , \qquad (4.18c)$$

where A, B and C are constants. They have the maximally helical property,

$$\underset{\sim}{\omega} \equiv \underset{\sim}{\nabla} \times \underset{\sim}{u} = (1/L)\underset{\sim}{u} \quad , \tag{4.19}$$

everywhere.

The two-dimensional case, A = 0, B = C, reduces after a suitable change of coordinates to the G.O. Roberts flow (4.2), with U proportional to ψ. As A is increased from zero, much of the flow remains on streamsurfaces but the outer surfaces, where in the two-dimensional case ψ = 0, begin to break up and new vortex structures begin to emerge. When A = B = C three distinct families of spiralling vortices each aligned roughly with the three principle axes, Ox, Oy, Oz, interlock to form a symmetric network. The small regions between the vortices, which contain the stagnation points, consist of particle paths which are largely chaotic (see Dombre et al., 1986).

The importance of the Beltrami flow in the dynamo context is its helical property. From the point of view of fast dynamo action, the existence of chaotic particle paths for which neighbouring particles can separate indefinitely at an exponential rate is also interesting. Furthermore, for the symmetric case, A = B = C, there are regions of the flow in which the Childress mechanism can operate. For example, the flow in the vicinity of the line connecting the stagnation points at x = y = z = -Lπ/4 and x = y = z = 3Lπ/4 is the same as the axisymmetric flow (4.7a) close to the axis, s = 0. Accordingly, the spatial periodicity of (4.18) and the symmetry with respect to the principal axes suggest an isotropic α-effect with

$$\underset{\sim}{\bar{e}} = \alpha\underset{\sim}{\bar{B}} \quad , \tag{4.20}$$

where α is of order L/T_c. The problem of evaluating the magnitude of α is very difficult largely because, unlike the cylinder, the streamsurfaces connected to the stagnation points reside in the chaotic regions and do not terminate, except at isolated stagnation points. Childress and Soward (1984) attempt an estimate and propose a value of α order unity. Recently, Galloway and Frisch (1984, 1986) have solved the dynamo equations numerically and find in the case of large magnetic Reynolds number most of the flux is concentrated in ropes consistent with the present model and our remarks at the end of Section 4.3.

REFERENCES

Alfvén, H., 1950, Tellus, 2, 74.
Batchelor, G.K., 1967, 'An Introduction to Fluid Dynamics'. Cambridge
  University Press.
Childress, S., 1979, Phys. Earth Planet. Int., 20, 172.
Childress, S. and Soward, A.M., 1984, 'Chaos in Astrophysics', NATO
  Advanced Research Workshop, Palm Coast, Florida, U.S.A.
Dombre, T., Frisch, U., Greene, J.M., Hénon, M., Mehr, A. and Soward,
  A.M., 1986, J. Fluid Mech., 167, 353.
Fearn, D.R., Roberts, P.H., and Soward, A.M., 1987, 'Energy,
  Stability and Convection', eds. Straughan, B. and Galdi, P.G.,
  Capri, Italy.
Galloway, D.J., and Frisch, U., 1984, Geophys. Astrophys. Fluid Dyn.
  29, 13.
Galloway, D.J., and Frisch, U., 1986, Geophys. Astrophys. Fluid Dyn.
  36, 53.
Krause, F., and Rädler, K-H., 1980, 'Mean-field Magnetohydrodynamics
  and Dynamo Theory', Akademie-Verlag and Pergamon.
Moffatt, H.K., 1978, 'Magnetic field generation in Electrically
  Conducting fluids', Cambridge University Press.
Moffatt, H.K., and Proctor, M.R.E., 1985, J. Fluid Mech., 154, 493.
Parker, E.N., 1979, 'Cosmical Magnetic Fields', Clarendon.
Roberts, G.O., 1972, Phil. Trans. R. Soc. Lond., A271, 411.
Soward, A.M., 1987, J. Fluid Mech., 180, 267.
Soward, A.M., and Childress, S., 1986, 'Analytic Theory of Dynamos',
  Adv. Space Res., 6, 8, 7, Proc. COSPAR meeting, Toulouse.
Vainstein, S.I., and Zel'dovich, Ya B., 1972, So. Phys. Usp., 15, 159.
Weiss, N.O., 1966, Proc. R. So. Lond. A293, 310.
Zel'dovich, Ya B., Ruzmaikin, A.A., and Sokoloff, D.D., 1983,
  'Magnetic fields in Astrophysics', Gordon and Breach.

THE PATTERN OF LARGE-SCALE CONVECTION IN THE SUN

C.A. Jones,
School of Mathematics, University of Newcastle upon Tyne,
Newcastle upon Tyne NE1 7RU, U.K., and
D.J. Galloway,
Department of Applied Mathematics, University of Sydney,
N.S.W. 2006, Australia.

ABSTRACT. The pattern of large scale convection in the Sun has generally been considered to be in the form of rolls fitted into a sphere and aligned with the rotation axis. This picture has been called 'cartridge belt' convection or 'banana cell' convection. The theoretical reasoning behind this pattern is that the Taylor-Proudman theorem for a rotating fluid constrains the fluid to be as near two-dimensional as is consistent with spherical geometry. This view is supported by calculations using weakly nonlinear theory and fully nonlinear numerical simulations of the solar convection zone.

Despite the theoretical evidence, there is little observational evidence to support this picture. Indeed, it has recently been argued that observations of surface velocities are more consistent with convection in the form of axisymmetric toroidal rolls than with 'cartridge belt' convection.

A possible explanation is that the magnetic field in the solar convection zone can influence the pattern of convection, breaking down the Taylor-Proudman constraint. It is argued that the relevant parameter measuring the strength of the magnetic field in this regard is the Elsasser number $\Lambda = \dfrac{B^2}{2\mu\rho\eta\Omega}$, η being the magnetic diffusivity. Because it is appropriate to use a turbulent value of η, the exact value of Λ is hard to estimate accurately: however, values generally used in the literature give Λ of order 1 for reasonable values of B and ρ, indicating that the magnetic field may play a role in pattern selection in large-scale solar convection.

Because the toroidal field strength varies during the course of the solar cycle, it is possible that Λ can vary above and below the threshold value: in that case it may be possible that the pattern of convection could vary during the solar cycle.

1. INTRODUCTION

The pattern of large-scale convection in the Sun is an important issue for our understanding of solar behaviour. It strongly affects the

F. R. Stephenson and A. W. Wolfendale (eds.),
Secular Solar and Geomagnetic Variations in the Last 10,000 Years, 97–108.
© *1988 by Kluwer Academic Publishers.*

dynamo activity which generates the solar magnetic field and the pattern of differential rotation. It may also lead to variations in solar radius and luminosity. Whereas the convection pattern near the solar surface can be observed in some detail (Bray, Loughhead and Durrant, 1984) no conclusive observational evidence exists concerning the global convection pattern. We have to rely on a combination of theory, numerical experiment, laboratory experiment and observational inference.

Photospheric granules with a typical horizontal length scale of 1000 km have a life-time comparable to their eddy turn-over time, of the order of 10 minutes. The supergranulation has a horizontal length scale of around 15,000 km and a life-time of order 24 hours. It is unlikely that either of these phenomena have a great deal to do with the generation of the solar magnetic field, which appears to be generated on a much larger length scale and on a much longer time scale.

It has been conjectured (Bumba, 1967; Simon and Weiss, 1968) that a scale of convection, known as giant cells, exists which extends over the whole depth of the convection zone. Estimates based on mixing length theory give around a month for the eddy turnover time for giant cells, and a typical velocity of circulation of around 100 m/s. These estimates are, of course, qualified by the usual difficulties surrounding mixing length theory. Although there are a number of arguments in favour of the existence of giant cells (Weiss, 1976), it should be noted that there is no clearcut explanation of why convection in the Sun should break down into three distinct length scales.

Despite the fact that the giant cells have been postulated for twenty years, hard observational evidence for their existence is still lacking (Gilman, 1986). There is some evidence for a sector structure in the apparent occurrence of spots at preferred ('active') longitudes, and also in coronal holes. However, Ribes, Mein and Mangeney (1985) have recently argued that analysis of sunspot motions indicates a meridional large scale velocity pattern, with a typical velocity of 100 m/s, about that expected for giant cells.

The theoretical question posed by these observations is whether the pattern of large-scale convection is predominantly in the form of 'cartridge belt' convection, with a strong sector structure, as in Figure 1, or in the form of toroidal rolls, as in Figure 2. The toroidal rolls picture is axisymmetric, and it is argued by Ribes et al. to be the best explanation of their sunspot motion data. The case for a 'cartridge belt', or banana cell pattern has been put forward by Busse (1970), Gilman (1986 and references therein), Glatzmaier (1984) and others. There is, of course, no certainty that either picture is exactly correct: at the high Reynolds numbers encountered in the Sun, convection is likely to be time-dependent rather than stationary, so the question is more whether toroidal or banana cells predominate: nevertheless, even if the question can only be given a statistical answer, it still has importance.

Pure toroidal rolls cannot produce poloidal field from toroidal field generated by differential rotation. This has been noted by Dogiel and Syrovatsky (1979) and Zeldovich, Ruzmaikin and Sokoloff (1983), who suggested that during the Maunder minima the apparent collapse of dynamo action may be associated with a change from nonaxisymmetric to

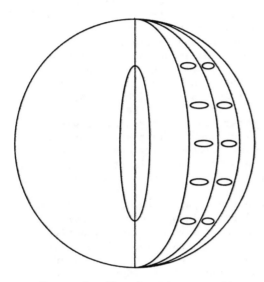

Figure 1.   Sketch of banana cells

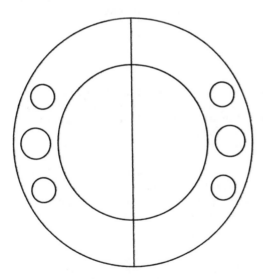

Figure 2.   Sketch of toroidal rolls

axisymmetric rolls. A similar point of view was expressed by Parker (1979).

The main arguments for the banana cell pattern are theoretical. Linear analysis of convection in the presence of rotation (Busse, 1970) indicates that a 'cartridge belt' pattern onsets at a lower Rayleigh number than an axisymmetric pattern. This behaviour can be understood in terms of the Taylor-Proudman theorem: in an inviscid fluid, slow, steady motions have to have the momentum density independent of the coordinate in the direction of $\underline{\Omega}$. For the astrophysical convection problem we do not assume the fluid is inviscid, but nevertheless motions which obey the Taylor-Proudman constraint are strongly preferred over motions which depend on viscosity for their existence.

The fully nonlinear calculations of Glatzmaier (1984) and Gilman (1979) also reveal a preference for 'cartridge belt' convection. These calculations also show that such motions give values of the variation of rotation rate with latitude which is consistent with observation, and a variation of rotation rate with depth which is consistent with current solar oscillation data (Duvall and Harvey 1984). Whether similar results could be obtained by axisymmetric motions is not known: one might conjecture, however, that axisymmetric motions will tend to mix angular momentum so that it becomes uniform over the convection zone: this would lead to a differential rotation that is faster at the poles than the equator, the opposite of what is observed. There are, however,

some theories with axisymmetric meridional motions which produce an
equatorial acceleration: for example, the anisotropic eddy viscosity
model (Kippenhahn, 1963; Kohler, 1974), or the latitude-dependent
turbulent heat transport model (Belvedere and Paterno, 1976).

There are a number of laboratory experiments which have been
performed on convection in a rotating system (e.g. Carrigan and Busse,
1983; Hart et al. 1986) and these experiments generally support the
view that the Taylor-Proudman constraint is effective, and hence that
the 'cartridge belt' pattern is to be expected in the solar convection
zone. In the Carrigan and Busse experiments, the radial gravity is
mimicked by a large centrifugal force, whereas in the Hart et al.
experiments the apparatus was put in a microgravity environment aboard
Spacelab 3. The radial gravity was simulated by electrostatic forces.
Flow visualization techniques indicated that in both experiments banana
cells occurred in many parameter regimes.

In view of both the theory and the experiments we conclude that
axisymmetric convection in toroidal rolls is unlikely to occur in a
rotating convecting shell unless there is some dynamically active
ingredient to counter the Taylor-Proudman constraint. A toroidal
magnetic field is such an ingredient, and the rest of this paper is
devoted to forming estimates of the strength of field required to make
toroidal rolls the preferred planform. The inclusion of both magnetic
field and rotation increases the complexity of the problem considerably,
so we restrict ourselves to the comparatively simple case of the linear
theory in the Boussinesq approximation.

2. THE MODEL PROBLEM

We adopt a Cartesian geometry, which approximates the behaviour near
the equator of the solar convection zone. We take the magnetic field
to be in the $\hat{x}$-direction, the rotation axis to be in the $\hat{y}$-direction,
and gravity to be in the $-\hat{z}$-direction (see Figure 3). This configuration has been considered by Hughes (1985) in connection with
magnetic buoyancy, and by Condi (1978).

We assume a Boussinesq fluid with thermal conductivity κ, kinematic viscosity ν and magnetic diffusivity η. The rotation is $\Omega\hat{y}$,
the unperturbed field $B_0\hat{x}$. We assume an initial unstable temperature
gradient $\beta\hat{z}$ in a layer of fluid of depth d.

The relevant nonlinear equations are then

$$\frac{Du}{Dt} + 2\underline{\Omega} \wedge \underline{u} = -\frac{1}{\rho}\nabla p + \frac{1}{\mu\rho}(\underline{\nabla} \wedge \underline{B}) \wedge \underline{B} + g\alpha\theta\hat{z} + \nu\nabla^2\underline{u} \quad (2.1)$$

$$\underline{\nabla}\cdot\underline{u} = 0, \quad \underline{\nabla}\cdot\underline{B} = 0 \quad (2.2)$$

$$\frac{\partial\underline{B}}{\partial t} = \underline{\nabla} \wedge (\underline{u} \wedge \underline{B}) + \eta\nabla^2\underline{B} \quad (2.3)$$

$$\frac{D\theta}{Dt} = (\underline{u}\cdot\hat{z})\beta + \kappa\nabla^2\theta \quad (2.4)$$

where $\underline{u}$ is the fluid velocity, p is the pressure and ρ the density. The temperature $T = T_0 - \beta z + \theta$.

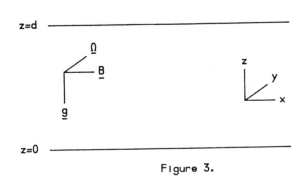

z=d

z=0

Figure 3.

We adopt the simplest boundary conditions, which are that $\underline{u} \cdot \hat{\underline{z}} = u_3 = 0$ and

$$\frac{d^2 u_3}{dz^2} = 0 \text{ on } z = 0 \text{ and}$$

d (stress-free boundary conditions), $\theta = 0$ on z = 0 and d (constant temperature boundaries) and $B_3 = \dfrac{d^2 B_3}{dz^2} = 0$

(perfectly conducting boundaries). These boundary conditions are chosen so that they avoid boundary layers at large R.

We now linearise the equations about the static solution $\underline{u} = 0$, $\underline{B} = B_0 \hat{\underline{x}}$, $T = T_0 - \beta z$ and we also non-dimensionalise the variables taking the unit of distance as d, of time d^2/κ, of field B_0 and temperature βd. We also assume that the x, y and t-dependence is

$$e^{\sigma t} . e^{ikx + i\ell y} \qquad (2.5)$$

so that k is the wavenumber in the field direction and ℓ the wavenumber in the rotation direction. We note at this stage that modes with k = 0 are independent of the coordinate in the field direction and modes with $\ell = 0$ are independent of the coordinate in the rotation direction. So the Taylor-Proudman constraint imposes $\ell = 0$: if the toroidal field is to significantly alter the direction of the convection rolls near onset, it is necessary that ℓ be increased from 0 until it is of the same order as k. Eliminating the pressure perturbation by taking the curl of (2.1), the linearised equations can be written

$$i\ell P^2 T u_3 = M(i\ell u_1 - iku_2) + PP_m Q(k^2 B_2 - k\ell B_1) \qquad (2.6)$$

$$i\ell P^2 T u_1 = M(Du_2 - i\ell u_3) + PP_m Q(i\ell DB_1 + k\ell B_3) - i\ell PR\theta \qquad (2.7)$$

$$L B_i = -iku_i \quad (i = 1,2,3) \qquad (2.8)$$

$$N\theta = -u_3 \qquad (2.9)$$

$$iku_1 + i\ell u_2 + Du_3 = 0, \quad ikB_1 + i\ell B_2 + DB_3 = 0 \qquad (2.10)$$

where $D \equiv \dfrac{d}{dz}$, $L \equiv P_m(D^2 - k^2 - \ell^2) - \sigma$, $M \equiv P(D^2 - k^2 - \ell^2) - \sigma$,

$$N \equiv (D^2 - k^2 - \ell^2) - \sigma \text{ are linear operators.}$$

Here $\underline{u} = (u_1, u_2, u_3)$ and $\underline{B} = B_0\hat{\underline{x}} + (B_1, B_2, B_3)$, and the dimensionless parameters are

$$P = \frac{\nu}{\kappa}, \quad P_m = \frac{\eta}{\kappa}, \quad T = \frac{4\Omega^2 d^4}{\nu^2}, \quad Q = \frac{B^2 d^2}{\mu\rho\eta\nu}, \quad R = \frac{g\alpha\beta d^4}{\kappa\nu}. \tag{2.11}$$

T, Q and R are the Taylor, Chandrasekhar and Rayleigh numbers respectively. P is the Prandtl number and P_m the magnetic Prandtl number.

The magnetic field perturbations can be eliminated using (2.8), the temperature perturbation using (2.9), u_2 using (2.10a) and u_1 using (2.7). A single 10th order equation for the vertical velocity u_3 results. Because the boundary conditions have been chosen appropriately, the required solution of this 10th order system is

$$u_3 = \sin\pi z. \tag{2.12}$$

B_3 and θ both also are proportional to $\sin\pi z$, the other perturbed quantities in equations (2.6) to (2.10) being proportional to $\cos\pi z$. This simple behaviour allows us to obtain the dispersion relation giving the eigenvalue σ as a function of k and ℓ. Although this dispersion relation is complicated, we state it here as it is crucial to what follows:

$$a\sigma^5 + a^2\sigma^4(1 + 2P') + \sigma^3[P^2T\ell^2 + P'a^3(2 + P') + 2PP_m aQ' - PRb]$$

$$+ \sigma^2[T\ell^2 aP^2(1 + 2P_m) + a^4P'^2 + 2PP_m a^2Q'(1 + P') - PaRb(P + 2P_m)]$$

$$+ \sigma[a^2P^2T\ell^2P_m(2 + P_m) + P^2P_m^2aQ'^2 + 2a^3PP_mP'Q' -$$

$$PbR(P_mP'a^2 + PP_mQ')]$$

$$+ P^2P_m^2a^3T\ell^2 + a^2P^2P_m^2Q'^2 - P^2P_m^2abRQ' = 0, \tag{2.13}$$

where $P' = P + P_m$, $a = \pi^2 + k^2 + \ell^2$, $b = k^2 + \ell^2$, $Q' = a^2 + Qk^2$.

This dispersion relation is fifth order in time, hence fifth order in σ. We cannot hope to solve (2.13) analytically, of course, so a program was written to evaluate the roots of (2.13) for a given imput of ℓ and k, and the parameters (2.11).

Numerical investigation concentrated on the case $P = P_m = 1$, which is usually considered appropriate if turbulent values of the diffusion coefficients are used. We note that there appears to be very little hard evidence to support the conjecture that the turbulent Prandtl numbers are unity when the molecular values are very far from unity (the molecular values of P and P_m are of order 10^{-6}), but the usual argument is that the same eddies which transport heat transport momentum and field.

3. ANALYSIS OF THE DISPERSION RELATION

In order to find the critical Rayleigh number at which convection occurs, we have to find that value of R at which one root of (2.13) has the real part of σ zero, and the other four roots of (2.13) have real part of σ less than or equal to zero. This value of R is denoted by R_c, and so the condition gives a relation

$$R(k^2, \ell^2, T, Q, P, P_m) = R_c. \qquad (3.1)$$

Now for fixed values of T, Q, P and P_m in an infinite plane layer any value of k or ℓ is possible. The selection criterion is that we choose those values of k and ℓ which minimise R_c. These criteria can be formally written as

$$\frac{\partial R}{\partial k} = 0, \quad \frac{\partial R}{\partial \ell} = 0, \qquad (3.2)$$

both derivatives being evaluated at $R = R_c$. In practice because of the complexity of (2.13) it proved simpler to use a general purpose minimization routine from the NAG (Numerical Algorithm Group) library to evaluate the k and ℓ which minimises R_c.

When this is done we end up with the minimum critical Rayleigh number, R_{min}, and k and ℓ are determined,

$$R(T, Q, P, P_m) = R_{min}, \quad k = k_{min}, \quad \ell = \ell_{min}. \qquad (3.3)$$

We now consider the question which we wanted to answer originally, are banana cells preferred to toroidal rolls or vice versa? If k_{min} is close to zero we are in the sector structure case. The natural borderline between these cases is

$$k_{min} = \ell_{min} \qquad (3.4)$$

the rolls here being aligned at 45° to both magnetic field and rotation axis. So we impose the constraint (3.4), and this gives a relation

$$f(T, Q, P, P_m) = 0 \qquad (3.5)$$

which must be satisfied. Since we set $P = P_m = 1$, this gives a relation between T and Q, which we can write as

$$Q = Q(T), \quad k_{min} = \ell_{min}. \qquad (3.6)$$

This gives the critical value of Q, and hence toroidal field B, as a function of T, and hence rotation rate Ω for the rolls to equally incline to both field and rotation axis. It turns out to be more convenient to plot Q against $T^{\frac{1}{2}}$ rather than T, and this is done in Figure 4.

We find, as expected, that if Q is greater than the critical value, $k_{min} \to 0$ rapidly is Q as increased, so the rolls rapidly align with the

104

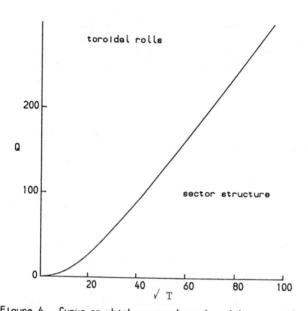

Figure 4. Curve on which wavenumbers k and ℓ are equal

magnetic field, and if Q is decreased below the critical value, $\ell_{min} \to 0$ rapidly so that the rolls quickly align with the rotation axis as the field strength falls below critical.

It transpires that if $P = P_m = 1$, the critical line (3.6) is obtained entirely from modes for which $\sigma = 0$ at critical. This is not the case if the Prandtl numbers are significantly below 1, as oscillatory modes can onset first in that case. Nevertheless, this enables us to make (for the case $P = P_m = 1$) a great simplification to the dispersion relation (2.12), which becomes

$$a^2 T\ell^2 + a(a^2 + Qk^2)^2 = Rb(a^2 + Qk^2).$$ (3.7)

This relation is sufficiently simple for some analysis to be possible. We look for the limit of large Q and T, so that R is also large. Then $a^2 + Qk^2 \sim Qk^2$, and (3.7) becomes further simplified to

$$R = \frac{a^2 T\ell^2}{Qk^2(\ell^2 + k^2)} + \frac{aQk^2}{\ell^2 + k^2} \text{ (large Q, T).}$$ (3.8)

Then the conditions (3.2) reduce to

$$(\pi^2 + 2k^2)(3\pi^2 + 2k^2)T = (\pi^2 + 4k^2)k^2Q^2,$$
$$(\pi^2 + 2k^2)(\pi^2 + 6k^2)T = \pi^2k^2Q^2$$ (3.9)

where we set $k = \ell$. These two equations give k in the large Q limit, the only acceptable root being

$$k = \ell = \frac{\pi}{\sqrt{6}}$$ (3.10)

Equations (3.9) give

$$Q^2 = 16T \quad \text{and} \quad R = \pi^2 Q \tag{3.11}$$

also valid in the large Q limit. The reason for plotting Q against $T^{\frac{1}{2}}$ in Figure 4 is now clear; as $T^{\frac{1}{2}} \to \infty$ the curve approaches the straight line $Q = 4T^{\frac{1}{2}}$.

The natural parameter we are led to consider is then the Elsasser number

$$\Lambda = \frac{Q}{T^{\frac{1}{2}}} = \frac{B^2}{2\mu\rho\Omega\eta} = \frac{(\text{Alfven speed})^2}{2\Omega\eta} \tag{3.12}$$

since the asymptotic value of this in Figure 4 is $\Lambda = 4$. This parameter Λ is of importance in discussions of the geodynamo; see e.g. Acheson, 1978; Eltayeb, 1972; Soward, 1978.

The main conclusion we draw from this analysis is that at large Elsasser number we expect convection to line up so that the axis of the rolls is parallel to the field, and at small Elsasser number convection lines up with the rotation axis.

The condition that the Elsasser number be of order unity can be seen to be equivalent to the balance of Lorentz and Coriolis force in the equation of motion (2.1). In the type of large scale convection we envisage, in (2.3) there is a balance of terms

$$\underline{\nabla} \wedge (\underline{u} \wedge \underline{B}) \sim \eta\nabla^2\underline{B}$$

so that

$$|\underline{u}| \sim \frac{\eta}{d},$$

where d is a typical length scale, here the depth of the convection zone. If we return this to (2.1), if the Coriolis and Lorentz forces are of the same order of magnitude,

$$|2\underline{\Omega} \wedge \underline{u}| \sim \left|\frac{1}{\mu\rho}(\underline{\nabla} \wedge \underline{B}) \wedge \underline{B}\right|,$$

or

$$\frac{2\Omega\eta}{d} \sim \frac{B^2}{\mu\rho d},$$

so

$$\frac{B^2}{2\mu\rho\Omega\eta} \sim 1, \tag{3.13}$$

so the Elsasser number is of order unity.

Order of magnitude analysis suggests that the terms $(\underline{u}\cdot\underline{\nabla})\underline{u}$ and $g\alpha\theta\hat{z}$ are both comparable to the Lorentz and Coriolis forces. The dynamics of the solar convection zone is extremely complicated!

4. CONCLUSIONS

An interesting (and perhaps surprising) feature of the parameter
$\Lambda = B^2/2\mu\rho\Omega\eta$ is that it depends on the diffusion coefficient, η. This
is in some ways unfortunate as it makes it rather hard to estimate
Λ well, as our knowledge of η is very uncertain. If we take a laminar
value of η, almost any sensible estimate of B gives Λ very large, so
that convection would line up with the field (toroidal rolls) rather
than the rotation. However, it seems inappropriate to take η as the
laminar value in the turbulent solar convection zone. More likely
is a value of η in the range 10^7 to 10^9 m^2s^{-1}. The lower figure comes
from the decay of long-lived spots, the upper value coming from the
mixing length approach. Priest (1984) argues that the best fits for
$\alpha-\omega$ dynamo models are given at $\eta \sim 10^8$ m^2s^{-1}.
    The density, of course, varies a great deal from the top of the
convection zone to the bottom: η is thought to be more constant. If
we use the values given from a mixing-length model of the convection
zone (Spruit, 1974) at three different levels, we obtain Table 1.

Table 1

Convection zone

	top	middle	bottom
Depth	10^3 km	10^5 km	2.10^5 km
Density	3.10^{-3} kg m^{-3}	60 kg m^{-3}	230 kg m^{-3}
Laminar η	3×10^2 m^2s^{-1}	0.6 m^2s^{-1}	0.2 m^2s^{-1}
Turbulent η (estimate)	$10^8 m^2s^{-1}$	$10^8 m^2s^{-1}$	$10^8 m^2s^{-1}$
Λ_{turb}	$5 \times 10^{-3}B_G^2$	$2.4 \times 10^{-7}B_G^2$	$6.4 \times 10^{-8}B_G^2$
Λ_{lam}	1600 B_G^2	40 B_G^2	32 B_G^2

[N.B. B_G is the field in gauss (1 gauss = 10^{-4} Tesla).]

    From this table we see that even for a field of 1 Gauss, which
is very low for a toroidal field, Λ_{lam} is large, but the issue is far
less clear-cut in the turbulent case, particularly in view of the
uncertainties in η in this case. Globally, a mean toroidal field of
100 G is probably reasonable, but of course this may vary with the
solar cycle. With this value Λ_{turb} is probably small throughout most
of the convection zone, a possible exception being near the top, where
the low value of ρ enhances Λ. We note that this is in accord with
Glatzmaier's results, which had B_G around 100 G and η around $10^8 m^2s^{-1}$,
and in which the convection pattern has the form of banana cells.
However, these results indicate that if η was reduced in the numerical
simulations toroidal rolls might result (we note, however, that reducing

η substantially increases the demand on computing time). We also note that the Glatzmaier calculations are fully nonlinear, so that it is possible that different solutions may be reached depending on in which basin of attraction the calculation is started off. The startup conditions used in simulations are that the convection is started without magnetic field, so that cartridge belt convection is obtained, and then the field is added: the seed field then grows in a way which fits into the existing convection pattern. If, on the other hand, a large toroidal field was initially present, it is possible that the final state would be predominantly in the form of toroidal motions, with a small non-axisymmetric component to maintain dynamo action.

Locally much larger fields than 100 G are possible. Galloway and Weiss (1981) have argued that a field of 10,000 G may fill a substantial volume near the base of the zone. This could well bring Λ_{turb} up to a significant value, and promote toroidal rolls in this neighbourhood. How a local preference for toroidal rolls would interact with the rest of the convection zone is a question that requires a fully compressible analysis, well beyond the scope of what we have done so far.

An interesting possibility that emerges from this analysis is the possibility that the magnetic field could, through pattern selection, exert a significant influence on the dynamics of the convection zone. If this is the case, we might expect that fluctuations of luminosity and radius will occur during the eleven year cycle (Spiegel and Weiss, 1980; Weiss and Spiegel, 1987).

The mechanism discussed above may also add something to the Dogiel and Syrovatsky (1979) idea that changes in convection pattern are responsible for Maunder minima, since it provides a feedback mechanism by which changes in the field could influence the convection pattern. This would give the solar dynamo the character of a nonlinear dynamical system (Weiss, Cattaneo and Jones, 1984; Jones, 1983; Ruzmaikin, 1983). The calculations done in this paper are based on a very simplified picture; we only model the equatorial belt of the convection zone: a calculation done in a sphere is likely to reduce the critical value Λ required for toroidal rolls, as the angle between gravity and the rotation vector decreases. In view of the considerable variation in Λ_{crit} across the convection zone, we expect that compressibility will be an important factor: our analysis is essentially local, and we cannot say for certain whether the global value of Λ_{crit} will depend more on conditions at the top of the zone or at the base of the zone. The effect of nonlinearity on pattern selection is also not clear: Knobloch, Rosner and Weiss (1981) have suggested that nonlinear effects may lead to the possibility of different states being realised at the same rotation rate, as a possible explanation of the Vaughan-Preston gap.

REFERENCES

Acheson, D.J., 1978. In 'Rotating Fluids in Geophysics', eds. Roberts,
    P.H. and Soward, A.M., p. 315, Academic Press.
Belvedere, G., and Paterno, L., 1978. Solar Phys. 60, 203.
Bray, R., Loughhead, R.E. and Durrant, C.J., 1984. 'The Solar
    Granulation' (2nd ed.), Cambridge University Press.
Busse, F.H., 1970, Astrophys. J. 159, 629.
Bumba, V., 1967. In 'Plasma Physics', ed. Sturrock, P.A., p.77.
    Academic Press.
Carrigan, C.R. and Busse, F.H., 1983. J. Fluid Mech. 126, 287.
Condi, F.J., 1978. In 'Notes of the Summer Study Program at Woods
    Hole Oceanographic Institute', p.18.
Dogiel, V.A. and Syrovatsky, S.I., 1979. Bull. Acad. Sci. U.S.S.R.
    43, p.35.
Duvall, T.L. and Harvey, J.W., 1984. Nature, 310, p.19.
Eltayeb, I.A., 1972. Proc. Roy. Soc. A., 326, p.229.
Galloway, D.J. and Weiss, N.O., 1981. Astrophys. J. 243, p.945.
Gilman, P.A., 1979. Astrophys. J. 231, 284.
Gilman, P.A., 1986, In 'Physics of the Sun, vol. 1', ed. P.A. Sturrock
    p.95.
Glatzmaier, G.A., 1984, J. Comput. Phys. 55, 461.
Hart, J.E., Toomre, J., Deane, A.E., Hurlburt, N.E., Glatzmaier, G.A.,
    Fichtel, G.H., Leslie, F., Fowlis, W.W. and Gilman, P.A., 1986,
    Science, 234, p.61.
Hughes, D.W., 1985. Geophys. Astrophys. Fluid Dyn. 34, p.99.
Jones, C.A., 1983. In 'Stellar and Planetary Magnetism', ed. A.M.
    Soward, p.159. Gordon and Breach.
Kippenhahn, R., 1963. Astrophys. J. 137, 664.
Kohler, H., 1974. Solar Phys. 34, 11.
Moffatt, H.K., 1978. 'Magnetic Field Generation in Electrically
    Conducting Fluids'. Cambridge University Press.
Parker, E.N., 1979. 'Cosmical Magnetic Fields', p.762, Clarendon Press.
Priest, E.R., 1984. 'Solar Magnetohydrodynamics'. Reidel.
Ribes, E., Mein, P., Mangeney, A., 1985. Nature, 318, 170.
Ruzmaikin, A.A., 1983. In 'Stellar and Planetary Magnetism', ed.
    A.M. Soward, p. 151. Gordon and Breach.
Simon, G.W. and Weiss, N.O., 1968. Z. Astrophys. 69, 435.
Soward, A.M., 1978. In 'Rotating Fluids in Geophysics', ed. P.H.
    Roberts and A.M. Soward, p. 409. Academic Press.
Spiegel, E.A. and Weiss, N.O., 1980, Nature, 287, 616.
Spruit, H.C., 1974. Solar Phys. 34, 277.
Weiss, N.O,, 1976. In 'Basic Mechanisms of Solar Activity', eds.
    Bumba and Kleczek. I.A.U. Symposium No. 71.
Weiss, N.O., Cattaneo, F. and Jones, C.A., 1984. Geophys. Astrophys.
    Fluid Dyn. 30, 305.
Weiss, N.O. and Spiegel, E.A. 1987. In preparation.
Zeldovich, Y.A., Ruzmaikin, A.A., and Sokoloff, D.A., 1983. 'Magnetic
    Fields in Astrophysics'. Gordon and Breach.

SOLAR VARIABILITY FROM HISTORICAL RECORDS

F.R. Stephenson
Department of Physics
University of Durham
Durham DH1 3LE, U.K.

ABSTRACT. The application of historical observations of (i) the Sun's diameter, (ii) the corona and (iii) sunspots to the question of long-term solar variability is considered. Although solar diameter measurements can be mapped in detail since about A.D. 1700, at present it is not possible to use these results to estimate changes in the solar output of energy with any degree of confidence. Coronal observations are too infrequent to prove of much value in the study of changes in solar activity. The telescope record of sunspots since A.D. 1700 is of paramount importance to this problem. However, it is not possible to use observations of sunspots made before this time to reliably delineate solar activity owing to the presence of spurious trends in the data.

1. INTRODUCTION

Historical observations which have proven potential to yield direct information on solar variability in the past can be grouped in three main categories. These are:

(i)   solar diameter measurements,
(ii)   descriptions of the form of the corona at total solar eclipses,
(iii)   sightings of sunspots.

The main aim of this paper is to discuss the applicability of observations in all three groups to the question of long-term solar variability - i.e. on the centennial to millennial time-scale. Only telescopic measurements of the solar diameter are available and there are few extant observations of the corona before the 17th century. However, sightings of sunspots cover a much longer time-scale, extending back to the 2nd century B.C. Hence in considering the data in this last category, special emphasis will be placed on the study of pre-telescopic records.

F. R. Stephenson and A. W. Wolfendale (eds.),
Secular Solar and Geomagnetic Variations in the Last 10,000 Years, 109–129.
© 1988 by Kluwer Academic Publishers.

2. SOLAR DIAMETER MEASUREMENTS

Direct measurements of the solar constant (S) which can be considered
in any way reliable are only available in the past twenty years or
so. In default of such data, changes in the solar diameter (D = 2R)
observed since about A.D. 1700 ought to be linked directly with changes
in the solar luminosity (L) since they alter the area of the radiating
surface. However, the relation between R and L is as yet not well
understood. Defining $W = \Delta(\log R)/(\log L)$, estimates of W at present
range over two orders of magnitude. Frohlich and Eddy (1984) obtained
a preliminary empirical value for W by comparing simultaneous changes
in S and R over the previous 15 years or so; these were obtained as
the result of compiling available measurements during the period from
1967 to 1983. However, their result of 0.078 ± 0.026 is much larger
than the theoretical upper limit of about 0.005 as derived by Gough
(1981). Hence at present the fairly accurate measurements (both direct
and indirect) of R which are available over the past 250 years or so
are of limited value in the study of changes in L. When satisfactory
accord between the observational and theoretical results for W is
eventually reached, the full potential of the existing measurements
of solar diameter as indicators of changes in the solar output may
be realised.

Observations of the following types have been used to determine
variations in D over the past few centuries: meridian transits;
timings of transits of Mercury across the solar disc; durations of
total solar eclipses; and observations of the size of the lunar shadow
cast during total solar eclipses. The stimulus for this field of
investigation was provided by Eddy and Boornazian (1979), who analysed
meridian transits of the Sun observed at both Greenwich (since 1836)
and Washington (since 1890). At each visible transit, it was the
practice to record the time of passage of the western and eastern limbs
of the Sun across the meridian (leading to a measurement of the
horizontal solar diameter D_h) and to determine the zenith distances of
the upper and lower limbs of the Sun (a measurement of its vertical
diameter D_v). From the many thousands of observations available,
Eddy and Boornazian inferred a mean secular decrease in D of as much
as 0.1 per cent per century over the last 150 years.

The use of meridian observations for determining changes in D
was, however, criticised by Parkinson et al. (1980). In particular,
they showed that observer bias and instrumental defects marred the
Greenwich data. Further, Stephenson (1981) commented on the great
disparity between the individual results obtained by Eddy and Boornazian
from the following sets of measurements: (a) Greenwich determinations
of D_h; (b) Greenwich D_v; (c) Washington D_h; (d) Washington D_v - see
Figure 1. Especially significant is the marked discord between the
observed rates of change of D_h and D_v. This is at variance with the
extremely precise measurements of Hill and Stebbins (1975) which
demonstrate that the Sun has an accurately circular outline - at the
0".01 level.

At each meridian transit, a single observer was responsible for
making all four measurements (times and declinations) within about

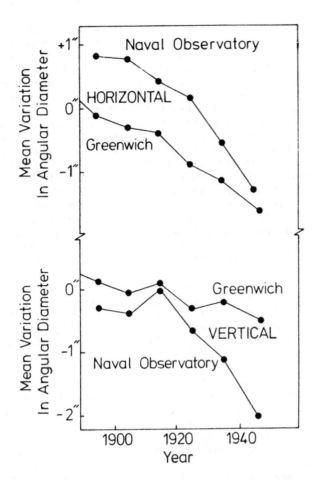

Fig. 1.   Changes in the mean horizontal and vertical diameter of
the Sun between about A.D. 1890 and 1950 as deduced by
Eddy and Boornazian (1979) from Greenwich and Washington
meridian observations.

two minutes - no easy task. Further because of the rapid apparent
(diurnal) motion of the Sun, an error in timing of only 0.1 sec would
lead to an uncertainty in the horizontal diameter of as much as 1.5
arcsec. Observations of higher sensitivity covering 250 years are
provided by transits of Mercury and total solar eclipses. Although
less numerous than meridian transits, these number some 2,000 individual
observations. A typical transit of Mercury lasts for about 7 hours.
This is more than two orders of magnitude longer than a meridian transit
so that if similar precision in timing could be realised, the Mercury
data would lead to greatly improved results for D. However, transits
of Mercury have two main disadvantages. Firstly, the contacts are

not easy to resolve owing to the so-called "black drop" effect, caused by diffraction at the solar limb. In consequence, astronomers making simultaneous observations at the same location may differ in the time of a particular contact by more than 10 sec. In practice, only internal contacts are useful and even these measurements reveal considerable scatter. Secondly, because of the lengthy duration of a transit, it is not often that one observer can time both the beginning and end of an event. Thus pairs of independent absolute timings - rather than durations - are usually necessary, and this requires the availability of fairly accurate clocks. Nevertheless, analysis of occultation timings by Morrison and Stephenson (1986) shows that even in the early 18th century, the standard deviation in timing a single event (based on the results of many observers) was only about 5 seconds.

Shapiro (1980) investigated observations of 23 Mercury transits since A.D. 1736 to study the variation in the solar diameter. He reported a secular decrease in the solar diameter of only about 0.3 arcsec per century, with > 90 per cent confidence. This result is almost an order of magnitude smaller than that obtained by Eddy and Boornazian (1979).

Total solar eclipses require only the measurement of a duration (typically two or three minutes) whose beginning and end are well defined (to about 0.1 sec). A total solar eclipse thus lasts for a similar length of time to a meridian transit. However, considerable gain in the precision of determining D is possible since the relative difference between the apparent diameters of the Moon and Sun is only about 3 per cent on average.

Parkinson et al. (1980) made a more extensive investigation than Shapiro, analysing timings of 30 transits of Mercury and six total solar eclipses since A.D. 1715 - see Figure 2. They demonstrated that both sets of data reveal negligible secular trend in D (-0.008 ± 0.007 per cent) throughout this period. Standard errors in the mean results for D shown in Figure 2 are small, ranging from 0".025 at A.D. 1700 to 0".1 in recent years. The only long-term variation in D which was detected was an 80-year periodic oscillation of amplitude approximately 0.02 per cent. This latter result was closely confirmed by Gilliland (1981) from a wider variety of data. He deduced a periodicity of 76 years and amplitude 0.02 per cent, with the last maximum occurring in 1911. This fluctuation was found to be approximately in antiphase with the so-called Gleissberg cycle (the apparent modulation of the sunspot cycle) but the reality of the latter is still in doubt - see Section 3. No satisfactory physical explanation of the observed 76-year fluctuation in D has so far been given.

Recently Ribes et al. (1987) have analysed meridian observations made at Paris between A.D. 1666 and 1719. This interval covered much of the Maunder Minimum (1645 to 1715) - a period when solar activity seems to have been at an unusually low level (Eddy 1976). The results of Ribes et al. imply that during this time the mean diameter of the Sun was some 4 arcsec (0.2 per cent) greater than at present. In response to this paper, Morrison et al. (1988) have discussed several rather precise observations made near the margins of the umbral zone at the eclipse of 1715 as well as measurements of the duration of

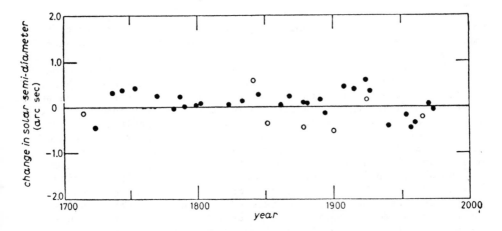

Fig. 2.   Mercury transit (shaded circles) and total solar eclipse
(open circles) determinations of the semi-diameter of
the Sun between A.D. 1715 and 1973 (Parkinson et al.,
1980).   Standard errors in the mean for each point plotted
range from about 0.25 arcsec near 1700 to 0.1 arcsec in
recent years.

totality of this same event.   These two independent sets of data are
all agreed that in the early 18th century the mean solar diameter
was within about 0.5 arcsec of the present value.   A fair proportion
of the measurements analysed by Ribes et al. were made within a few
years either side of 1715 so that they are in direct conflict with
the careful eclipse observations made in that year.   A possible
explanation has been suggested by O'Dell and van Helden (1987), who
have argued that Ribes et al. underestimated the effects of optical
imperfections in early telescopes on the image size of the Sun.   Thus
the evidence for a significantly larger Sun during the Maunder Minimum
must be regarded as highly marginal.

3.   DESCRIPTIONS OF THE FORM OF THE CORONA AT TOTAL SOLAR ECLIPSES

The general form of the corona provides a useful indication of the
level of solar activity. Since the 1950's, when improvements were made
to Lyot's coronagraph (which was first developed around 1930) it has
been possible to map the form of the corona in white light without
the need for an eclipse.   However, before that time only total eclipses
provided a viable opportunity to view the corona so that only sporadic
observations could be made.   About 70 total eclipses are visible somewhere on the Earth's surface during a typical century but many of these
occur in thinly populated regions or are clouded out.   Before the mid-
19th century, eclipse expeditions were rare and rather limited in scope.

In consequence, historical records of the corona cannot compare in
consistency with those of sunspots or aurorae. Despite this, it is
important to summarize existing data during and prior to the Maunder
Minimum.

There are two components of the visible corona, known as K and
F. Both are caused by the scattering of light from the photosphere:
the first by electrons which are contained within the corona; the second
by solid particles which lie between the Earth and the Sun. The form
of the K corona is a function of solar activity since it traces out
the lines of force of surface magnetic fields. However, the form of
the F corona is unrelated to solar activity since it is merely an
indication of the distribution of dust in the interplanetary medium.

When the Sun is active, the K corona dominates. This is rather
bright and has a white glow. A striking feature of the K corona near
sunspot maximum is the numerous tapered streamers, often extending
out to several solar radii. In contrast, the K corona around sunspot
minimum is almost devoid of structure and is relatively dim except
near the edge of the Sun. The zodiacal light, although fairly feeble
in intensity, then tends to dominate. This produces a faint reddish
glow which is mainly of uniform width, but has extensions in the plane
of the ecliptic. In early accounts, it is often difficult to decide
whether writers were describing this F (or false) corona or the actual
corona of the Sun. However, if streamers or irregular structures are
described, we can be fairly sure that it is the true K corona.

Because total eclipses were rather infrequent in the more
scientifically developed areas of Europe during much of the 16th and
17th centuries, it was not until the early 18th century that the corona
was widely observed. Four total solar eclipses visible in Western
Europe in A.D. 1706, 1715, 1724 and 1733 provided suitable opportunities
to view it. Interesting descriptions - if a little untechnical - were
recorded in 1706, 1715 and 1733. Many of these observations have been
compiled by Ranyard (1879). Although the eclipses of 1706 and 1715
occurred towards the end of the Maunder Minimum, by this time the
sunspot cycle was beginning to be detectable again. Unfortunately,
both eclipses took place near the minima of individual cycles - see
Figure 3, which is a histogram of mean annual sunspot frequency
(McKinnon, 1987). Hence the various coronal descriptions, which are
somewhat indicative of a quiet Sun, are of little value. The eclipse
of A.D. 1733 also took place around sunspot minimum. Total eclipses
were rather rare in Europe during the mid-18th century and it was not
until as late as 1778 that we find the first identifiable account
(including illustrations) of the K corona in all its splendour.

Only four clear reports of total eclipses - all from Europe -
have survived from the 16th and 17th centuries: in the years 1560,
1567, 1605 and 1652. All but the earliest of these records do mention
the corona, but the descriptions provide insufficient detail to clearly
distinguish between an active or quiet Sun. Thomas Wyberd, observing
the eclipse of 1652 at Carrickfergus in N. Ireland (Wing, 1669), wrote:
"There was a corona of light around the moon which had a uniform breadth
of about half a digit or a third of a digit at least. It emitted a
bright and radiant light and it was concentric with the Sun and Moon

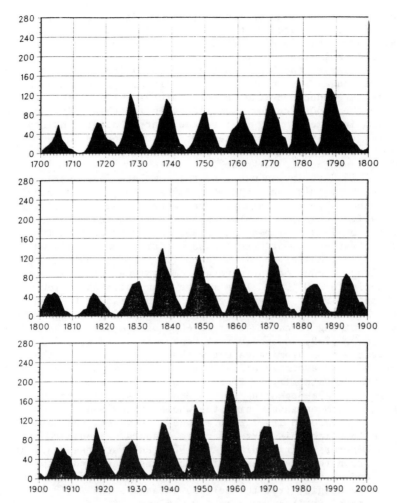

Fig. 3. Variation in the annual mean sunspot numbers from A.D. 1700 to 1975, as depicted by D.C. Wilkinson (McKinnon, 1987).

when in conjunction". A <u>digit</u> had represented one-twelfth of the Sun's apparent diameter since ancient Greek times, so that the breadth of the corona was evidently about one arcmin. In 1605, Kepler (1606) reported the following observation made in Southern Italy: "The surface of the moon appeared quite black but around it shone a brilliant light of a reddish hue and of uniform breadth; this occupied a con- siderable part of the sky".

In 1560 and 1567, Clavius was fortunate to observe by chance two total eclipses, the first at Coimbra in Portugal, the latter at Rome (Clavius, 1593). In 1567, Clavius appears to have witnessed the bright

inner corona near the edge of the photosphere since the angular
diameters of the moon and Sun were almost identical. He noted that
"a certain narrow circle was left on the Sun surrounding the whole
of the Moon on all sides". However, seven years before - when the
apparent diameter of the Moon was significantly greater than that of
the Sun - he had made no mention of the corona, concentrating instead
on the effects produced by the intense darkness. He wrote: "The Moon
... covered the whole Sun for a considerable length of time and there
was darkness in some manner greater than that of night. Neither could
one see very clearly where one stepped; stars appeared in the sky and
(miraculous to behold) the birds fell down from the sky to the ground
in terror of such horrid darkness".

This vivid description mentions the complete disappearance of
the Sun, but is mainly concerned with emphasising the accompanying
darkness rather than the residual appearance in the solar vicinity.
Such a negative attitude is unfortunately all too characteristic of
the numerous medieval European and Arabic accounts of totality which
are still extant - see for example the descriptions translated by
Newton (1972) and Muller and Stephenson (1975). I am only aware of
a single clear account of the corona from medieval times - from
Constantinople in A.D. 968. This may be translated as follows (Muller
and Stephenson, 1975): "Everyone could see the disc of the Sun without
brightness, deprived of light, and some dull and feeble glow like a
narrow band shining round the extreme edge of the disc". At the total
eclipse of A.D. 1239, an observer at Cesena in Italy recorded a fairly
clear reference to a prominence but without any mention of the corona.

There is only a single allusion to the corona in ancient European
writings, by Plutarch in the first century A.D. In his dialogue
entitled On the Face Appearing in the Moon's Disc, Plutarch stated:
"Even though the Moon should hide the whole of the Sun .. a peculiar
radiance is seen around its circumference". However, the date of the
observation to which this text refers is obscure. Around the same
time, when Emperor Domitian died (A.D. 96), Philostratus mentions that:
"A certain crown (corona), resembling the Iris, surrounded the disc
of the Sun and obscured his light". This phenomenon must have been
at atmospheric origin; from the eclipse atlas of Ginzel (1899), no
eclipse was total in Italy between A.D. 75 and 174.

Wang and Siscoe (1980) attempted to identify sightings of the
corona from ancient and medieval Chinese history. However, in many
cases it could be argued that the phenomena described in the works
which they consulted originated in the Earth's atmosphere rather than
in the solar neighbourhood itself. For the dated texts translated
by Wang and Siscoe, comparison with an atlas of solar eclipses for
China (e.g. Stephenson and Houlden, 1986) shows that in or near none
of the suggested years - namely 720, 713 and 488 B.C. and A.D. 1292
and 1639 - was a total eclipse visible in China. In A.D. 1292 the
text specifically mentions an eclipse but modern computations show
this was only annular, no more than 93 per cent of the Sun's diameter
being covered by the Moon. Such a small diminution of light would
not enable the corona to be seen. The only direct reference to the
corona discovered by Wang and Siscoe is in the form of an astrological

omen and dates from some time during the period 77 to 33 B.C. This
reads as follows: "If the ruler does not share his fortune with his
subjects ... there will be a total eclipse with the Sun being black
and its light shooting outward". This is a fairly clear allusion to
the K corona, which must have been observed on some previous occasion,
but the eclipse responsible cannot be identified.

4. SIGHTINGS OF SUNSPOTS

Sunspot observations are the most systematic and direct indices of
solar activity over the historical period. Following the discovery
of the sunspot cycle by Schwabe (1843), Wolf (1856, 1868) developed
the present-day criterion for expressing sunspot frequency. The monthly/
annual mean sunspot number is a measure of the frequency of separate
spots which can be effectively discerned with a telescope of 8 cm
aperture - as employed by Wolf himself.

Figure 3 shows the variation in the annual mean sunspot number
since A.D. 1700 as shown by D.C. Wilkinson in the paper by McKinnon
(1987). The cycle of solar activity which is indicated by the varying
frequency of sunspots has a period of about a decade - half the magnetic
period. Over the last three centuries or so the mean interval between
two successive sunspot maxima has been 11.1 years but the actual interval
has been far from constant, ranging from about 9 to 14 years. The
mean proportion of the visible solar hemisphere covered by sunspots
at the time of maximum activity averages around 0.003 but may reach
0.005, as occurred in June 1957 (Willis and Tulunay, 1979).

The first telescopic observations of sunspots were made independently by Galileo and others around A.D. 1610. However, it was not
until the mid-19th century, that Wolf initiated a systematic monitoring
of the Sun for spots. Wolf eventually established a network of observing
stations. He also made a careful search of published literature and
observatory annals for historical records of sunspots reaching back
to the beginning of the telescopic era. This search was so extensive
that few additional telescopic observations made before Wolf's time
have since been uncovered. There would appear to be considerable scope
for a renewed search for sunspot records from this early period.

Based on descriptions and drawings of the Sun, Wolf was able to
reconstruct an almost complete series of daily sunspot numbers going
back to 1818. Before that date, the available information proved of
lesser quality. Wolf was still able to derive monthly means between
1749 and 1817 but only annual means between 1700 and 1748. Prior to
1700, he made no attempt to deduce even annual mean sunspot numbers.
Eddy (1976) has described the results obtained by Wolf as "poor"
between 1700 and 1748, "questionable" between 1749 and 1817, "good"
between 1818 and 1847, and "reliable" since 1848. It is in this context
that we must attempt to interpret the telescopic record of solar
behaviour down to the mid-19th century.

Figure 4, which is taken from Eddy (1980), is produced from similar
data to Figure 3 but contains estimates of the annual mean sunspot
number back to the early 17th century. This exhibits little trace

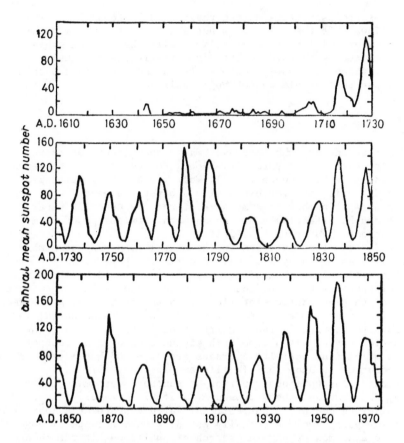

Fig. 4.   Variation in the annual mean sunspot number since the
beginning of the telescopic era (Eddy, 1980).

of periodicity before A.D. 1700.   Just how much of this apparent lacuna
is due to an inactive sun (e.g. during the Maunder Minimum) and how
much is the result of data selection effects cannot be ascertained
satisfactorily.   Clear evidence for a lengthy period of weak solar
activity between about 1645 and 1715 is provided by the witness of
several astronomers of the time - such as Cassini and Flamsteed - who
commented on the scarcity of sunspots (Eddy, 1976).   In particular,
Cassini wrote in 1671 that about 20 years had elapsed since astronomers
had seen "any considerable spots on the Sun".   However, because of
the relative inaccessibility of astronomical records of any kind
during the 18th century (compared with more recent times), it is not
possible to objectively quantify the level of solar activity in this
time period with any degree of confidence from sunspot records alone.
      This difficulty is exemplified by the sampling of data relating
to other celestial phenomena from similar sources to those recording
sunspot observations.   Sightings of solar eclipses and occultations

since the early 17th century have been extensively studied because
of their importance in investigating variations in the Earth's period
of rotation. Unlike for sunspots, the frequency of such observations
is not expected to show any real causal trends. Hence the trends which
are apparent must be of artificial origin. Figure 5 illustrates the
frequency of telescopic timings of solar eclipses (usually no more
than one event per year) made between A.D. 1620 and 1800. These observations are taken from the catalogue compiled by Morrison et al.
(1981), which was based on a fairly comprehensive literature search.
The diagram reveals considerable scatter depending on a variety of
factors: weather, eclipse magnitude, observer interest, relative
accessibility of records, etc. - some of which might be expected to
affect the frequency of sunspot sightings. In addition, over the 180
year interval there is a gradual trend towards increased frequency
of observations - as indicated by the dashed line in Figure 5. The
eclipse catalogue of Morrison et al. terminated at A.D. 1806 because
after this time there was declining interest in the use of solar eclipse
timings to determine geographic longitude. Hence we have a situation
in the early 18th century which is much the antithesis of that for
sunspots at a similar period.

Figure 6 shows the annual frequency of telescopic timings of
occultations of stars by the Moon between A.D. 1620 and 1860. This
diagram is from the paper by Stephenson and Morrison (1984), which

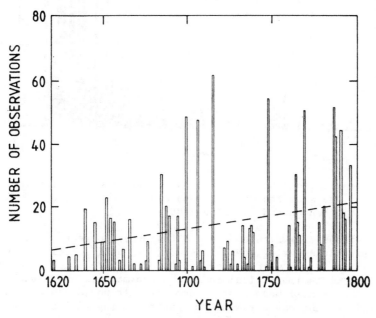

Fig. 5.   Frequency of telescopic timings of solar eclipses (usually
no more than one event per year) made between A.D. 1620
and 1800 - from data compiled by Morrison et al. (1981).

120

is again based on the catalogue of Morrison et al. (1981). Figure
6 reveals a dramatic rise in the frequency of observation around A.D.
1800 and a lesser but still conspicuous rise near 1670. Such changes
are partly the result of a greater number of astronomers making regular
observations in the later periods. However, a further important factor
is the relative ease of accessibility of their records. By the late
17th century, many observations were communicated to the leading
scientific journals of the time; whereas earlier material is often
scattered in unpublished papers or books whose circulation was very
limited.

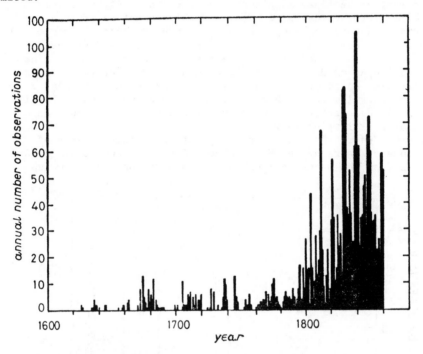

Fig. 6. Annual frequency of telescopic timings of occultations
of stars by the Moon between A.D. 1620 and 1860
(Stephenson and Morrison, 1984).

It is evident from Figures 5 and 6 that the apparent frequency
of sunspot sightings during the 17th century is artificially depressed
by a significant factor which is probably both variable and indeterminate. Hence it is impractical to attempt normalisation of such
data in terms of annual mean sunspot numbers. It would appear that
the true scale of the Maunder Minimum cannot be properly assessed from
sunspot data alone; proxy records are also necessary.

The total number of known sunspot observations during the Maunder
Minimum is by no means small: some 700. Hence the discovery of six
further naked-eye sightings from China during the Maunder Minimum (in
addition to 15 sightings in the early part of the 17th century) by

Xu and Jiang (1982) has little bearing on the extent of the Maunder Minimum. Eddy (1983) noted that some of these six sunspots in fact coincided with spots reported in Europe. Where Chinese and other East Asian sunspot records are of importance is in the period prior to the telescopic era, during which there are scarcely any known sightings from Europe.

The sunspot record since A.D. 1700 illustrated in Figure 3 reveals some indication of modulation on the centennial time-scale. This is the so-called "Gleissberg Cycle" (Gleissberg, 1965). Possibly the Maunder Minimum may be interpreted as a continuation of this pattern - see Figure 4. However, the reality of the Gleissberg Cycle is open to question, particularly on account of its obviously variable length. The interval between successive maxima ranges between about 50 and 100 years. At present, the possibility cannot be ruled out that the various modulations are random in nature. This would give a false impression of periodicity since only a restricted time-span is available (Barnes et al., 1980). The accumulation of future observations for several more decades - not a very encouraging prospect to present-day solar physicists - may be necessary before the existence of the Gleissberg Cycle can be proved or disproved.

There may be an analogy here in the apparent motion of the Moon. During the 19th and early 20th centuries it was believed that the lunar mean longitude contained a fairly smooth sinusoidal term of unknown origin (Newcomb, 1878; Brown, 1919). This variation, of approximate amplitude 10 arcsec and period 250 years was called the "Great Empirical Term" - see Figure 7. However, marked variations in the

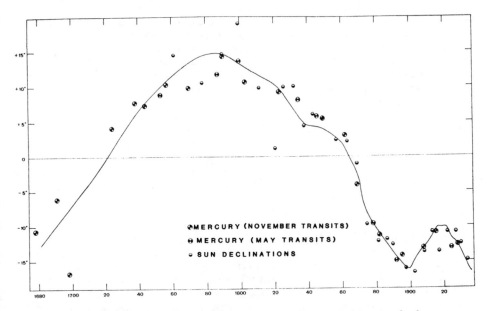

Fig. 7.   Apparent fluctuations in the lunar mean longitude between A.D. 1680 and 1930 (Spencer Jones, 1939).

Earth's rate of rotation (at the millisecond level) which occurred
at the beginning of the present century led to the demonstration by
Spencer Jones (1939) that the Great Empirical Term was merely the
reflection in the lunar motion of irregularities in the Earth's rate
of rotation. These random fluctuations in the length of the day had
followed a rather smooth trend for more than two centuries but around
1900 the pattern became more complex.

Although the earliest known sunspot sighting dates from 165 BC
(Schove, 1950), less than 200 observations of sunspots - averaging
about one per decade - are extant from the whole of the pre-telescopic
period. This is far fewer than the number of telescopic sightings
during the 70 years of the Maunder Minimum. The pre-telescopic record
from Europe is very disappointing, totalling only about seven separate
observations (Wittmann and Xu, 1987). Some of these were made not
long before the invention of the telescope - in 1590 and 1607. The
others date from medieval times - A.D. 807, 1139, 1365 and 1371. No
ancient European records are known.

Four observations, reported as transits of Mercury or Venus by
Arabic astronomers, were shown by Goldstein (1969) to probably relate
to sunspots. These observations date from A.D. 840, c. A.D. 1000,
c. 1080, and c. 1100, only the first date being exact. Mercury is
actually much too small (apparent diameter no more than 15 arcsec)
to be discernible by the unaided eye when in transit. However, Venus
(apparent diameter fully 1 arcmin) is fairly readily visible when
projected on the solar disc. Using the list of Venus transits computed
by Meeus (1958), Goldstein ruled out the possibility that the supposed
transits of A.D. 840, c. 1080, and c. 1100 were real. The date of
the event around A.D. 1000 was not accurate enough to make a definite
decision. In inferring transits, the medieval Arabic astronomers
were presumably influenced by the Aristotelian theory of a faultless
Sun. This notion may have been partly responsible for the lack of
other identifiable references to sunspots in Arabic works and ancient
European writings. However, in medieval Europe there may be a much
simpler explanation for the paucity of sightings. After the close
of the Classical Age, until the Renaissance, there was little European
interest in all but the most spectacular astronomical phenomena
(notably eclipses, comets and aurorae). Such a refined pursuit as
the observation of sunspots would arouse minimal interest. In this
context, it seems relevant that the two known medieval Russian sightings
of sunspots - in A.D. 1365 and 1371 - took place when smoke from forest
fires dimmed the Sun (Vyssotsky, 1949).

In order to find anything approaching a systematic record of
sunspots during the pre-telescopic period, it is necessary to examine
the chronicles of East Asia. This region may be taken to include
China, Korea, Japan and Vietnam. Catalogues of sunspots from oriental
sources have in recent years been compiled by the "Ancient Sunspots
Records Research Group" (1977), Clark and Stephenson (1978), Wittmann
and Xu (1987), and Yau and Stephenson (1988). Almost all of the observations in these catalogues are from Chinese and Korean history. An
extensive search of the astronomical records of Japan from earliest
times down to the 17th century by Kanda (1935) revealed only a single

identifiable sunspot observation (in A.D.851). The few Vietnamese sunspot reports (in A.D. 1276, 1593 and 1603) are similar to contemporary Chinese or Korean accounts and thus may not be original.

The principal sources of astronomical observations (including sunspots) from both China and Korea are the treatises on astronomy in the various dynastic histories. These works are largely compilations from the records of the court astronomers and contain references to all kinds of celestial phenomena which can be observed with the unaided eye. Usually the exact date of an observation is given and this can be readily converted to the Julian or Gregorian Calendar using tables. The earliest Chinese treatise - in the Han-shu ("History of the Former Han Dynasty") - covers the last two centuries before the Christian Era. From then on, similar works are continuously available until after the invention of the telescope. An additional source of Chinese data from the 14th century is the Ming-shih-lu, a chronicle of the Ming Dynasty (A.D. 1358 to 1644). Independent Korean observational records probably do not begin until about A.D. 1000 but by A.D. 1100 they are of comparable frequency to those of China. The earlier Korean observations (down to A.D. 1392) are reported in the astronomical treatise of the Koryo-sa, the official history of the period. However, later material is found in the Wangjo Sillok, a detailed chronicle of the Yi Dynasty, which extends down to the 19th century.

Most sunspots in Chinese and Korean history are described either as "black spots" (hei-tzu) or "black vapours" (hei-ch'i) seen on the Sun. However, the oriental astronomers did not always make distinctions between sunspots and atmospheric phenomena. Thus we occasionally find reference to black vapours "beside" the Sun, possibly relating to clouds viewed obliquely near sunrise or sunset. In order to attempt a definitive classification, it is better to restrict attention to those records which describe black vapours or spots specifically "within" the Sun unless the sunspot nature is obvious - e.g. when a duration of several days was noted. Such criteria were used by Clark and Stephenson (1978) in compiling their catalogue of oriental sunspot sightings and also by Yau and Stephenson (1988) in producing an enlarged version of this work.

There is little evidence that the oriental astronomers regularly used artificial aid to dim the Sun. In most cases atmospheric diminution of sunlight - e.g. during a dust storm or near sunrise or sunset - enabled spots to be seen. Thus it is frequently stated that the Sun appeared unusually yellow or red when spots were visible on it. The investigation by Willis et al. (1980) demonstrates that the frequency of sunspot observations in East Asia shows a high spring maximum - a time when dust storms are prevalent.

The East Asian record of 170 separate sunspot sightings in the whole of the pre-telescopic period (Yau and Stephenson, 1988) is remarkable by the standards of the rest of the world at that time. However, this total must represent no more than a minute proportion of the number of spots actually visible to the unaided eye over the same time interval. Eddy (1983) noted that if during the pre-telescopic period the Sun has been active as it was over the last century, the efficiency

124

of the oriental astronomers in recording visible sunspots was as low
as 0.1 per cent. This deduction was based on the assumption that
a sunspot of angular diameter 1 arcmin or more is detectable by the
unaided eye. Systematic observations by Mossmann (1988) in 1981-1982,
near the last solar maximum, indicate that the resolution of the eye
is rather better than this; sunspots of angular diameter approximately
0.5 arcmin were discernible. The implications are either that (i)
most oriental observations of sunspots are no more than chance sightings
or (ii) if the Sun was indeed scrutinized fairly regularly very few
sunspot sightings were actually recorded (Eddy et al. 1988). Neither
alternative is in keeping with the remarkable degree of completeness
of most East Asian astronomical records such as of eclipses (Wylie,
1897) and comets (Ho Peng Yoke, 1962). However, this conclusion seems
unavoidable. In this context, it is interesting to note that as many
as 25 per cent of oriental sunspot sightings occurred on the first
day of the lunar month - a time when the astronomers would be
scrutinizing the Sun for possible eclipses (Yau and Stephenson, 1988).

Under these circumstances it is debatable whether the meagre
pre-telescopic sunspot record is of value in detecting the solar cycle
except in occasional periods of unusually frequent reports such as
around A.D. 1400 - see Figure 8. The use of these data in the study
of long-term solar variability would seem to be a more viable prospect
but even this is subject to limitations.

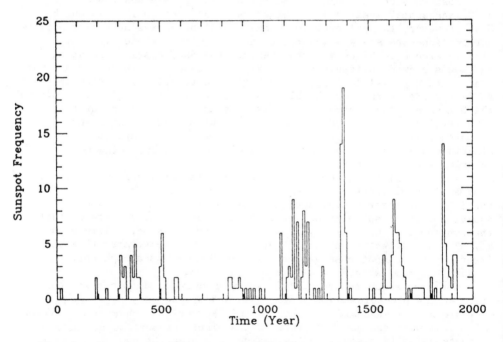

Fig. 8. Frequency of sunspot sightings per decade recorded in
oriental history from the beginning of the Christian Era
to the early 20th century (Yau, 1988).

Figure 8 is a plot of the frequency of sightings of sunspots per decade reported in East Asian history from the beginning of the Christian Era to the early 20th century (Yau, 1988). There are only four allusions to sunspots in more ancient times: in 165, 43, 32 and 28 B.C. Although the diagram extends to relatively modern times, only the pre-telescopic period (before A.D. 1610) will be of concern here. Figure 8 is characterised by several marked peaks and lengthy gaps, but it is difficult to determine how much of this variation is a reflection of real changes in solar activity. Thus the gap between about A.D. 600 and 800 is probably largely due to the effects of a major rebellion in China around A.D. 760. During this insurrection, most of the astronomical records dating back to about A.D. 610 were destroyed. Hence there is no valid reason for supposing an unusually inactive Sun around this time.

As an illustration of just how variable the record of astronomical phenomena in a single dynasty can be, Figure 9 shows a plot of the annual frequenty of astronomical events of all kinds recorded in the Koryo-sa (Yau, 1988). This history covers the entire period between A.D. 918 and 1392 but no astronomical records are preserved until after A.D. 1000. Scarcely any of the astronomical events noted (e.g. eclipses, comets, lunar and planetary conjunctions) would be expected

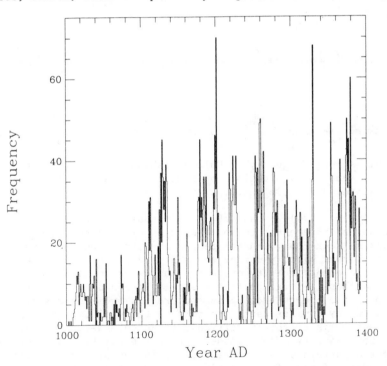

Fig. 9. Annual frequency of astronomical observations of all kinds reported in Korea during the Koryo Dynasty - A.D. 918 to 1392 (Yau, 1988).

to show obvious trends in the frequency of occurrence so that most
of the pattern visible in Figure 9 must be of artificial origin. Some
of the gaps can be readily explained by historical considerations.
Thus when Korea was invaded by the Mongols in A.D. 1233, the king
and his court fled from the capital to an island refuge where they
remained for many years. This event is largely responsible for the
low frequency of astronomical records around this time. Varying rates
of preservation of historical observations as well as changing
attitudes towards celestial portents (Park, 1977) must account for
much of the distribution shown in Figure 9 but it is clear that the
histogram is much too complex to be explained in detail.

The existence of similar spurious trends in the record of one
type of celestial phenomenon is shown in Figure 10. This covers the

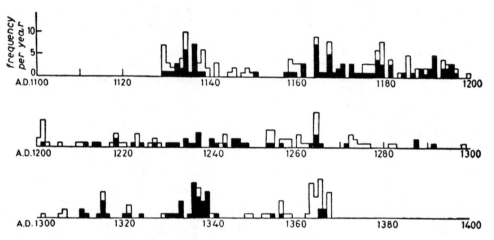

Fig. 10. Annual frequency of Chinese (shown shaded) and Korean
(unshaded) sightings of atmospheric phenomena involving
the Sun between A.D. 1129 and 1367.

interval between A.D. 1129 and 1367 and is based on an annual count
by the author of references to solar changes of atmospheric origin
– e.g. haloes, mock Suns, unusual reddening of the Sun – in both
Chinese and Korean history during the selected period. The precise
interval was chosen since it covers two Chinese dynasties (Southern
Sung and Yuan) and a large part of a single Korean dynasty (Koryo).

How much the pre-telescopic sunspot record from the Orient (Figure
8) is affected by similar data artefacts to those illustrated in
Figures 9 and 10 is extremely difficult to judge. Nevertheless, it
would be surprising if strong biases were not present. In order to
interpret the pattern of solar variability in the pre-telescopic period
with any degree of confidence, it is necessary to compare the sunspot
record with other independent evidence, particularly C-14 measurements.
Figure 11 is taken from Stuiver and Quay (1980) and is based on C-14
measurements in tree rings between about A.D. 1000 and 1900. This

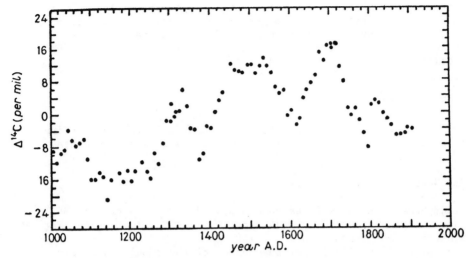

Fig. 11.   Variations in atmospheric C-14 (parts per million) between
           A.D. 1000 and 1900 from tree ring measurements.   The
           long-term trend caused by changes in the intensity of
           the geomagnetic field has been removed.   Counting errors
           (1 standard deviation) range from about 2 parts per
           million around A.D. 1000 to roughly half of this amount
           since A.D. 1800 (Stuiver and Quay, 1980).

diagram shows clear evidence of increased C-14 production (i.e.
decreased solar activity) around the time of the Maunder Minimum and
again around A.D. 1500 and 1300 - the Sporer and Wolf Minima.   Reference to Figures 4 and 8 shows that there is a good inverse
correlation with both the telescopic and pre-telescopic sunspot records.
However, in view of the critical comments made above, at least some
of the accord in the pre-telescopic period may be coincidental; the
earlier sunspot record is tantalisingly incomplete.

5.   CONCLUSION

Historical observations enable changes in the diameter of the Sun
on the centennial time-scale to be detected since A.D. 1700 but at
present these cannot be confidently related to variations in the solar
output.   Observations of the corona at total eclipses are too infrequent to effectively supplement the sunspot record as an index
of solar activity except in relatively recent times.   The cycle of
solar activity can be mapped with fair reliability during the 18th
century and in detail since the early 19th century using sunspot
observations.   However, before A.D. 1700, it is only possible to use
sunspot records to trace the long-term pattern of activity in an
essentially qualitative way.   Spurious trends due to data artefacts

may be of large magnitude in these earlier periods, at least partially masking any real evidence of long-term solar variability. It is concluded that at any time prior to A.D. 1700, proxy records such as C-14 and Be-10 are more reliable indicators of solar activity than historical observations.

ACKNOWLEDGEMENTS

I wish to thank Dr. J.A. Eddy of UCAR, Boulder, U.S.A. for reading this paper and offering several valuable criticisms and comments.

REFERENCES

Ancient Sunspots Records Research Group, 1977, Chin. Astr., 1, 347.
Barnes, H.A., Sargent, H.H. and Tryon, P.V., 1980, in 'The Ancient Sun', Pergamon Press, New York, 159.
Brown, E.W., 1919, 'Tables of the Motion of the Moon', Yale Univ. Press, New Haven, Vol. 1.
Clavius, C., 1593, 'In Sphaeram Ioannis de Sacrobosco Commentarius', Lugduni, 508.
Clark, D.H. and Stephenson, F.R., 1978, Q. J. R. Astr. Soc., 19, 387.
Eddy, J.A., 1976, Science, 192, 1189.
Eddy, J.A., 1980, in 'The Ancient Sun', Pergamon Press, New York, 119.
Eddy, J.A., 1983, Solar Phys., 89, 195.
Eddy, J.A. and Boornazian, 1979, Bull. Am. Astr. Soc., 11, 437.
Eddy, J.A., Stephenson, F.R. and Yau, K.K.C., 1988, submitted to Q. J. R. Astr. Soc.
Frohlich, C. and Eddy, J.A., 1984, Adv. Space Res., 4, 121.
Gilliland, R.L., 1981, Astrophys. J., 248, 1144.
Ginzel, F.K., 1899, 'Spezieller Kanon der Sonnen-und Mondfinsternisse', Berlin.
Gleissberg, W., 1965, J. Br. Astr. Assoc., 75, 227.
Goldstein, B.R., 1969, Centaurus, 14, 49.
Gough, D.O., 1981, in 'Variations of the Solar Constant', Washington, D.C., 185.
Hill, H.A. and Stebbins, R.T., 1975, Astrophys J., 200. 471.
Ho Peng Yoke, 1962, Vistas in Astronomy, 5, 127.
Kanda Shigeru, 1935, 'Nihon Tenmon Shiryo', Tokyo.
Kepler, J., 1606, 'De Stella Nova in Pede Serpentarii', Prague, 116.
McKinnon, J.A., 1987, 'Sunspot Numbers: 1610-1985', World Data Center A, Boulder.
Meeus, J., 1958, J. Br. Astr. Assoc., 68, 98.
Morrison, L.V., Lukac, M.R. and Stephenson, F.R., 1981, Royal Greenwich Observatory Bull., No. 186.
Morrison, l.V. and Stephenson, F.R., 1986, in 'Earth's Rotation: Solved and Unsolved Problems', Reidel, Dordrecht, 69.
Morrison, L.V., Stephenson, F.R. and Parkinson, J.H., 1988, Nature, 331, 421.

Mossmann, J.E., 1988, submitted to Q. J. R. Astr. Soc.

Muller, P.M. and Stephenson, F.R., 1975, in 'Growth Rhythms and the History of the Earth's Rotation', Wiley and Sons, London.

Newcomb, S., 1878, Washington Observations for 1875, Appendix II, 1.

Newton, R.R., 1972, 'Medieval Chronicles and the Rotation of the Earth', Johns Hopkins Univ. Press, Baltimore.

O'Dell, C.R. and Van Helden, A., 1987, Nature, 330, 629.

Park, S.R., 1971, 'Portents and Politics in Early Yi Korea 1392-1519', Ph.D. Thesis, University of Hawaii.

Parkinson, J.H., Morrison, L.V. and Stephenson, F.R., 1980, Nature, 288, 548.

Ranyard, A.C., 1879, Mem. R. Astr. Soc., 41.

Ribes, E., Ribes, J.C. and Barthalot, R., 1987, Nature, 326, 52.

Schove, D.J., 1950, J. Br. Astr. Soc., 61, 22.

Schwabe, H., 1843, Astr. Nacht., 20, 283.

Shapiro, I.I., 1980, Science, 208, 51.

Spencer Jones, H., 1939, Mon. Not. R. Astr. Soc., 99, 541.

Stephenson, F.R., 1981, J. Br. Astr. Assoc., 91, 345.

Stephenson, F.R. and Morrison, L.V., 1984, Phil. Trans. R. Soc. Lond. A, 313, 47.

Stuiver, M. and Quay, P.D., 1980, Science, 207, 11.

Wang, P.K. and Siscoe, G.L., 1980, Solar Phys., 66, 187.

Willis, D.M., Easterbrook, M.G. and Stephenson, F.R., 1980, Nature 287, 617.

Willis, D.M. and Tulunay, Y.K., 1979, Solar Phys., 64, 237.

Wing, V., 1669, 'Astronomica Britannica', London, 356.

Wittmann, A.D. and Xu Zhen-tao, 1987, Astr. Astrophys. Suppl. Ser., 70, 83.

Wolf, R., 1856, Astr. Mitt. Zurich, 1, viii.

Wolf, R., 1868, Astr. Mitt. Zurich, 24, 111.

Yau, K.K.C., 1988, 'An Investigation of some Contemporary Problems in Astronomy and Astrophysics by way of Early Astronomical Records', Ph.D. Thesis, University of Durham.

Yau, K.K.C., and Stephenson, F.R., 1988, Q. Jl. R. Astr. Soc. (in press).

THE BEHAVIOUR OF SOLAR ACTIVITY AS INFERRED FROM SUNSPOT OBSERVATIONS
165 BC TO 1986

A.D. Wittmann
University Observatory, Goettingen, West Germany.
Z.T. Xu
Purple Mountain Observatory, Nanjing, China.

ABSTRACT. We have studied the behaviour of solar activity on the basis
of a catalogue of non-telescopic sunspot observations covering the period
165 BC to 1684 (Wittmann and Xu, 1987), to which recently has been
appended a catalogue of naked-eye sunspots of the period 1764 to 1986.
Our main results are as follows: between AD 302 and 1980 there are 44
conspicuous maxima which may be expressed as: Year (Max.) =
$4.0 + 11.116 \times N$ (where $N = 178$ for the maximum of 1981, i.e. $N =$
Zürich cycle + 157). The residuals (O-C) are less than 3 years in most
cases, and only one (in 1790) slightly exceeds 4 years. The average
period of the sunspot cycle is $P = 11.116 \pm 0.007$ years, with
individual periods ranging from 7.5 to 14.5 years (30% are between
10.5 and 11.7 years; 70% are between 9.9 and 12.3 years). While both the
'modern' and the 'reliable' ancient spots show a bimodal distribution
of O-C values with a distinct concentration around zero, the residuals
for the whole sample of ancient spots show only weak indications of a
maximum and are almost evenly distributed (which we interpret in terms
of a statistically incomplete sample, rather than as evidence for a
strange attractor model of solar activity). Whereas the Oort Maximum
of AD 1130, the Wolf Maximum of AD 1380, the Spoerer Maximum of AD 1600
and the associated minima preceding them by about 90 years (viz. the
Oort Minimum 1010-1050, the Wolf Minimum 1282-1342, and the Spoerer
Minimum 1416-1534) are conspicuous in our data (as are several other
intervals characterized by a scarcity of observations), there is little
or no indication of the Maunder Minimum (1654-1714) from our data alone.
The Maunder Maximum of AD 1780 is again conspicuous, but this time in
the modern sunspot record.

1. INTRODUCTION

Solar activity is traditionally measured in terms of relative sunspot
numbers, which approximately correspond to the total number (or, more
precisely, to the total area) of all spots on the disc. The sunspot
relative number is closely correlated with the solar radio flux at
10.7 cm and - as we have learned from the ACRIM experiment aboard the

131

F. R. Stephenson and A. W. Wolfendale (eds.),
Secular Solar and Geomagnetic Variations in the Last 10,000 Years, 131–139.
© *1988 by Kluwer Academic Publishers.*

Solar Maximum Mission Satellite - with minute fluctuations of the
solar constant. Proxies of solar activity, such as aurorae, geomagnetic
storms, tree ring widths, lake sediments, etc. are, however, more
severely biased by terrestrial influences, and do not in general show
a very close correlation with the sunspot number. We have, therefore,
restricted our investigation to sunspot observations alone. By
updating, correcting and merging previously published catalogues of
large sunspots visible to the naked eye (i.e. sunspots with angular
diameters in excess of about 105", or with areas in excess of about
1500 millionth of a hemisphere), and by adding a substantial
amount of new data, we have produced a machine-readable catalogue of
naked-eye sunspot observations 165 BC - 1684. Our catalogue has been
published   elsewhere (Wittmann and Xu, 1987), and the present paper
will, therefore, concentrate on additional aspects and on the main
results obtained so far.

2. OBSERVATIONAL DATA

References to previous catalogues of large sunspots (including the
modern telescopic era, are summarized in Table i.

Table i

Author(s)	Period	Reference
Williams (1873)	301 - 1205	Monthly Not. 33, 370.
Fritz (1882)	28 BC - 1607	Sirius 10, 227.
Hosie (1879)	28 BC - 1617	Nature 20, 131.
Hirayama (1889)	188 - 1638	The Observatory 12, 217.
Moidrey (1904)	28 BC - 1638	Bull. Astron. 21, 59.
Kanda (1932)	28 BC - 1743	Annals Obs. Tokyo 5, 1.
Zhu (1933)	28 BC - 1638	Tian Wun Kao Gu Lu.
Richardson (1937)	1882 - 1937	Publ. Astron. Soc. Pac. 49, 87.
Spencer Jones (1955)	1874 - 1954	Greenwich Sunspot Geomag. Data.
Newton (1955)	1874 - 1954	J. Brit. Astron. Assoc. 65, 225.
Chen (1957)	43 BC - 1638	J. Nanjing Univ. 4, No. 4.
Gnevysheva (1972)	1955 - 1969	Solnechnye Dannye 7, 76.
Kopecký/Kotrč (1974)	1955 - 1964	Bull. Astron. Inst. Czech. 25, 171.
ASRR Group (1977)	43 BC - 1638	Chin. Astron. 1, 347.
Clark/Stephens. (1978)	28 BC - 1604	Quart. J. R.A.S. 19, 387.
Stephenson/Cl. (1978)	28 BC - 1604	Appl. of Early Astron. Rec., Ch. 4.
Wittmann (1978)	467 BC - 1638	Astron. Astrophys. 66, 93.
Xu/Jiang (1982)	1603 - 1665	Chin. Astron. Astrophys. 2, 84.
Xu (1983)	165 BC - 1684	Kunming Workshop, Vol. 1, p. 108.
Xu (1985)	32 BC - 1684	cf. Wittmann/Xu (1987).
Jiang/Xu (1986)	1402 - 1597	Astrophys. Space Sci. 118, 159.
Wittmann/Xu (1987)	165 BC - 1684	Astron. Astrophys. Suppl. in press.

Our catalogue now contains 235 entries for the period 165 BC to 1684 (the ancient record) and 210 entries for the period 1764 to 1986 (the modern record). The latter have mainly been compiled from the sources of Table i. Due to the lack of regular observations before the commencement of the (now abandoned) Greenwich series, they are necessarily incomplete before 1872. As is well known, sunspots were − at least occasionally − also observed during the Maunder Minimum: for instance by Flamsteed around 1676 (cf. Flamsteed, 1725), and by Kirch in 1678-1688 and 1700-1718 (cf. Landsberg, 1980), and these will be included as time permits. There remain, of course, the facts that a few, isolated spots do not make a maximum, and that many experienced observers (among them Cassini and Flamsteed) have distinctly noted the absence of sunspots for extended periods of time (for details cf. Eddy, 1983). It should be borne in mind, however, that the 11 year cycle had not yet been discovered and that observers − familiar with books like Scheiner's "Rosa Ursina sive Sol" (1630) − probably expected an abundance of sunspots to be a regular and permanent feature of the sun.

Table ii lists the central meridian passage (CMP) dates of naked-eye sunspots 1764-1986 (note that there were no such spots in 1985-1986, and that the table is still somewhat incomplete).

Figure 1 is a plot of the distribution in time of all known naked-eye sunspots of the period 32 BC to 1986, where it has been assumed that each catalogue entry corresponds to only one spot (although in some cases multiple spots have been described), and where all spots occurring in the same calendar year have been plotted on top of each other (note the high concentration of large spots in 1981).

Whereas the sunspot relative number (or the sunspot record in general) is known with some confidence since AD 1700, little is known before that date (cf. e.g. Figure 1 of Eddy, 1977). Our Figure 1 shows that the record of giant spots is highly incomplete (undersampled) even before, say, 1830. On the other hand, Figure 1 clearly shows several conspicuous maxima (such as, for instance, that in AD 1382, cf. Section 5) and also shows − or at least suggests − that the spacings between individual maxima tend to be multiples of 10-11 years (this has, of course, already been observed by most of the authors quoted in Table i, and can further be substantiated by mathematical scrutiny of the data).

Power spectra of the sunspot relative number (e.g. Cohen and Lintz, 1974; Wittmann, 1978; Lomb and Andersen, 1980; Otaola and Zenteno, 1983; Sonett, 1983b) show significant peaks at periods of 8.4 years (relative power 0.1), 10.0 years (0.4), 11.1 years (1.0), 12.2 years (0.1), 56.0 years (0.2), and 92.5 years (0.3). Whereas the secondary peaks close to the main peak of 11.1 years are caused by a multiplet structure of the basic maxima (cf. Section 4), the last-mentioned period (92.5 years) corresponds to the 80 year Gleissberg cycle (Gleissberg, 1952), which manifests itself in an amplitude modulation of the 11 year Schwabe cycle (see also Yoshimura, 1979). It seems, however that − contrary to Gleissberg (1955) − the Gleissberg cycle is probably transitory in nature and persisted only during 1720-1960;

Table ii

1764 APR 15	1848 JUL 19	1858 OCT 16	1884 APR 01	1925 DEC 29	1946 FEB 06
1779 --- --	1848 JUL 31	1858 NOV 09	1892 FEB 12	1926 JAN 25	1946 JUL 27
1801 SEP --	1848 SEP 22	1858 NOV 29	1892 JUL 10	1926 MAR 04	1946 DEC 17
1828 MAY 24	1848 NOV 16	1859 JUL 25	1893 AUG 07	1926 SEP 20	1947 FEB 11
1828 SEP --	1848 DEC 25	1859 AUG 31	1894 OCT 08	1928 SEP 27	1947 MAR 10
1829 APR --	1849 JAN 22	1859 OCT 28	1896 SEP 17	1929 NOV 30	1947 APR 07
1836· JUL --	1849 APR 06	1861 MAR 29	1897 JAN 09	1935 DEC 02	1948 DEC 24
1839 AUG 30	1849 MAY 06	1861 APR 19	1898 SEP 09	1937 JAN 31	1949 JAN 23
1839 OCT --	1849 MAY 20	1861 MAY 26	1903 OCT 12	1937 APR 25	1949 FEB 05
1842 JUL 01	1849 JUN 20	1861 AUG 01	1905 FEB 04	1937 JUL 29	1950 FEB 20
1846 DEC 06	1849 SEP 14	1861 SEP 25	1905 MAR 08	1937 OCT 04	1950 APR 14
1847 JUN --	1849 NOV 05	1864 JAN 09	1905 JUL 17	1938 JAN 18	1951 APR 19
1847 JUN 14	1850 JUL 12	1864 AUG 13	1905 OCT 20	1938 JUL 15	1951 MAY 16
1847 JUL --	1852 DEC 30	1864 OCT 23	1907 FEB 12	1938 OCT 12	1951 JUN 18
1847 AUG --	1854 NOV 14	1864 OCT 31	1907 JUN 20	1938 NOV 11	1956 JAN 21
1847 AUG --	1857 DEC 11	1864 NOV 07	1908 AUG 31	1939 SEP 01	1956 FEB 17
1847 SEP --	1858 MAR 15	1864 NOV 28	1917 FEB 10	1939 SEP 10	1956 SEP 12
1847 NOV --	1858 JUN 04	1882 APR 17	1917 AUG 10	1939 SEP 28	1956 DEC 10
1848 MAY --	1858 AUG 28	1882 APR 19	1920 JAN 28	1940 JAN 06	1957 JAN 03
1848 JUN 27	1858 SEP 30	1882 NOV 19	1920 MAR 22	1941 SEP 17	1957 MAY 12

1957 JUN 21	1960 NOV 12	1978 DEC 13	1981 APR 15	1982 JUL 15
1957 JUN 23	1967 FEB 27	1979 JUN 12	1981 APR 21	1982 NOV 15
1957 JUN 25	1967 MAY 26	1979 NOV 09	1981 MAY 11	1982 NOV 20
1957 SEP 19	1967 JUL 28	1980 FEB 05	1981 MAY 18	1982 DEC 02
1957 OCT 18	1968 JAN 31	1980 APR 07	1981 MAY 20	1983 FEB 02
1957 DEC 25	1969 APR 03	1980 APR 08	1981 JUN 26	1983 FEB 06
1958 MAR 07	1969 OCT 26	1980 MAY 26	1981 JUL 01	1983 JUN 05
1958 MAY 03	1969 NOV 22	1980 JUL 17	1981 JUL 22	1983 AUG 03
1958 JUN 10	1970 NOV 14	1980 OCT 11	1981 JUL 24	1983 OCT 13
1959 JAN 11	1971 JAN 22	1980 OCT 24	1981 JUL 28	1984 AUG 31
1959 JAN 23	1971 AUG 23	1980 NOV 06	1981 AUG 10	
1959 MAR 18	1972 AUG 05	1980 NOV 11	1981 SEP 10	
1959 MAR 31	1972 OCT 30	1980 DEC 18	1981 OCT 14	
1959 APR 23	1974 JUL 04	1981 FEB 11	1981 OCT 19	
1959 JUN 23	1974 OCT 12	1981 FEB 22	1981 NOV 05	
1959 AUG 01	1974 NOV 21	1981 FEB 29	1982 FEB 01	
1959 AUG 29	1976 MAR 19	1981 MAR 03	1982 FEB 10	
1959 DEC 02	1978 MAY 01	1981 MAR 12	1982 MAY 22	
1960 MAR 31	1978 MAY 28	1981 MAR 20	1982 JUN 08	
1960 JUL 04	1978 DEC 12	1981 MAR 24	1982 JUN 18	

Gleissberg's period of 78.8 years indeed gives a better fit during the past, but is not in accordance with recent power spectra (e.g. Cole, 1973) and would, for instance, predict a maximum amplitude in 1939 (and not after 1960, as has been observed).

3.   EPOCH ANALYSIS

A total of 44 conspicuous maxima may be identified in our catalogue, of which 17 are based on the ancient (non-telescopic) record.   Many additional maxima may be tentatively identified, but we have used only the most prominent and well-established ones in our analysis.   From progressive least-squares fits back in time we get in terms of Julian Day number (JD):

$$JD(Max) = (1722516 \pm 365) + (4060.1 \pm 2.5) \times N \qquad (1)$$

where N is an arbitrary cycle number (N = 178 for the maximum of
1980.83, i.e., N = Zürich cycle number + 157). With practically no
loss of accuracy, equation (1) may also be written as:

$$Year(Max) = (4.0 \pm 1.0) + (11.116 \pm 0.007) \times N \qquad (2)$$

Hence the mean period of the sunspot cycle (i.e. half the period of the
22 year Hale cycle) is P = 11.116 ± 0.007 years, which is in good
agreement with the results of other authors (e.g. Wolf, 1861; Newcomb
1901; Kimura, 1913; Nicolini, 1952; Nicolini, 1976; Dicke, 1979).

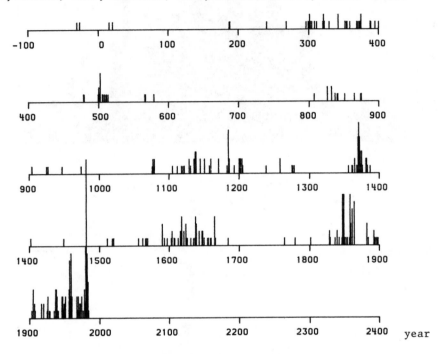

Fig. 1.   Distribution in time of naked-eye sunspots (i.e.
           sunspots with areas in excess of about 1500 millionth
           of a hemisphere). Data are from the catalogues
           referenced in Table i.

In Figure 2 we have plotted the relative phase (years from the
nearest maximum) of the 360 most reliable spots of the period AD 299 –
1985 with respect to Equation (2). Figures 3 and 4 show the same
diagram for the modern record (i.e., for the 153 largest spots of the
period 1872 – 1985) and for the ancient record (i.e., for the 235 spots
or 'entries' of the period 165 BC – 1684).

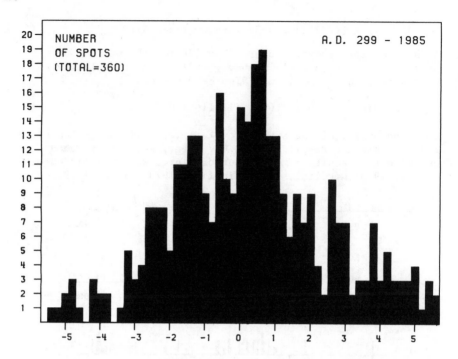

Fig. 2.   Relative phase (O–C with respect to the nearest
          maximum) histogram for the 360 most reliable spots
          of the period AD 299 – 1985.

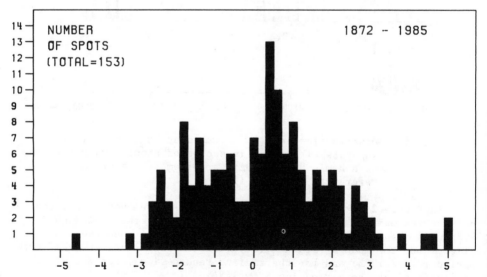

Fig. 3.   Same as Figure 2, but for the 153 largest spots of the
          period AD 1872 – 1985.

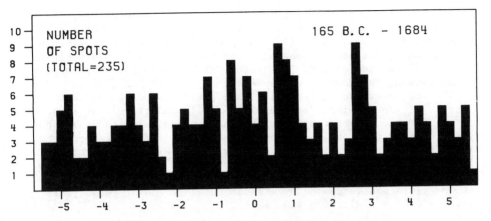

Fig. 4.   Same as Figure 3, but for the 235 entries of our
catalogue (i.e., for the period 165 BC - AD 1684).

From Figures 2 and 3 it is seen that large spots are not distributed at random, but concentrate around the maximum epochs. It thus
seems that, although some transient phase irregularities have occurred
(such as, for instance, that in 1790; cf. Sonett, 1983a), in the long
run the solar cycle exhibits a fairly good phase stability and tends
to recover from irregularities after a few cycles (compare Fairbridge
and Hameed, 1983, who find recurrent patterns after 16 cycles).   This
may be interpreted as evidence against the strange attractor hypothesis
of solar activity (Ruzmaikin, 1981; cf. Stix, 1984).   Figure 3 also
shows the Gnevyshev gap (with peaks at about -1.4 and +0.6 years from
'maximum', and a minimum in between; for details cf. section 4).
Figure 4, on the other hand, exhibits considerable noise, and only
marginally shows a few peaks (at -1 year, +1 year and +2.7 years)
which might have something to do with the Gnevyshev gap.   From Figure
4 we conclude that our catalogue does not represent a statistically
clean sample (i.e., the data are highly undersampled, and some of the
entries may either be incorrectly dated or otherwise obscure).   We
stress, however, that some of the entries e.g. the observations of AD 807,
or Kepler's observation of AD 1607) are obviously (viz. for good
reasons) more reliable than others, and that these epochs - selected
exclusively on the basis of their 'reliability' and not on their
goodness of fit - show a greatly reduced amount of scatter.

4.   THE GNEVYSHEV GAP

From Figures 2 and 3 it is also seen that the bimodal distribution of
large sunspots discovered by Gnevyshev (1963), with two maxima at
phases of about -1 and +1 years and a minimum - the Gnevyshev gap -
in between, is much more pronounced in the modern record (and even
more pronounced if the area limit is raised to 1900 millionths instead
of 1500 millionths, cf. Figure 1 of Wittmann, 1978).   As the

Gnevyshev gap tends to be filled by smaller spots, this seems to
indicate that not necessarily only large, singular spots, but also
conglomerates (groups) of smaller spots were observed with the naked
eye. The Gnevyshev double peak has also been found in the distribution
of proton flares and coronal line intensities (5303 Å); an early plot
already showing the double maxima is Figure 1 of Newton (1955).

The Gnevyshev gap is generally interpreted as being due to the
existence of two distinct distributions (or 'populations') of sunspots,
with the first one occurring at higher heliographic latitudes and about
2 years earlier than the second one (Gnevyshev, 1977). It seems that
this multiplet structure of the activity maxima also causes the
multiplicity of peaks in the power spectrum near periods of 11 years.
For instance, Cole (1973) has suggested that a 190 year phase modulation exists and is caused by a fundamental period of 10.45 years
(viz. 18x10.45 = 17x11.06 = 188 years).

5. THE MAUNDER MINIMUM

The terrestrial Carbon-14 record (i.e., the decay-corrected abundance
of Carbon-14 in dateable tree-rings), shows a pronounced anticorrelation with the envelope of the sunspot number curve (not the
curve itself, because the 11 year cycle is not resolved; cf. e.g.
Stuiver and Quay, 1980; Stuiver and Grootes, 1980). Several long-
term minima and maxima are clearly discernible, the strongest maxima
during the pre-telescopic era being those of AD 1130 (Oort Maximum),
1380 (Wolf Maximum), and 1600 (Spoerer Maximum). The first strong
maximum during the telescopic era is that in AD 1780 (Maunder Maximum);
soon thereafter the C-14 curve is spoiled due to human activity (viz.
fossile fuel combustion),

All of these maxima are also seen in the sunspot record, and in
particular the pre-telescopic maxima are in accordance with our data.
Of the envelope minima (which precede the above-mentioned maxima by
about 90 years), the Oort Minimum (1010-1050), the Wolf Minimum
(1282-1342), and the Spoerer Minimum (1416-1534) are conspicuous
in our data (as are several other intervals characterized by a
scarcity of observations), whereas from the naked-eye sunspot
observations alone there is no indication of the Maunder Minimum
(1654-1714).

ACKNOWLEDGEMENTS

Most of the programmed data processing for the present investigation
has been made with our PDP-11/73, which was financed by the Deutsche
Forschungsgemeinschaft (DFG) and will soon be installed at
Observatorio del Teide/Tenerife.

REFERENCES

Dicke, R.H., 1979, Nature 280, 24.
Cohen, T.J., Lintz, P.R., 1974, Nature, 250, 398.
Cole, T.W., 1973, Solar Phys. 30, 103.
Eddy, J.A., 1977, Climatic Change 1, 173.
Eddy, J.A., 1983, Solar Phys. 89, 195.
Fairbridge, R.W., Hameed, S., 1983, Astron. J. 88, 867.
Flamsteed, J.,'Historia Coelestis Britannica', Vol. 1, Lib. I, Pars VI.
Gleissberg, W., 1952, 'Die Haeufigkeit der Sonnenflecken', Akademie-
    Verlag. Berlin.
Gleissberg, W., 1955, Publ. Univ. Obs. Istanbul, No. 57, 1.
Gleissberg, W., 1972, J. Interdiscipl. Cycle Res. 3, 391.
Gnevyshev, M.N., 1963, Astron. Zhurn. 40, 401.
Gnevyshev, M.N., 1977, Solar Phys. 51, 175.
Kimura, H., 1913, Monthly Not. Roy. Astron. Soc. 73, 543.
Landsberg, H.E., 1980, Arch. Met. Geophys. Biokl., Ser. B, 28, 181.
Lomb, N.R., Andersen, A.P., 1980, Monthly Notices Roy. Astron. Soc. 190,
    723.
Newcomb, S., 1901, Astrophys. J., 13, 1.
Nicolini, T., 1952, Oss. Napoli Coll. Misc. I, 51.
Nicolini, T., 1976, Sci. Fis. Rend. Acad. Mat. Napoli, Ser. IV, 42, 1.
Otaola, J.A., Zenteno, G., 1983, Solar Phys. 89, 209.
Ruzmaikin, A.A., Comments Astrophys. 9, 85.
Sonett, C.P., 1983a, Nature 306, 670.
Sonett, C.P., 1983b, in McCormac, B.M. (Ed.), 'Weather and Climate
    Responses to Solar Variations', Colorado Associated Univ. Press,
    Boulder, p. 607.
Stix, M., 1984, Astron. Nachr. 304, 215.
Stuiver, M., Quay, P.D., 1980, Science 207, 11.
Stuiver, M., Grootes, P.M., 1980, in Pepin, R.O., Eddy, J.A., Merrill,
    R.B. (ed.) 'Proc. Conf. on The Ancient Sun', Pergamon Press, New
    York, 1980, p.165.
Wittmann, A., 1978, Astron. Astrophys. 66, 93.
Wittmann, A.D., Xu, Z.T., 1987, Astron. Astrophys. Suppl., 70, 83.
Wolf, R., 1861, Astron. Nachr. 54, 343.
Yoshimura, H., 1979, Astrophys. J. 227, 1047.

SOLAR, GEOMAGNETIC AND AURORAL VARIATIONS OBSERVED IN HISTORICAL DATA

Joan Feynman
Jet Propulsion Laboratory
4800 Oak Grove Drive
Pasadena, CA 91109

ABSTRACT. The use of historical data sets to investigate solar and
solar-terrestrial variability was demonstrated dramatically by J.A.
Eddy when he described the Maunder Minimum. The historical record
can also be examined for evidence of less dramatic changes such as
those associated with the 88 year cycle in auroral activity. Power
spectral analyses of historical auroral data from 450 AD to 1450 AD
show a strong line at 88.4 ± 0.7 years verifying the existence of the
Gleissberg cycle in auroras during that 1,000 years. The relation
of these results to observations recorded since 1700 will be discussed.
Emphasis will be placed on developing a description of the changes
in solar and solar terrestrial phenomena during the 1900 and 1815
minima in sunspot cycle amplitude and auroral and geomagnetic disturbance levels.

INTRODUCTION

Many different solar and solar terrestrial phenomena have been studied
to investigate the variability of the sun on time scales longer than
one or two solar cycles. These phenomena give information on differing
aspects of solar variability and there is no a priori reason that they
should all vary in the same way or in phase with one another. Furthermore there is no a priori reason why the parameter changes that take
place during the 11 year solar cycle should be analogous to the changes
taking place during other cyclic and non-cyclic variations, for example
the Maunder Minimum or Gleissberg cycle. In this paper we will describe
several of these data sets used in studies of solar variation and the
relations among them. We will then discuss studies which have
established the reality of the 88 year Gleissberg cycle. Finally we
describe the changes that took place in solar and solar-terrestrial
phenomena during the 1811 and 1901 minima, which are widely believed
to be examples of Gleissberg minima. These minima are evident in solar,
solar-terrestrial and heliospheric phenomena.

F. R. Stephenson and A. W. Wolfendale (eds.),
Secular Solar and Geomagnetic Variations in the Last 10,000 Years, 141–159.
© 1988 by Kluwer Academic Publishers.

HISTORICAL DATA SETS

Table 1 lists the data types that are often used to discuss solar
variability and the region of space to which each data type refers.
Variations evident in any one of the data types should be distinguished
carefully from variations observed in any other data type since they
refer to different phenomena. This point is emphasized because in
some earlier studies this distinction was not made. For example, it
was sometimes assumed that auroral observations could be used as proxy
data for sunspot numbers and if the number of auroras seen was known
then the sunspot number could be implied. However, as we shall show,
this is not the case.

In Table 1 we list several types of data that refer to the "solar
surface", i.e. the photosphere, chromosphere and lower corona. These
include the sunspot number, solar flares and prominences. The sunspot
number is defined as proportional to the number of spots plus 10 times
the number of spot groups. With this definition it seems surprising
that it correlates strongly with any other parameters. However, it
does and the sunspot number is often used as the standard of solar
variation against which all other variations are measured. There does
not appear to be any theoretical or observational justification to
assume the sunspot number is more basic to solar variability than say
the solar wind flux or the size and extent of coronal holes.

TABLE 1

DATA SETS

TYPE	REFERS TO
SOLAR	
SUNSPOTS	SOLAR SURFACE
OTHER (FLARES ETC.)	
GEOMAGNETIC	SOLAR WIND AT
aa (Kp)	EARTH
SUDDEN COMMENCEMENTS	
AURORAS	
GALACTIC COSMIC RAYS	SOLAR WIND THROUGHOUT
(^{14}C)	HELIOSPHERE
SOLAR COSMIC RAYS	SUN AND INTERPLANETARY
	MEDIUM

The annual mean sunspot number since 1610 is shown in Figure 1 (from Eddy, personal communication). There is very little sunspot data for any earlier periods. In Figure 1 the Maunder Minimum is evident, especially between 1645 to 1700. The increased sunspot number around 1716 was accompanied by the reappearance of auroras in Europe (Siscoe, 1980). Notice also the dramatic drop in sunspot cycle amplitude between 1800 and 1820 and the much less dramatic drop to relatively small amplitude cycles between 1880 and 1930. The second type of variation listed in Table 1 is geomagnetic. These variations have to do with the state of the solar wind at Earth. They include the Kp and aa variations that measure the strength of magnetospheric disturbances seen at mid-latitudes during 3 hour intervals (see Allen and Feynman, 1980 for a brief description and Mayaud, 1980 for a complete discussion). Kp and aa essentially measure the same variations but the network of stations used to derive the values is different and aa is linearly related to the range of variation (in nT) whereas Kp is quasi-logarithmically related.

The second item listed under geomagnetic variations, i.e. sudden commencements, are sudden increases in the horizontal component of the magnetic field at mid and low latitudes. They are usually followed within a few hours by an increase in Kp (and all other 3 hour mid-latitude indices). Sudden commencements are caused by shocks in the solar wind (Smith 1986) which in turn are caused by sudden emissions of high speed solar wind from the corona. Studies of the past

Annual Mean Sunspot Number, A.D. 1610-1975

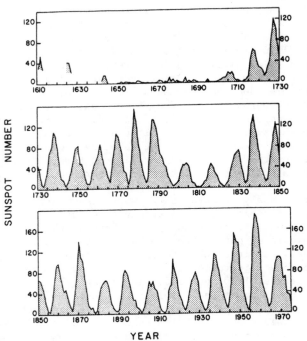

Fig. 1. Annual mean sunspot number for 1610-1975 (from Eddy, personal communication). Note the Maunder Minimum in the 17th century and the series of low amplitude cycles at the beginning of the 19th and 20th centuries. The most recent cycle (not shown) maximized at about 120.

behaviour of sudden commencements and the aa index have been enormously
facilitated by the work of Mayaud, who scaled the original magnetograms
from observations in Great Britain and Australia and produced a 100
year (1868-1967) record of the aa index and sudden commencements. There
is only some sporadic data on geomagnetic variations earlier than 1868.
The aa record can be extended to the present by using the am values.
The am values are the same as the aa except that a more extensive
network of stations is used.

The third type of data set that refers to the solar wind at Earth
is auroral. Since auroras are very dramatic manifestations of
magnetospheric disturbance we have records of them occurring several
hundred years B.C. See Siscoe (1980) for a very interesting discussion
and a compilation of the early auroral record. For more modern auroral
observations, the data set from Sweden compiled by Rubenson for the
period from 1720 to 1876 has been successfully used in several studies
of auroral variations and its main features are in accord with other
data independently compiled in the United States (Silverman and Feynman,
1980). More modern auroral data sets have been compiled by Silverman
and Blanchard (1983). Aurora and aa are closely related and the
correlation between the annual average aa given by Mayaud and the
number of auroras seen in Sweden given by Rubenson was 0.97 (Silverman
and Feynman, 1980) for the nine years of overlap between the two data
sets.

The relation between in situ observations of the solar wind at
Earth and the sunspot number is shown in Figure 2 (Feynman, 1983).
The differences between the sunspot number and the velocity of the
solar wind are striking. They appear to have little to do with one
another. The average log of the magnitude of the interplanetary
magnetic field (log B) is somewhat more closely related to sunspot
number but it is still markedly different. Log B is used in this comparison rather than B because B is distributed log normally (King,
1979). Now it has been shown that magnetic disturbances are driven
by the velocity of the solar wind and the southward component of the
magnetic field (Hirsberg and Colburn, 1967). The third panel in this
figure shows the north-south component of the field and this quantity
is also not proportional to the sunspot number. Although there is
still strong controversy as to what function of the velocity and
southward magnetic field drives the geomagnetic activity on time scales
of a few minutes, for annual averages the activity is proportional
to $V^2 B_Z$ (Crooker et al. 1977). It is clear from Figure 2 that this
driving function will not be closely related to the sunspot number.

The relation between the sunspot cycle and geomagnetic activity
can be seen by comparing the annual average sunspot number and the
annual average aa over 100 years of data. See Figure 3. The aa has
a cyclic variation of 11 years but it is not the same as that of the
sunspot cycle. The most obvious difference can be seen by comparing
the minimums of the cycles between 1900 and 1950. The sunspot cycle
returns to nearly the same low value at each minimum but there is a
steady rise in the value of aa at sunspot minimum. This rise has been
interpreted as a rise in the values of solar wind velocity and/or
southward field for the same sunspot number, and as an expression

SOLAR CYCLE VARIATION

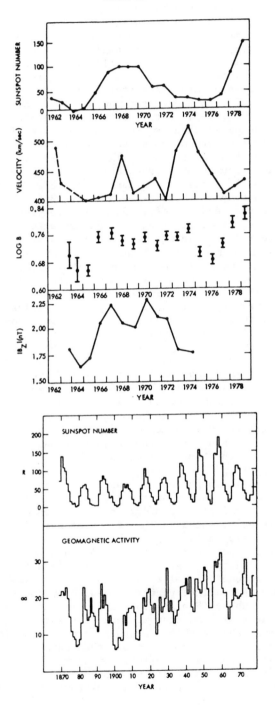

Fig. 2. A comparison between the sunspot number cycle and solar wind parameters. The parameters that drive geomagnetic activity (velocity, 1B$_Z$1) do not mimic the sunspot number. (Adapted from Feynman, 1983).

Fig. 3. A comparison between sunspot number cycles and geomagnetic activity cycles as measured by the mid-latitude 3 hour range index, aa. Although they are both accurately measured they do not follow one another, (from Feynman, 1982).

of the Gleissberg cycle (Feynman and Crooker, 1978).

It has been known since early in this century that there are at least two kinds of magnetic storms, sudden commencement storms and recurrent storms. The sudden commencement storms begin with the sudden increase in surface magnetic fields as mentioned earlier. They have no tendency to reappear 27 days later, i.e. after the sun has rotated once around. The number of sudden commencement storms per year has a correlation coefficient of 0.85 with the sunspot number (Mayaud, 1973), and so they are strongly related. Note that the relationship is between the number of sudden commencement storms, not the intensity of disturbance. The second type of storm has no sudden commencement but it does recur with each rotation of the sun. These recurrent storms rise from coronal holes such as those seen during the Skylab mission (Neupert and Pizzo, 1973). They appear during the decreasing or minimum phase of each sunspot cycle (Newton and Milson, 1954). Sargent (1979) has quantified the tendency of these storms to recur. He calculates the correlation coefficient of each 27 day rotation with the next, takes a 13 rotation running average, and multiplies by 100. He assigns the resulting number to the middle 27 day period of the 13 rotations used in the running average. His results are shown in Fig. 4 for the period from 1868 to 1980. Ohl (1971) and Sargent (1978) have developed a remarkably successful method of predicting the amplitude of the sunspot cycle from the intensity of magnetic disturbance during the period of high recurrence in the preceding cycle. That is, geo-magnetic activity in one cycle is used to predict sunspot number in the next. In some sense, then a sunspot cycle begins with the formation of the coronal holes from which the recurrent storms arise. Further-more it appears that coronal holes are as basic to the solar cycle as sunspots, and theories of solar variability that do not include the formation of polar coronal holes are doomed to failure.

The third type of data in Table 1 is the data on galactic cosmic rays. As these particles enter the heliosphere and propagate through it, they are modified by their interaction with the solar wind. When they reach the Earth the size of the Earth's field and the state of disturbance of the magnetosphere influence their ability to enter the Earth's atmosphere. Once in the atmosphere they can indirectly produce ^{14}C. This process and the results of the ^{14}C anomaly analysis are described by Stuiver (this volume). This data set has been one of the most useful in unequivocally establishing the existence of solar-terrestrial and heliospheric variations.

The final solar-terrestrial data set to be described are the solar cosmic rays. Particles in the energy range from 10 MeV/nucleon to hundreds of MeV/nucleon are produced in solar flare events. These particles then propagate through the interplanetary medium, where they may be further accelerated by shocks (Hewish 1986) and some reach the vicinity of the earth. At Earth, if the energies and fluxes are high enough, the particles can enter the Earth's atmosphere along magnetic field lines that lead into the polar caps. They can then cause additional ionization in the ionosphere, interfering with radio com-munication. These events are called polar cap absorptions (PCA). The historical record on the number of PCA events goes back to 1938

RECURRENCE INDEX
AND
SUNSPOT NUMBER

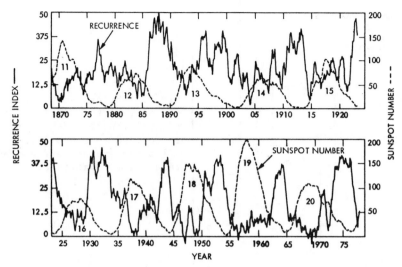

Fig. 4.   A comparison between sunspot number cycle and the recurrence index. See text for a definition of recurrence
index. (From Sargent, 1979).

(Svestka, 1966). On the fluxes of protons in space the data begins
in 1956. Very large events of this kind are important for spacecraft
design and would be lethal to an astronaut in EVA, and so there has
been considerable interest in being able to predict the event occurrence
rate. Very large events occurred frequently during the solar cycle
that maximized in 1957 but since that time only one large event occurred,
that of August 1972. This suggests the possibility that we may be
seeing a variation in event frequency that is part of the Gleissberg
cycle. This point will be discussed further below. We now have 30
years of data on the proton fluxes in solar energetic particle events.
The early data, 1956 to 1961, was observed by riometers supplemented
by high altitude and space observations (Fichtel et al. 1962). The
later data is from a tape compiled by Armstrong et al. (1983) from
observations in space. Although the early data was not from spacecraft
the large events were studied carefully and the fluxes were estimated
to be within a factor of two of the true fluence (Fichtel et al., 1962).
    In Figure 5 we show the solar cycle variation of the yearly inte-
grated flux (yearly fluence) of protons with energies greater than
30 MeV. We use a superimposed epoch technique to display the solar
cycle variation. The 12 month period centered on sunspot maximum is
plotted as year zero. Sunspot maximum is defined correct to 0.1 year
(Heckman, personal communication). The other "years" are defined in
an analogous way. The observations for the same phase relative to
maximum are joined with a line. There is a clear sunspot cycle
variation such that the yearly integrated fluxes are greater during

YEARLY FLUENCES (>30 MeV)

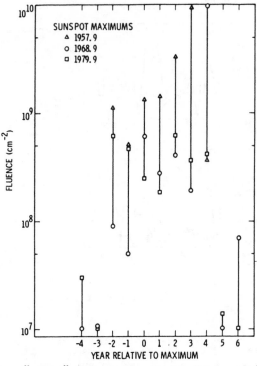

Fig. 5. Solar cycle variation of solar proton yearly fluence (annual flux). Sunspot maximum is defined to the 0.1 year. All other "years" are defined relative to that time.

the "years" between 2 years before and 4 years after maximum. There is only one exception to this rule. Some studies of earlier polar cap absorption events (Svestka, 1966, Hakura, 1974) found a tendency for the number of PCA's to be somewhat lower at solar maximum than during the years immediately before or after maximum. There does not seem to be an analogous tendency present in this 30 year data set of yearly fluences. If anything there seems to be a gradual increase in fluence size until year +4, but it is not clear that this rise is statistically significant.

Figure 6 shows the annual integrated solar cosmic ray flux for each year of the data set. The data is plotted on probability paper. The abscissa is ruled so that if the data is distributed log-normally, the plot of cumulative percentage will be a straight line. The abscissa gives the percentage of years in the data set for which the annual integrated flux is equal to or smaller than that of the year being plotted. For example, in this data about 3% of the years have annual fluxes exceeding 10^{10} particles/cm^2 above 30 MeV energy because one year had such a flux. The "years" are defined as they were in Figure 5. The points do not define a single straight line but seem to be arranged into 3 lines as shown. The lowest fluence line fits 12 or 13 points, all but two of which come from solar minimum years. The highest fluences line is fit to the 6 or 8 points above 7×10^8 particles/cm^2. Five of the 6 highest fluence years occur in the first

ANNUAL FLUENCE DISTRIBUTION (E >30 MeV)
(1956-1985)

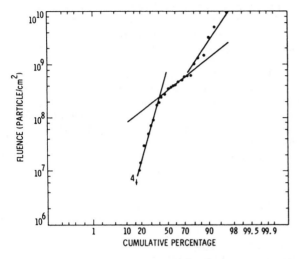

Fig. 6. Distribution of solar proton yearly fluence (annual flux) for the three most recent sunspot cycles.

of the solar cycles. The 6th year contains the great August 1972 event.

Although strictly speaking there is no a priori reason to expect that the event fluences should have a log normal distribution, the 3 lines invite the speculation that there are 3 different processes. The causes of the highest flux line is most interesting from the point of view of solar variability in the last 10,000 years. There are several hypotheses that could be involved. For example, the distinction between the highest flux line and the mid-flux line could be caused by a break or non-linearity in the distribution of the log of the flare fluence size at the sun or it could be due to a log normal distribution at the sun for the solar active years followed by further acceleration in space. It is widely suggested that the particles emitted from the sun in August of 1972 were accelerated in space by interacting with a series of shocks in the solar wind. These shocks were caused by other transient events from the same activity center. The 1972 event was quite complex, consisting of many flares of differing brightness and "importance" occurring in a short period of time. Indeed, almost all major solar particle events are complex (Malitson and Webber, 1960) and it is not unreasonable to suggest that interplanetary particle acceleration is a common feature of these events.

The line fit to the highest fluence years also allows speculations to be made about the largest events expected in time periods longer than 30 years. If we assume that the line can be extended indefinitely we find that the largest event expected in 10,000 years would have a fluence of about 2×10^{12} particles cm^{-2}. The largest in 100,000 years would have about 10^{14} particles cm^{-2}.

THE 88 YEAR CYCLE

Gleissberg (1965), through an analysis of sunspot number and auroral data suggested there was a variation in solar-terrestrial phenomena with a period of about 87 years. The existence of that period was in doubt for many years but recent work has shown that it is real. Stuiver (this conference) showed new power spectral analyses of high accuracy ^{14}C data that have the Gleissberg line. It had also been found (Feynman and Fougere, 1985) in a power spectral analysis of the number of auroras reported per decade at midlatitudes in China and Europe during the years from 450 A.D. to 1450 A.D. (Siscoe, 1980). Figure 7 shows the power spectrum found by Feynman and Fougere. The stability of this spectrum was tested by using 2 different power spectral techniques and by adding a high level of noise to the data set. The spectrum shown in Figure 7 is typical of the results. The best estimate of the period was 88.4 ± 0.7 years. The amplitude was 2.2 auroral observations per decade.

POWER SPECTRUM
REPORTED AURORA, 450 AD – 1450 AD

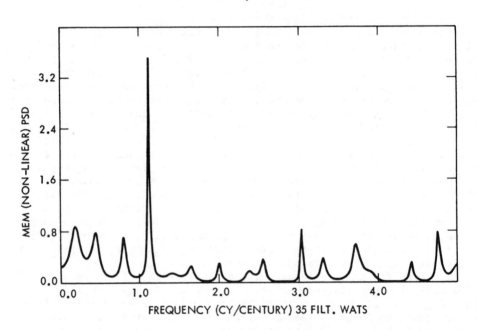

Fig. 7.   A power spectrum of the number of auroras recorded as having been sighted per decade from 450 A.D. to 1450 A.D. (From Feynman and Fougere, 1985).

The establishment of the reality of the Gleissberg cycle is important in investigations of solar variability because the changes in solar and solar-terrestrial parameters during the Gleissberg cycle may be different from the changes during the 11 or 22 year cycle. If so (or even if not) a description of the changes taking place during the Gleissberg cycle will constitute an important constraint on theories of solar and stellar variability. That is, theories of solar variability should be simultaneously able to predict the sunspot number cycle, the formation of coronal holes and the changes taking place in the Gleissberg cycle. With Stuiver's latest result the last important data set that did not show the 88 year cycle is now in agreement that the cycle is real.

The most straightforward way to get information on the changes taking place during the Gleissberg cycle would be to identify a modern Gleissberg minimum and analyze the data from that period. The minima in sunspot amplitude and auroral frequency that occurred early in the 19th and 20th centuries immediately suggest themselves, (Feynman and Crooker, 1978, Feynman and Silverman, 1980) since they occur about 87 years apart. Although we will describe the solar and solar-terrestrial changes that take place across these two minima it is not altogether established that these are Gleissberg type minima and before we describe them we will give the arguments for and against their identification as Gleissberg minima. The arguments for the minima near the beginnings of the 19th and 20th century being Gleissberg minima are strong. First and foremost, the time interval between them is of the right length. Auroral activity was low during both the candidate periods and a ^{14}C anomaly occurred in conjunction with the 19th century minimum. The increase of CO_2 in the atmosphere that began at the end of the 19th century distorted the ^{14}C data so that no conclusion can be drawn concerning an anomaly in the 20th century. The arguments against these minima being of the Gleissberg type are less strong but cannot be altogether neglected. Sonett (1982) ran a power spectral analysis on the post 1720 sunspot number cycles shown in Figure 1 and found the expected line at 87 years. This analysis, in our opinion, does not establish that the minima in the early part of the 19th and 20th centuries are Gleissberg minima because the data set was arbitrarily cut at 1720. This could be justified if we did not have information on sunspots from   earlier data, but we do. Figure 1 shows that 1720 is not a relative minimum but part way into a steep increase that occurs as the sunspots reappear after the Maunder minimum. The most disturbing argument however against the minima early in the 19th and 20th centuries being Gleissberg minima concerns the comparison between the cycles seen in 450 to 1450 AD and the modern data. The analysis of the medieval auroras permitted the Gleissberg phase to be determined quite accurately because there are 11 eighty-eight year cycles in the thousand years of data. The phase was such that the last minimum occurred between 1403 and 1413 AD. Extending this phase forward there should have been a minimum between 1665 and 1680, which is in agreement with observation. The next minimum should have been between 1750 and 1770, the next between 1838 and 1858, and the most modern one should have been between 1936 and

1950. The only one of the expected post Maunder Minimum periods which shows a decrease in activity is the 1750-1770 period. In fact the observed minima around 1810 and 1900 are close to 90° out of phase with the extrapolated phase. We conclude that either the 88 year cycle changed phase post-Maunder Minimum or the early 19th and 20th century periods of small amplitude sunspot cycles are not Gleissberg minima.

With these caveats in mind we now describe the changes that took place during the early 19th and 20th centuries. The 19th century minimum was sharply marked in sunspot number amplitude (Figure 1). Two sunspot cycles with amplitudes less than 50 occurred between 1800 and 1820. The 20th century minimum was less severe and less well marked. There were no sunspot cycles with amplitudes less than 50 during the 20th century minimum period but there was a series of 5 relatively small amplitude cycles, the first and third of which were below 70.

Sunspot number changes have been analysed more closely by Kopecky (1967). He considered the sunspot number at any time to be a function of two variables, f_0, the number of spot groups formed and T_0, the average lifetime of the groups. His results are shown in Figure 8. The number of spot groups is highly periodic and shows the 22 year nature of the cycle very clearly in that each alternate cycle has a relatively small amplitude. There is no clear change in the f_0 cycles as the early 20th century solar-terrestrial minimum is crossed. In fact the 15th solar cycle shows the most spot groups. On the other hand the lifetime of the spots shows a general fall and rise over the 75 years with much less distinct minima corresponding roughly to f_0 minima for each 11 year cycle. If the data corresponding to the 11 year cycle minima are deleted, the overall lifetime minimum seems to occur some time between 1912 and 1922.

SUNSPOT PARAMETERS

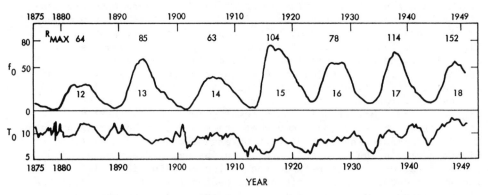

Fig. 8. A comparison of sunspot parameters. The top panel shows f_0, the number of spot groups formed per solar rotation. The bottom panel shows the mean lifetime of the group. The R_{max} numbers across the top give the maximum annual sunspot number for each cycle. (Adapted from Kopecky, 1967).

Figure 9 gives data on the position of sunspots on the sun. The sun usually has a larger sunspot area on one hemisphere than on the other and the figure shows a measure of that tendency. Although each cycle has some years of each dominance, one hemisphere predominates in most cycles. The southern hemisphere predominates until 1913, but the northern hemisphere predominates for at least the next two cycles (see also Waldmeir, 1957). Other sunspot parameters also changed in the first two decades of the 20th century (see review by Kopecky, 1967), Altolini et al., (1985) give evidence for a change in period and a 90° phase change in the second harmonic.

SUNSPOT AREAS
· N–S ASYMMETRY

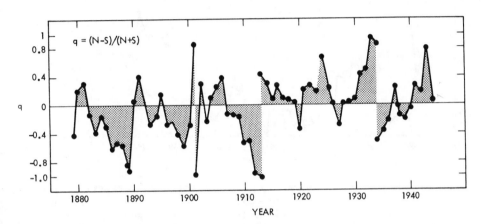

Fig. 9.   A comparison of the sunspot number in the two hemispheres of the sun. A value of +1 would indicate all the sunspots were in the northern hemisphere and −1 would indicate all were in the southern hemisphere.

Auroral and geomagnetic activity also mark the minimums. The geomagnetic changes taking place during the 20th century minimum have already been described in connection with Figure 3. Note that the geomagnetic changes define a time of minimum much more sharply than did the sunspot number. No geomagnetic data is available for the 19th century minimum. During both the 19th and 20th century minima the aurora retreated northward (Feynman and Silverman, 1980). An auroral minimum took place in Sweden and New England about 1809–1815. Between 1880 and 1920 the aurora in north Sweden underwent particularly interesting changes as shown in Figure 10. Haparanda and Karesuando are both stations in the north, with Karesuando (65°N corrected geomagnetic latitude) being north of Haparanda (at 62.6N corrected geomagnetic latitude). The auroral count at Haparanda becomes very low between 1900 and 1915, as would be expected. The behaviour at Karesuando is

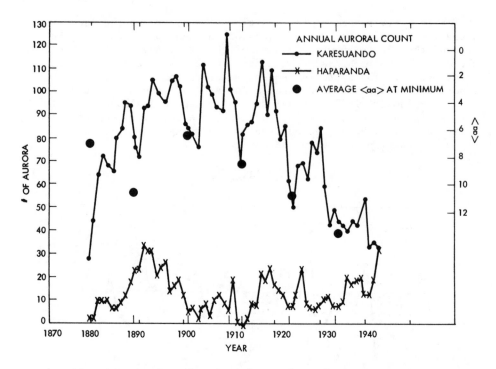

Fig. 10. A comparison between the number of auroræ reported for
two northern Swedish stations. Also given are the annual
average value of aa, mid-latitude geomagnetic index
(see text). Note the aa scale on the right increases
downward. (From Silverman and Feynman, 1980).

quite different. There is a general increasing trend which maximizes
between 1900 and 1915. Superposed on this trend is a solar cycle
variation in which a larger number of auroras are seen with high sunspot
number. We have emphasized this behaviour by plotting the annual
average aa index at minimum for each solar cycle. Note that the aa
scale increases downward. We therefore have the curious situation
in which the number of aurora increases with aa <u>increase</u> within each
of the 11 year solar cycles but also increases with a <u>decrease</u> of aa
in a Gleissberg cycle. This indicates that the changes taking place
in the solar wind at Earth during an 11 year cycle are not the same
as those taking place during a Gleissberg cycle (assuming the minima
are Gleissberg minima).
    Figure 11 shows a composite of the data on auroræ and geomagnetic
activity over the 250 years since the reappearance of the aurora in
1720. The left hand side of the figure shows an 11 year running
average of Swedish auroral sightings. There are several deep minimums
including the sharp minimum in 1811 which has been discussed here in

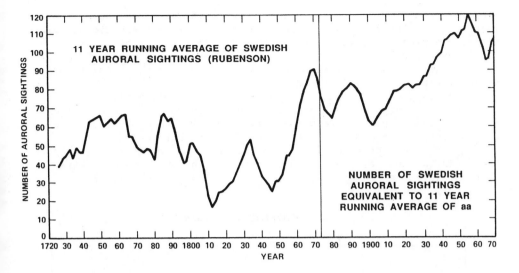

Fig. 11. The 11 year running average of the number of auroras seen
in Sweden from 1726 to 1872 and the number of sightings
that would be equivalent to the 11 year running average
of the aa index of geomagnetic activity. The equivalence
is based on the relation between annual sightings of
auroras and annual average aa indices during the nine years
of overlap between the two data sets.

detail. There is also a minimum about 1847, which time is well within
the bounds of the minimum expected by extrapolating the phase of the
medieval 88 year cycle to the post Maunder Minimum period. The right
hand side of the figure gives the number of Swedish auroras that would
be equivalent to the 11 year running average of aa. The equivalence
is determined from the relationship between Swedish auroras and aa for
the 9 year period of overlap mentioned earlier. Although this
equivalence is correct for that period it is not clear that it is
correct for the entire period of both data sets. For example it is
frequently claimed that the increase in auroral sightings between 1850
and 1870 was due to increased interest in auroras during the Victorian
period. The rise between 1900 and 1957 is real since the geomagnetic
activity was certainly measured well. Unfortunately these data cannot
be re-calibrated because scientific visual auroral observations
stopped in the 1940's. There are several relative minima in the equivalent data, i.e. 1880, 1901 and 1968. The 1901 minimum is the one
identified for study here. Note that the 1811 and 1901 minima are
so sharp that a single year can be selected for both of them.

     Several other interesting pieces of information on the changes
in the sun across the 1901 minimum can be deduced from geomagnetic
activity. In Figure 4 we showed a comparison between the sunspot
number cycle and the recurrence index. The recurrence during the

decline and minima of cycles 13 and 14 was just as strong as the
recurrence during the other cycles. We also know that the geomagnetic
activity during those recurrent periods was low. Apparently, stable
polar coronal holes formed much as in other cycles and remained as
long. However, the solar wind that issued from them had a low velocity
and/or southward magnetic field.

The top panel in Figure 12 shows the solar cycle variation of
the number of large sudden commencements per year. As argued earlier,
these data can serve as proxy for major coronal mass ejections. The
data show that large CME's were relatively rare but they did occur
during the 20th century minimum. In fact the largest storm (as measured
by aa) in the hundred years of data occurred in 1909. Auroras were
seen in Singapore and other low latitude cities (Botley, 1957).

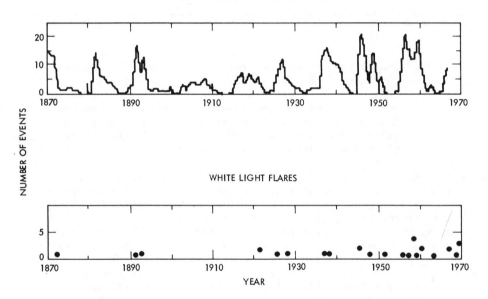

Fig. 12. The top panel shows the number of large sudden commencements
per year. Sudden commencements are equivalent to shocks
in the solar wind. The bottom panel shows the yearly
number of white light flares (from Neidig and Cliver,
1983). Note the paucity of both phenomena between 1895
and 1920.

The lower panel of Figure 12 shows the number of white light
flares observed per year from Neidig and Cliver (1983). White light
flares are very energetic events and have been observed to be related
to the 1956, 1958, 1960 and 1972 major solar cosmic ray events. Note
that no white light flares were reported between 1895 and 1920, i.e.

during the period of the early 20th century minimum. This could either have been due to lack of observation or to lack of events. (The lack of flares in the 11 year cycle beginning in 1881 is probably due to lack of observation.) Further examination of the post 1920 data shows that minima in the sudden commencements for 11 year cycles correspond closely to gaps in the white light flare reports. This suggests that the 1895 to 1920 gap may be real and part of the early 20th century variation and we speculate that both white light flares and major solar proton events may show a Gleissberg cycle variation.

SUMMATION

In the body of this paper we have discussed a large number of data sets that varied across the 1811 and 1901 minima. For convenience all our results will be listed here.

Solar and solar-terrestrial changes across the minima included:

o Both minima showed several low amplitude sunspot cycles.
o In both periods the auroras moved northward and were rarely seen at midlatitudes (although an aurora was seen in Singapore in 1909). The behaviour of auroral frequency across the 1901 minimum was different than it was in the 11 year cycle, indicating that the changes occurring in solar wind parameters were different for the two variations.
o Major geomagnetic storms are rare but can occur. The largest storm on record accompanied the 1909 aurora.
o Geomagnetic activity (in the year 1901) and auroral activity (in the year 1811) minimized sharply compared to sunspot amplitude.
o Strongly recurrent solar wind streams continued (early 20th century) but were weakly geo-effective.
o Major sudden commencements continued (20th century) but at a decreased rate.
o Sunspot lifetime minimized (20th century).
o No apparent effect on number of spots formed (20th century).
o ^{14}C anomaly occurred (19th century).
o No reported white light flares (20th century).
o Change of dominant solar hemisphere (20th century).

These changes can be referred back to the sun to describe changes taking place there in both solar activity and solar wind.

The following changes took place in solar activity:

The decrease in the sunspot number cycle amplitude was due to a change in lifetime of the groups rather than the number of spots formed. The dominant solar heliosphere changed (1901). Major coronal transients appear to decrease in frequency but when they do occur they can be surprisingly large. There is some evidence that white light flares were relatively rare, if they occurred at all. This suggests that major cosmic ray proton events would also be relatively rare.

158

The following changes took place in the solar wind:

The solar wind was weak and minimized sharply, probably due to a decrease in both velocity and magnetic field. Polar coronal holes continued to form and to maintain themselves for as many rotations as during other periods. The relatively mild solar wind from them suggests a low magnetic field at the base of the holes. The changes in solar wind parameters across these two "Gleissberg" minima was different from the changes taking place during the eleven year cycle.

This description of the changes across the 19th and 20th century relative minima is an attempt to bring together observations of Gleissberg minima to begin to describe this phenomena. The purpose of this exercise is to develop input for theories of solar and stellar variability. It is especially important that this be done now because it is about one Gleissberg cycle since the 1901 minimum and we can hope to see a repetition of some of these phenomena. We should keep in mind, however, that just as the depth of the 1811 minimum was greater than that of the 1901 period, so the depth of any current minimum would not necessarily be as great as the earlier ones. In fact it may be that the current period is already a minimum period that began after the sunspot cycle that maximized in 1957. If so, all of the data we have collected from space refers to a minimum period, and we can expect interesting and exciting changes to take place in solar and solar-terrestrial phenomena in the future, perhaps the near future.

ACKNOWLEDGEMENTS

I thank Dr. Paul Robinson for his interest and encouragement. The research described in this publication was carried out by the Jet Propulsion Laboratory, California Institute of Technology, under a contract with the National Aeronautics and Space Administration.

REFERENCES

Allen, J.H., and Feynman, J., 1979, 'Solar Terrestrial Predictions Proceedings', Vol. II, edited by R.F. Donnelly, NOAA, Boulder, Colorado.

Armstrong, T.P., Brungardt, C., and Meyer, J.E., 1983, 'Weather and Climate Responses to Solar Variations', ed. B.M. McCormac, University of Colorado Press.

Attolini, M.R., Galli, M., Cini Castagnoli, G., 1985, Solar Physics, 95, 391.

Botley, C.M., 1957, J. Brit. Astron. Assoc., 67, 188.

Crooker, N.U., Feynman, J., and Gosling, J.T., 1977, J. Geophys. Res., 82, 1933.

Feynman, J., 1983, Rev. of Geophys. and Space Phys., 21, 338.

Feynman, J., and Fougere, P., 1984, J. Geophys. Res., 89, 3023.

Fichtel, C.E., Guss, D.E., and Ogilvie, K.W., 1962, 'Solar Proton Manual', ed. Frank B. McDonald, NASA Goddard Space Flight Center, X-60-62-122, Greenbelt, MD.

Gleissberg, W., 1965, J. Brit. Astron. Assoc. 75, 227-231.

Hakura, Y., 1974, Solar Physics, 39, 493.

Hewish, A., and Bravo, S., 1986, Nature, 324, 44-46.

Hirshberg, J., and Colburn, D.S., 1969, Planetary and Space Science, 17, 1183.

King, J.H., 1979, J. Geophys. Res., 84, 5938.

Mailitson, H.H., and Webber, W.R., 1962, 'Solar Proton Manual', ed. Frank B. McDonald, NASA Goddard Space Flight Center, M-611-62-122, Greenbelt, MD.

Mayaud, P.N., 1983, IAGA Bulletin #33.

Mayaud, P.N., 1980, Geophysical monograph 22, AGU, Washington, D.C.

Neidig, D.F., and Cliver, E.W., 1983, AFGL-TR-83-9257, Air Force Geophysics Lab.

Neupert, W.M., and Pizzo, V., 1974, J. Geophys. Res., 79, 3701.

Newton, H.W., and Milson, A.S., 1954, Terrestrial Magnetism and Atmospheric Electricity, 59, 203.

Ohl, A.I., 1971, Geomagnetism and Aeronomy, 11, #4, 549.

Rubenson, R., 1882, Svenska: Vetenskaps - Akademeins Handlingar, 18, 1, 216.

Sargent, H.H., 1978, '28th IEEE Vehicular Technology Conference'.

Sargent, H.H., 1979, 'Solar Terrestrial Physics and Meteorology: Working Document III', World Data Center A. Boulder, CO.

Silverman, S.M., and Blanchard, D.C., 1983, Planet. Space Sci., 31, 1131.

Silverman, S.M., and Feynman, J., 1980, 'Exploration of the Polar Upper Atmosphere', eds. JC. S. Deehr and J.A. Holtet.

Siscoe, G.L., 1980, Rev. of Geophys. and Space Phys., 18, 647.

Smith, E.J., Slavin, J.A., Zwickl, R.D., Bame, S.J., 1986, 'Solar Wind - Magnetosphere Coupling', 345, ed. Y. Kamide and J.A. Slavin.

Sonett, C.P., 1982, Geophys. Res. Lttrs, 9, 1313.

Svestka, Z., BAC, 17, 262.

Waldmeir, M., 1957, Z. Ap., 43, 149.

ANALYSIS OF PRE-TELESCOPIC AND TELESCOPIC SUNSPOT OBSERVATIONS

Kevin K.C. Yau
Department of Physics
University of Durham
South Road
Durham DH1 3LE, U.K.

ABSTRACT. Sunspot data from the RGO series covering the period from 1874 to 1954 are analysed to show the asymmetric distribution of sunspot areas north and south of the solar equator. A detailed cata- logue of naked-eye sunspots from Far Eastern annals is compiled from both the pre-telescopic and telescopic periods. Analysis of these fragmentary records indicates a mean period of about 10 years.

1.1   INTRODUCTION

The earliest alleged reference to a sunspot in Western literature is found around 350 BC; it was seen by Theophrastus of Athens, who was a pupil of Aristotle (Bray and Loughhead, 1964). The Aristotelian view of a perfect Sun without blemish plus the Orthodox Christian theological teaching about an uncorruptible heaven in the Middle Ages prevented the potential recognition of sunspots in Europe. Most of the few accounts of European sightings of sunspots that are known were due to misidentification of other phenomena than a specific observation of sunspots. One of the well known examples is found in Einhard's Life of Charlemagne, in which a spot on the Sun around AD 807 was wrongly interpreted to be a transit of Mercury (Goldstein, 1969). A 14th century Russian chronicle - the Niconovsky Chronicle - recorded descriptions of dark spots on the Sun as seen through the haze of forest fires in the years 1365 and 1371 (Vyssotsky, 1949). Also the Carraras (father and son) in Italy were known to have observed sunspots in 1457. However, the doctrine of a perfect Sun persisted down to the dawn of telescopic observation. Even Kepler himself mistook a spot he had seen on May 18, 1607 to be a transit of Mercury (Sarton, 1947).
        In the Islamic world, Abu-1-Fadl Ja'far ibn al-Muktafi (AD 906-977) recorded that the philosopher al-Kindi observed a spot on the Sun in May 840, which was mistaken for a transit of Venus (Goldstein, 1969). There was also Ibn Rushd who mentioned the sighting of a spot on the Sun in 1196. However, these are only isolated events. In East Asia, one finds a long series of sightings of sunspots - almost

161

F. R. Stephenson and A. W. Wolfendale (eds.),
Secular Solar and Geomagnetic Variations in the Last 10,000 Years, 161–185.
© 1988 by Kluwer Academic Publishers.

exclusively from China and Korea – going back two millennia with the earliest recorded spot in 165 BC from China. As with most of the other types of astronomical records, these sightings are preserved in various astronomical treatises of the official histories.

It seems that before the invention of telescope in 1609, the existence of sunspots was virtually unknown in Europe. It was largely due to the publication in 1613 by Galileo of his telescopic sunspot observations in Istoria e Dimonztrazioni intorno alle Macchie Solari e loro Accidenti (Mascardi, Rome, 1613) that they gradually started to attract a wide interest. Although contemporary observations had also been made by Harriot in England, Scheiner in Germany and Goldsmid in Holland, Galileo was credited with the discovery that the spots were phenomena associated with the surface of the Sun (which he thought to be similar to clouds on Earth). On the contrary, others like Scheiner initially thought they were due to the passage of planets across the Sun's disc or small satellites of the Sun.

Before the full understanding of sunspot phenomena about a century ago, there had been much speculation about the nature of sunspots. When a spot is situated near the limb of the Sun, the penumbra on the side nearest to the limb is relatively broad compared with the side furthest from the limb, an effect due to perspective and foreshortening. This led Wilson (1774) to suspect sunspots to be depressed regions on the solar surface. His idea was generalised by Herschel (1795) who preferred them to be openings in the luminous solar cloud and from which one could glimpse the exposed cool surface of a solid Sun.

The existence of the sunspot cycle was not known until Schwabe of Dessau in 1843 showed from 17 years observations of sunspots a periodicity of approximately 10 years (Schwabe, 1843). About twenty years later, Carrington (1863) derived an accurate rate of rotation of the Sun with a mean length of 27.2753 days from the apparent motions of sunspots. Thereafter, the Sun's synodic rotation has been numbered with the Carrington Rotation Number which began on November 9th 1853. Carrington later also demonstrated the variation of sunspot latitude and distribution over the solar surface during the course of a spot cycle. It is possible that the quasi-random distribution of spots on the surface of the Sun was one of the major causes for not recognising their periodic nature earlier. As sunspot records are the only direct indicators of the long term behaviour of solar activity, their importance in our understanding of the Sun need not be more emphasized.

1.2 THE FORMATION OF SUNSPOTS

After the discovery of the differential rotation of the Sun by Carrington it was generally accepted that this effect was largely attributed to a greater acceleration of convection currents at the solar equator. In the following years many attempts were made to interpret the process of sunspot formation. Faye (1865) regarded sunspots to be places where the ascending currents of gas were particularly strong – such that they blew away the particles forming the photosphere, thus exposing the deep interior of the Sun. As the

gases were too hot to emit any visible radiation, this resulted in a
dark spot. In the same year, De la Rue et al. (1865) reckoned that
the currents of gas were descending as opposed to ascending into
the solar interior and cooler than the surroundings. Other interpretation of sunspots was largely based on terrestrial analogy. For
example, Herschel believed some spots were due to meteoritic collisions
while De la Rue et al. (1865) hypothesized that the frequency of
formation was due to an alignment of two or more planets.

However, from spectroscopic observation Lockyer (1886) was able to
show that currents of gas were descending inside sunspots and
different spectral lines appeared in them were resulted from a lower
temperature. This led to suggestions that sunspots were caused by
eruption of material from the solar surface. The material was thrown
up from the edge of a ring, cooled and fell back into the middle of
the ring forming a central depression. (Secchi, 1870 and Schaeberle,
1890).

At the end of the 19th century, the only common consensus among
researchers was that sunspots were depressions in a gaseous solar
surface. Further progress was not possible without one essential
element, namely the magnetic field. In 1908, George Hale proved
spectroscopically the existence of a magnetic field within sunspots
and measured its field strength via the Zeeman effect (Hale, 1908).
He was able to show that the magnetic fields of sunspots were intense
and that the north and south sunspot belts were in opposite sense
to each other and further were reversed with each 11-year cycle.
Also shortly afterwards, systematic motions of the material within
sunspots were observed by Evershed (1909).

Sunspots possess magnetic fields of about 300 mT for medium
to large size spots, which is much larger than the average solar
magnetic field of 0.1 mT (cf. geomagnetic field $\sim$ 0.05 mT). It can
be shown empirically that the sunspot magnetic field, B, can be
approximated by (Bray and Loughhead, 1964):

$$B = 370A/(A + 60) \text{ mT}$$

where A is the mean area of sunspots in millionths of the Sun's hemisphere. Most spots exist in pairs of opposite polarities, with the
larger, preceding spots (or leader spots) moving a few degrees in
latitude closer to the solar equator. It has been shown from an
analysis of the Mount Wilson white light plates, for the years 1917-
1983, that leader spots rotate faster than the follower spots by
$\sim$ 0.1 deg per day (Gilman and Howard, 1985). Individual sunspot
magnetic fields are a part of a more fundamental region in the photosphere known as the bi-polar magnetic region. Sometimes, a single
spot occurs instead of a pair; however, its associated bi-polar magnetic
region of opposite polarity can still be identified.

The principal energy transfer at the solar visible surface is
believed to be by way of convection columns. When these columns
of plasma carrying the energy from within the Sun rise to the surface,
they spread outwards before falling back to the surface again. The
appearance of a localised strong magnetic region like a sunspot

apparently inhibits this process, as first pointed out by Biermann
(1941). Under the influence of a strong magnetic field, a plasma
can still move freely along the field lines but only diffuses slowly
across them. As a result, the rising plasma of the individual convection columns is constrained from moving across the sunspot field
lines. However, the motion of the plasma along the radial sunspot
field lines is unaffected. The effective energy transfer due to
convection is then suppressed because the horizontal flow of the
plasma is hindered. It is now generally accepted that the coolness
or darkness of a sunspot   results   from a reduction in the amount
of energy convected upwards from the solar interior. The temperature
of umbrae is about 4300 K and penumbrae about 5700 K, as compared
with the photospheric temperature of about 6050 K (Allen, 1976).

So far, the origin of the intense sunspot magnetic field is
not yet fully understood despite many attempts at a theoretical solution.
Cowling (1934) was the first to put the magnetic field of a sunspot
in magnetohydrodynamic terms. The classic model of Babcock (1961)
suggested that the differential rotation of the Sun intertwined its
own field lines to such an intensity as to create a significant
magnetic pressure within the plasma near the surface. This upsets
the hydrostatic equilibrium and reduces the normal gas pressure within
the plasma at that intense flux region. As the plasma density is
reduced, it rises like a 'bubble' carrying field lines which have
been frozen into the plasma. Eventually the subsurface tube of field
lines penetrates the surface at the centre of activity and forming
an arch above the surface with its two ends still embedded in the
photosphere (Figure 1). Often a loop prominence will form along these

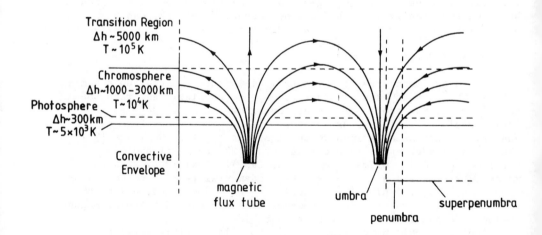

Fig. 1.   A diagram showing the surface layers of the Sun and the
          sunspot magnetic field configurations.

field lines, and the two ends linking the photosphere are the pair of sunspot regions. Parker (1979) explained that the subsurface magnetic field of a sunspot consists of a dynamical clustering of many separate flux tubes. At the visible surface, this loose cluster of flux tubes are pressed together to form a single large flux tube.

For the complex sunspot groups, the mechanism is much more com-plicated by the processes of compressing, breaking and reconnecting of magnetic field lines. There have been many theories proposed to explain the formation of sunspots or the underlying solar activity (see reviews by Parker, 1979; Moore and Rabin, 1985). The formation of sunspots or the cause of the sunspot cycle is still not clearly known; what is certain is the interplay between three factors - a large-scale relatively weak poloidal solar magnetic field beneath the photosphere, differential rotation and convection.

1.3 TELESCOPIC OBSERVATIONS

Following Galileo's telescopic rediscovery of sunspots in AD 1610, European astronomers took an enthusiastic interest in observing this phenomenon. However, there do not seem to have been many observational records from the second half of that century. The lack of reports of sightings can possibly be attributed to a real scarcity of sunspots during this period - the Maunder Minimum (Eddy, 1976). It was not until the early 18th century that fairly frequent observations are to be found again.

Since AD 1818, almost daily sunspot records are available but prior to that date little more than monthly means (between 1749 and 1817) or even annual means (1700 to 1748) can be deduced. Before AD 1700, data are relatively scarce and much more difficult to interpret. In Table 1, I have compiled a list of some of the principal observers of sunspots since Galileo and their respective period of observations. The list covers the period down to the 1890s. In 1874, RGO began a daily photographic patrol of sunspots and results were annually published as Greenwich Photoheliographic Results.

The most extensive compilation of the RGO series of observations covering 1874-1954 is summarised in Sunspot and Geomagnetic Storm Data 1874-1954 (RGO, 1955), compiled under the direction of Sir Harold Spencer Jones. The original data were recorded by the photo-heliograph on glass plates with 4" solar diameters which are now preserved in the plate archives at RGO in Herstmonceux. The RGO photo-heliographic plate measurements terminated after 1976 and the observatory at Debrecen in Hungary agreed to continue this work but so far has not yet published any data. The 1955-1976 series is only available in annual databooks.

Conventionally, the sunspot number is arbitrary defined by an index, R, as follows:

$$R = k(10g + f)$$

where g is the number of groups of spots, f is the number of individual spots and k is a correction factor taking into account the different

TABLE 1. Some of the principal observers of sunspots
since Galileo and their respective period of
observations

Observer	Period	References
G. Galileo	1611-12	Opera (1615) daily drawings ˜ 4" diameter
C. Scheiner	ca 1625-1627	Rosa Ursina (1630) in Latin - daily drawings ˜ 8" diameter
J. Hevelius	1642 - 44	Selenographia (c.1647) in Latin - daily drawings ˜ 8 inches in diameter
C.H. Adams	1819 - 23	
S.H. Schwabe	1825 - 67	39 Vols (MN, xxxvi, 297-99; xIi, 180)
R. Wolf	1610-1715	Historical Reconstructions (1856)
T.J. Hussey	1826 - 37	
H. Lawson	1831 Aug - 1832 Aug	
J. Herschel	1836 Dec - 1837 Oct 1826, 1836 1856 - 58 1865 - 71	176 diagrams
C. Shea	1847 - 65	5 Vols
T. Chevallier	1847 - 49	2 Vols
J.H. Griesbach	1850 - 65	
R.C. Carrington	1853 - 61 1870	3 Vols (MN, xxxvi, 249-50)
F. Howlett	1859 - 92	8 Vols (MN, xxxvi, 297; xxxvii,364; 1v, 73-6)
G.L. Bernaerts	1870 - 79	13 Vols

sizes of telescopes and observing conditions. The index R was first
established by Wolf at Zurich (1856). Appropriately, the international
sunspot number index used to come from the Swiss Federal Observatory
at Zurich; it was responsible for normalising measurements from various
observatories and with the determination of the subjective correction
factor k. Since 1981 the Sunspot Index Data Centre in Brussels has
taken over the task for world wide distribution of sunspot numbers.

1.4 FAR EASTERN NAKED-EYE SUNSPOT CATALOGUES

For the pre-telescopic period several compilations of Far Eastern
naked-eye sunspot sightings have appeared in various journals over the

years. The earliest of these catalogues was compiled by Williams
(1873) and was followed by Turner (1889). The more extensive compilations of the present century include the catalogues of Kanda (1933),
Schove (1950), Keimatsu (1976), Yunnan Observatory (1976), Clark and
Stephenson (1978), and Chen and Dai (1982). Recently, Wittmann and Xu
(1987) have compiled a list numbering some 235 entries including a few
observations from European sources. For the period since 1610, naked-
eye observations were first brought to note by Xu and Jiang (1979)
who argued against the existence of the Maunder Minimum. In addition,
there is also a list of naked-eye sunspots from Chinese local gazettes
given in a book by Chen (1984), which is extracted from the yet unpublished A Union Table of Ancient Chinese Records of Celestial
Phenomena. Table 2 gives a comparison of the above mentioned catalogues.

Table 2. A comparison of the various sunspot catalogues.

Authors	Period Covered	Total Number of Entries
Kanda (1933)	28 BC - AD 1743	142
Yunnan Group (1976)	43 BC - AD 1638	112
Clark and Stephenson (1978)	28 BC - AD 1604	139
Xu and Jiang (1979)	AD 1603 - AD 1684	33
Chen and Dai (1982)	165 BC - AD 1648	162
Chen (1984) (i)	28 BC - AD 1640	125
(ii)	165 BC - AD 1918	109
Xu and Wittmann (1987)	165 BC - AD 1684	235
The Present Work	165 BC - AD 1918	235

Despite the numerous attempts at a complete compilation, these
catalogues suffered in one form or another, most often by omissions,
dating errors or the inclusion of apparently spurious data. I have
compiled a new catalogue of naked-eye sunspots which aims at correcting
the mistakes made in earlier catalogues and incorporates several new
records. This investigation is based on a detailed study of dynastic
histories and other sources. For the period from earliest times to the
late 14th century I have consulted principally records contained in the
Astronomical Treatises of the various official Far Eastern dynastic
histories. The use of late secondary sources for this period is restricted as most of these works copied records directly from the earlier

dynastic histories.

For the period from the late 14th and mid-17th century I have consulted astronomical records in the <u>Ming-shih-lu</u> ('Veritable Records of the Ming Dynasty') for sunspots observed in China. This has been made easier by a recent compilation of all astronomical records in the <u>Ming-shih-lu</u> by Ho and Chiu (1986). For the period since 1644, no dynastic record of sunspots from China is found. From the late 14th century onwards up to about the mid-18th century, there are a number of Korean sightings of sunspots from the <u>Sillok</u> ('Veritable Records') of the Korean kings. Throughout the time span (164 BC - AD 1918) under consideration, there are only three Vietnamese and one Japanese records of sunspots.

The present catalogue (Yau and Stephenson, 1988) also extends into relatively modern times; the last entry is in AD 1918. The majority of the more recent data (since the 17th century) come from the Chinese local gazettes, of which Durham Oriental Library has a sizeable collection. The total number of entries contained in the present catalogue is 235 with 157 pre-telescopic and 78 post-telescopic records.

1.5 INVESTIGATION OF TELESCOPIC DATA

1.5.1 Sunspot Cycles Since 1610 –

I have reconstructed annual mean sunspot numbers (Figure 2) based on

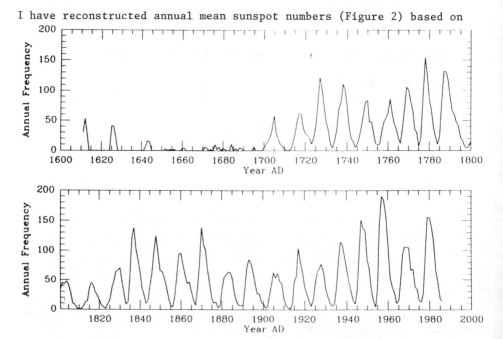

Fig. 2. The annual mean sunspot numbers from AD 1610 to 1986.

Eddy (1976) and Waldmeier (1961) for the period 1610 to 1931 and the
UK World Data Centre series for the period 1932 to 1986. As can be
seen, the nominal 11-year cycle is clearly depicted since 1700. Prior
to the 18th century, the 11-year period is obscure. Eddy (1976)
claimed this was due to a period of suppressed solar activity between
1645 and 1715, which following the original suggestion by Maunder
(1890) he called the 'Maunder Minimum'.

Several contemporary astronomers had commented on the paucity of
sunspots during this period. Concerning a sunspot in 1684, Flamsteed
comments: "These appearances, however frequent in the days of Scheiner
and Galileo, have been so rare of late that this is the only one I have
seen in his face since December, 1676" (Maunder, 1894). In 1705,
Cassini and Maraldi recorded that they had never seen a spot in the
northern hemisphere of the Sun.

It seems difficult to quantify the level of solar activity during
the Maunder Minimum on account of data selection effects. Stephenson
(1988) expected that imperfect sampling would cause spurious trends
in addition to real solar fluctuations. He illustrated this by the
sudden increase in the number of reports of occultations and eclipses
in the second half of the 17th century (Figure 3). Apart from more
observers, more astronomers communicated observations to the newly
established research journals, whereas formerly much material was
scattered in unpublished papers or books having a limited circulation.

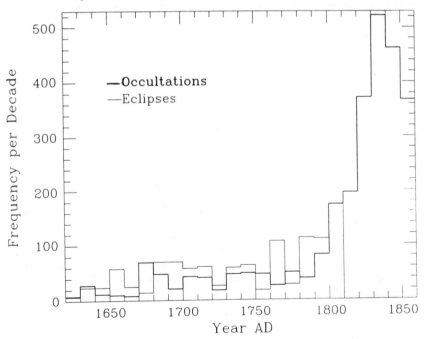

Fig. 3. The decade distribution of the number of reports of
solar eclipses (between AD 1620 to 1810) and occultations
(AD 1620 to 1860).

As is evident from Figure 2 the sunspot record since the early
18th century shows some indication of modulation on the centennial
time-scale. This is the so-called 'Gleissberg Cycle' of about 80
years (Gleissberg, 1944). It is tempting for one to interpret the
Maunder Minimum as a continuation of this pattern. However, the problems
relating to data selection have already been noted. In addition, the
reality of the Gleissberg Cycle is still in doubt, particularly
because of its obviously variable length. It is likely that these long
term cycles are caused by the random fluctuations of the 11-year cycles.

The amount of solar activity may vary considerably between
two consecutive sunspot cycles. As can be seen from Figure 2, the
onset and termination of one cycle can be markedly different from the
rest. The rise time to the sunspot maximum is approximately 4 or 5
years and the time for the cycle to decline is about 6 or 7 years.

The phase of the sunspot cycle also closely determines the mean
heliographic latitude of all groups. At minimum the first groups of
the new cycle appear at ± 30 to ± 35 deg. Thereafter the latitude
range moves progressively towards the equator, until by the next
minimum the mean latitude is around ± 7 deg. Then, while the equatorial
groups are fading, those of the succeeding cycle begin to appear in
their characteristically higher latitudes. Groups are seldom seen
farther than 35 deg. or closer than 5 deg. from the equator. This
latitude-time relation for the progression of sunspots was first
illustrated by Sporer (1889) and subsequently became known as Sporer's
Law. A graphical representatin of this law is depicted by the
'butterfly diagram' which is obtained by plotting the mean heliographic
latitude of individual groups of sunspots against time.

1.5.2 Asymmetric Distribution of Spot Areas

The size of a sunspot or sunspot group is conventionally measured in
units of millionths of the area of the Sun's visible hemisphere.
Sunspots may be morphologically classified into one of nine classes
of the Zurich system of classification (Bray and Loughhead, 1964).
The typical lifetime of a spot in days is approximately 1/10th of the
spot's maximum area in millionths of the solar hemisphere. In order
to show the distribution of relative sunspot sizes, I have analysed
the sunspot areas given per rotation in the RGO series covering the 80
year period between 1874-1954. Figure 4 shows the distribution of
sunspot areas for both the northern and southern hemispheres of the Sun.
Of all of the sunspot group areas in this period, about 22% have an
average area greater than 500 millionths of the Sun's hemisphere. It
can be seen from Figure 5a that sunspot areas for the two hemispheres
were not equally represented. I have calculated the anisotropy index,
A_s, for the two hemispheres by:

$$A_s = \frac{N - S}{N + S}$$

where N is the total mean areas in the northern hemispheres and S

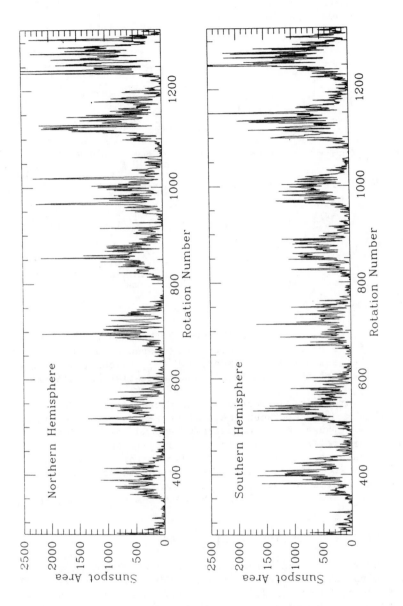

Fig. 4. The distributions of the sunspot areas per rotation from AD 1874 to 1954 for both the northern and southern hemispheres.

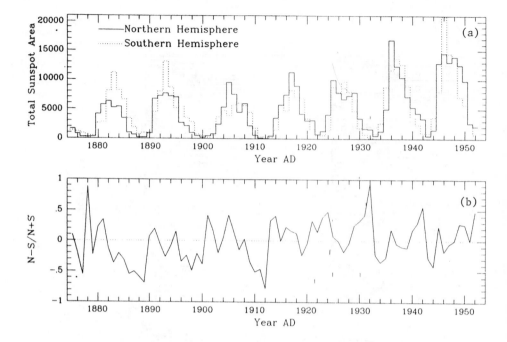

Fig. 5a.   A comparison between the annual total sunspot areas
           of the northern and southern hemispheres of the Sun.
    5b.   The area asymmetric index, N-S/N+S, calculated for the
           areas of the northern and southern solar hemispheres.

is the total mean areas in the southern hemisphere. It is worth
noting the marked deviation between the hemispheres, around the minima
of sunspot cycles (Figure 5b). In contrast, around the maximum the
two hemispheres are relatively equally covered by sunspots. The
deviation can be taken to signify that at least on the century time-
scale there is a systematic difference between the magnetic fluxes
from the two hemispheres during sunspot minimum, whereas near maximum
the magnetic fluxes are nearly balanced between the two hemispheres.

1.6   INVESTIGATION OF NAKED-EYE SUNSPOT RECORDS

Watching the Sun was a regular practice in China which may be traced
back in time to the pre-Han period (prior to 220 BC). It is generally
believed that the mythology of depicting a crow on a reddened Sun may
have originated from earlier sightings of unusually large sunspots.
Solar observation is likely to have been carried out routinely by
astronomers of the Imperial Astronomical Bureau. Throughout the
entire Chinese history down to the fall of the last dynasty in AD 1911,
watching the Sun for unusual occurrence was vigorously practised. In
later centuries, the same tradition was adhered to by astronomers in

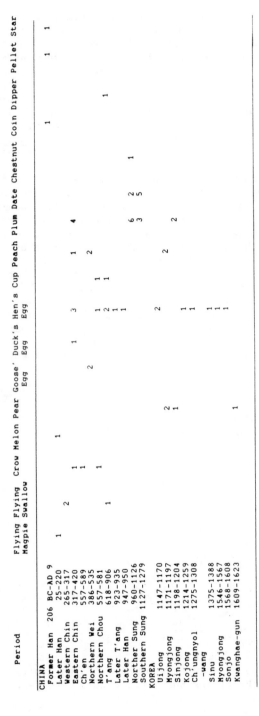

Table 3. A comparison of sunspot shapes described in Far Eastern texts.

Korea and Japan. The type of phenomena noted by these astronomers seems to include virtually anything on or near the Sun. These phenomena range from sunspots, solar haloes, parhelia to 'auspicous vapours' in the vicinity of the Sun (Ho Peng Yoke, 1966). They were regarded as portents and it thus can be understood that the Sun played an imortant role in astrlogical prognostications.

The general description of a sunspot record from the Far East is that of "A black spot (hei-tzu) or vapour (hei-ch'i) within the Sun". Some of the spots are compared with the shape of common objects like hen's eggs, plums, chestnuts and so on (Table 3). It can be seen that the comparison of spots with such objects is very much period dependent on and subject to the observers of that time. Beside the comparison of the shapes of spots, the duration of visibility of a spot was sometimes mentioned.

1.6.1 Observational Criteria

In order for the eye to look at the Sun directly, it is necessary for the brightness of the Sun to be sufficiently dimmed. Adequate dimming of the Sun's brightness may be gained in several ways:

natural aids include atmospheric haze, dust or sand storms, severe
atmospheric absorption near sunrise or sunset and reflection from
still waters. In addition, artificial aid may be used. Needham (1959)
thought that Oriental astronomers used pieces of semi-transparent jade,
mica or smoky rock crystal for looking at the Sun, but there does not
seem to be any early literature supporting this claim. On the other
hand, there are mentions that solar eclipses were observed by looking
at the reflection of the Sun in a basin of water blackened with Chinese
ink (Chu Wen-hsin, 1934 and Wang & Siscoe, 1980). In addition, I have
found a quote in the 12th century encyclopedia the <u>Wen-Hsien T'ung-k'ao</u>
saying that a basin of oil was used for observing an eclipse in the
Sung Dynasty (AD960 - 1279).

Provided viewing conditions are favourable, a large whole spot
or compact spot group of area 500 millionths of the Sun's hemisphere
(about one arc minute) when near the centre of the solar disk should
be just visible to the naked-eye (Newton, 1955). There are larger
spots with areas up to six times this size but they are less frequent.
A spot with a mean area of 1500 millionths of the Sun's hemisphere is
regarded as a 'giant spot' and should be readily observable by the
naked-eye under suitable conditions. Figure 6 shows the distribution
of the larger sunspots (areas greater than 500 millionths of the Sun's
hemisphere) together with the mean annual sunspot numbers in the
interval 1874-1954. It is clear that the larger sunspots do follow
closely the pattern exhibited by the 11-year cycles. Hence it may
be readily deduced that the probability of sighting a naked-eye sunspot
during the peak years of a sunspot cycle is much higher than at any
other times within that cycle.

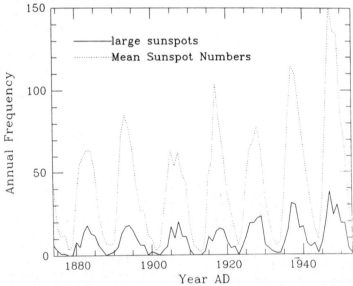

Fig. 6. A comparison between large sunspots (with area greater
than 500 millionths of the Sun's hemisphere) and yearly
mean sunspot numbers.

If the astronomers of the Far East kept a consistent watch of the sky, the extremely small sample of extant naked-eye sunspot records (about one per decade) is not easily explained. According to the RGO (1955) catalogue, there was a total of 761 large whole spots or compact spot groups (areas greater than 500 millionths of the Sun's hemisphere) observed during the 80 years from 1874 to 1954. Based on these statistics, one can readily infer that approximately 100 spots per decade should be visible to the naked-eye during normal solar activity. Comparing the number of naked-eye sunspots with modern observations, it seems that only about 1% of the theoretically observable sunspots were observed. Even if all occasions unfavourable for observing sunspots in the past were allowed for we would still have expected many more sightings. One may argue that this was due to inattention on the part of astronomers or due to some kind of astrological selection factors. Other reasons may be due to the attitude of the imperial court, both the rulers and their astronomers. A possible explanation is that these astronomers tended to observe at a specific time. Figure 7 shows a plot of the frequency of sunspots against the lunar age. The single marked peak on the first day of the month can be understood as due to accidental sightings when the Sun was being examined for possible eclipses. Hence it seems likely that the court astronomers did not keep a regular watch for spots; otherwise we would have a lot more records today. Sunspots observed on days other than the first were purely on a casual basis.

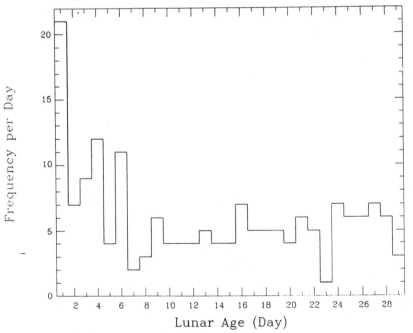

Fig. 7. The frequency of naked-eye sunspots plotted against lunar age.

176

1.6.2 Seasonal variation

Oriental sunspot sightings show a marked seasonal variation. The cause
of this is the general climatic conditions over much of central China
and Korea. During the late winter and whole of the spring, both dust
and sand storms originating from the Takla Makan, Gobi and Ordos
Deserts and the Chinese loesslands frequently dim the Sun considerably.
As a result, these natural geographical factors provide a favourable
viewing condition and enable direct observation of the Sun to be made
readily. Consequently, there would have been a higher probability of
observing sunspots during the spring and winter times than any other
seasons. On the other hand, the chance of directly looking at the Sun
in summer is limited due to a prevalent clear sky.

The variation of sunspots with seasons was first suggested by
Kanda (1933) and was brought into attention again by Clark and
Stephenson (1978). A subsequent detailed analysis by Willis et al.
(1980) demonstrated the existence of a marked seasonal variation in the
Oriental sunspot records. With the help of satellite photographs they
were able to show the large extent of dust storms over much of East
Asia. Figure 8 shows the seasonal variation of sunspot sightings with
the addition of all new data from the present catalogue. The histogram
clearly indicates a marked variation of sunspots with seasons; furthermore the characteristic seasonal variation as shown by Willis et al.
(1980) has been retained.

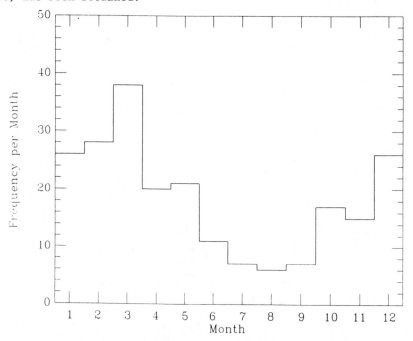

Fig. 8. The seasonal variation of naked-eye sunspots.

1.6.3 Secular Variation

By studying early naked-eye sunspot observations, it is hoped that one
may be able to obtain an indication to the long term activity levels
of the Sun in the past. It is assumed that a close correlation between
the appearance of large sunspots and a higher level of solar activity
existed. Due to the small number of recorded naked-eye sunspots, it is
very much doubted that if one could ever retrace accurately the 11-year
cycle. Furthermore, the 11-year cycle may well be described as
'quasi-regular' since the actual interval between successive maxima
may vary by several years. It seems likely that at best, one might just
be able to recover longer trends similar to that of the Maunder Minimum
type of long depression intervals. It is difficult now to assess the
true scale of the sunspot cycle before the telescopic period from the
scanty naked-eye sunspot data alone.

Figure 9 shows a histogram of the decade distribution of naked-eye
sunspot records from far Eastern histories covering the period 165 BC
to AD 1918. For the first millenium AD only a few features can be seen.
Due to the preservation of the records, one can at most claim that
these peaks confirm the Sun was active to a certain degree during
these periods. Whereas the absence of sunspot records do not
necessarily mean the Sun was inactive due to the problem with data gaps.

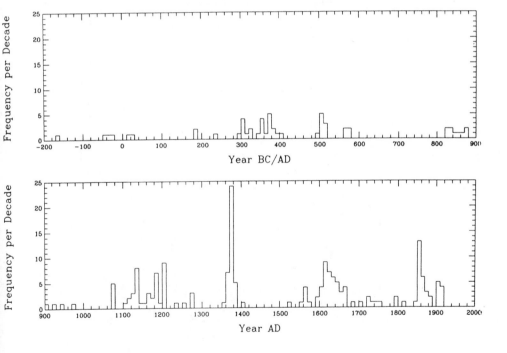

Fig. 9. The decade distribution of naked-eye sunspots from
165 BC to AD 1918.

178

Some of the gaps exhibited by historical records may be attributed
to the chaos caused by the downfall of a dynasty in China, or wars
in Korea or Japan. Presumably, during these difficult times, the
astronomers were hindered from carrying out their normal observational
routines. In addition, it is also difficult to distinguish between
the changing attitudes towards the value of celestial portents and the
amount of preserved materials available today. The problem can best
be illustrated by a comparison with other solar related phenomena,
e.g. the solar haloes (Figure 10). The various peaks shown in this
figure are likely due to random fluctuations caused by the varying
number of preserved records. The peaks appearing in the sunspot
distribution around the 12th century may be similarly explained. The
strongest peak of the sunspot distribution is around 1370; it is
plausible that this represents a period of relatively higher solar
activity judging from the size of the feature.

For the post-telescopic period, the features in the first half
of the 17th century could be due to selection effects. The lack of
naked-eye sunspot records in the telescopic period prompted researchers
to look for sightings recorded in other sources than the official
dynastic histories. Sunspot records in the Fang-chih ('Local
Gazettes') were first drawn to attention by Xu and Jiang (1979). They

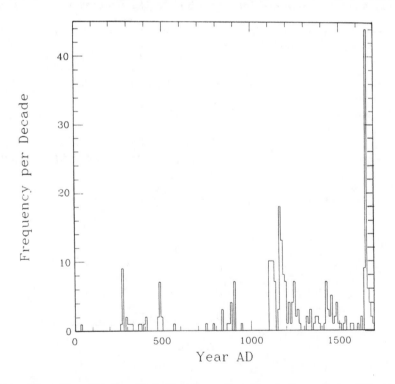

Fig. 10.   The decade distribution of solar haloes from AD1 to
           1700.

compiled a list of 21 records in the 17th century from these local
gazettes. Based on these records Xu and Jiang proposed that the Sun
did not lower its sunspot production during the Maunder minimum period
as favoured by Eddy (1976). They believed that the 11-year sunspot
cycles were present at these times and the Maunder Minimum was due to
a deficiency in the data.

In a reassessment of the Maunder Minimum, Eddy (1983) pointed out
that only six of the Chinese sightings actually occurred during this
interval, and some of these coincided with spots reported in Europe
at the time. Hence these isolated occurrences - only a tip of a large
iceberg - represent only a minute fraction of the total number of
sunspots seen telescopically in Europe at this period.

The lack of naked-eye sunspot records during the last three
centuries in China had been interpreted as either the Chinese astronomers
were not interested in sunspots or the records were simply lost. It is
possible that the introduction of the telescope around the mid-17th
century into China might have brought an end to the omen value of sunspots. Cullen (1980) argued that sunspot records could have been lost
at the time when the Peking Observatory was occupied by Allied troops
during the Boxer Uprising in 1900. However, from the state of preservation of other astronomical records in the early Ch'ing period
down to AD 1800,   Cullen's conclusion may not be entirely correct.
During this period, there is a profusion of other records such as
eclipses, planetary conjunctions and so on. If sunspots were recorded
they would have been preserved together with other contemporary
astronomical records at least until AD 1800.

The absence of Japanese sunspot records in the Maunder Minimum
period and other times is curious. Only one sighting in the year AD 851
is reported for the entire period since the first Japanese observation
in the 7th century. It is possible that they too, like their
European counterparts, believed in an unblemishable Sun. It is of
interest to note that the insignia of Japan is none other than the Sun.

Figure 11 compares the distributions of telescopic and naked-eye
sunspots for the period from AD 1600 to 1920. The naked-eye records
for this period originated mainly from local gazettes. Due to incomplete telescopic data, not much can be inferred from the distributions
before about AD 1700. But from about AD 1700 onwards, a number of
interesting features are shown by the distributions. It can be seen
that most of the naked-eye sunspots lie within the maximum envelopes
of the 11-year cycle. Only in a few cases did the naked-eye sunspots
coincide with the cycle minima. Thus, the assumption that the
observations of naked-eye sunspots represent times when the solar
activity is at a higher level appears essentially correct.

1.6.4 Period Search on Naked-eye Sunspot Records

(1) Methods of Period Analysis

One of the reasons for studying early sunspot data is to see whether
the 11-year sunspot cycle persisted in the pre-telescopic past.
Several available methods can be performed which will search for the

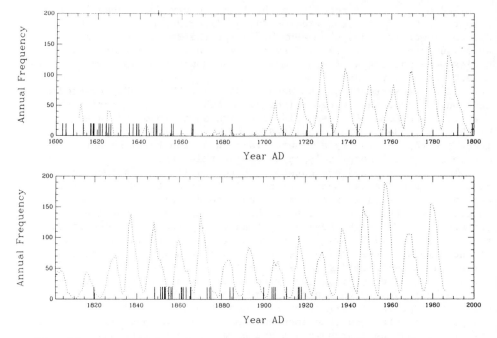

Fig. 11. A comparison between the number of naked-eye and
telescopic sunspots for the period AD 1600 to 1910.

underlying periodicities of the data. One of the better known methods
is that of   power spectral analysis. This type of analysis has
already been performed on the telescopic sunspot series by a number of
researchers (Cole, 1973; Cohen and Lintz, 1974; Wittmann, 1978; Otoalo
and Zenteno, 1983). Although there is a considerable disparity in the
periods obtained, from 5 to 180 years, there seems to be a general
agreement on the existence of an 11-year period. Beside the power
spectral method and its variant methods - see details in Berry (1987) -
two other methods worth mentioning are the cyclogram method (Attolini
and Cecchini, 1984) and epoch folding method (Wittmann and Xu, 1987).
    The criteria for using the above methods vary considerably. For
instance, both the power spectral analysis and the cyclogram method
assume that the input data are equally spaced, continuous and homogeneous; whereas the epoch folding method requires a relatively large
sample and fairly accurate knowledge of the search period. I have
initially analysed the naked-eye sunspot data (and the telesopic data
as a check) for periodicities by the power spectral analysis employing
a Parzen lag window (NAG, 1984 and Bloomfield, 1976). Figure 12a shows
the results of the power spectral analysis on the telescopic sunspot
data. It seems that in addition to the primary period of 11 years,
there is a secondary period of 10 years. At lower periods the spectrum
becomes noisy, it is difficult to draw any concrete conclusion on the
existence of other periods. The results of the analysis on the naked-

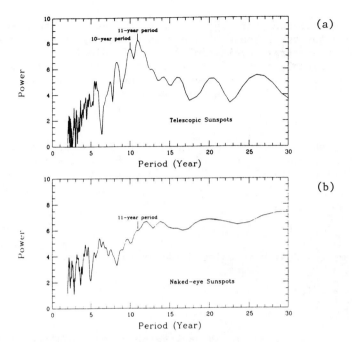

Fig. 12 (a) The spectrogram obtained from a power spectral analysis
of the telescopic sunspot data. (b) The spectrogram
obtained from a power spectral analysis of the naked-eye
sunspot data.

eye sunspot records is shown in Figure 12b; no obvious feature is
present at either the 11 or 10 year period.

In order to examine the data in another fashion to see whether it
is the power spectral method that failed to pick out the underlying
periods or the 10/11-year cycles are really absent from the naked-eye
observations, I have adopted a different approach. The method I
have selected here, which I shall call the pseudo-spectral method, is
based on a similar method used by Stothers (1979) in the investigation
of the Greco-Roman records of aurorae.

(2) Pseudo-Spectral Method

Several peculiarities of the naked-eye sunspot records need to be taken
into consideration when selecting a suitable method of period analysis.
The naked-eye sunspot records: (i) have large data gaps; (ii) are not
expected to coincide exactly with the maximum of the 11-year cycles
and (iii) constitute only a relatively small sample for each 11-year
cycle. In addition, the naked-eye sunspot data set is not expected
to follow a true sinusoidal pattern as can be inferred from the rather
irregular 11-year cycles exhibited by the telescopic data. The ability
to incorporate large data gaps and not to involve a sine wave term

are intrinsic features of the pseudo-spectral method which makes it more suitable for the purpose of analysing naked-eye data than methods like the power spectral, cyclogram or epoch folding.

First a suitable array of trial periods is set up. Then a continuous sequence of predicted times of maximum for each combination of trial period and trial epoch is calculated from:

$$t_{max} = t_o + nP$$

where t_{max} is the time of the nth maximum, t_o is the initial epoch and P is the trial period. Both P and t_o are to be determined simultaneously. The residual index R_{index} is then estimated from:

$$R_{index} = \frac{\sigma_c - \sigma}{P}$$

where the rms residual σ is calculated for each sequence by:

$$\sigma = \left(\sum_{i=1}^{N} \frac{d_i^2}{N} \right)^{\frac{1}{2}}$$

Here d_i (i = 1, 2, ..., N, for N observations) is the difference between the predicted maximum and the observed time. The expected rms residual, σ_c assuming a rectangular distribution is given by:

$$\sigma_c = \frac{P}{N} \left(\frac{N^2 - 1}{12} \right)^{\frac{1}{2}}$$

Figure 13a shows the results of a search for periodicities in the naked-eye sunspot records. There is only one broad peak at a period of about 10 years. The results obtained here are compatible with the secondary feature seen in the telescopic data (Figure 12a). It appears that the early sunspot cycles were more like 10 years rather than the dominant 11-year cycles seen in recent times. If this interpretation is correct, it would be worth looking at the mechanisms that caused the slow-down of the sunspot cycle. Figure 13b shows the resulting spectrum of a similar search carried out on the solar halo records. As is expected, no visible peak is seen in this case.

1.7 SUMMARY

The results obtained from an analysis of the available sunspot records from both the pre-telescopic and telescopic periods are in general agreement with prevous investigations. For the telescopic period, the analysis of sunspot areas from the RGO series showed an asymmetric distribution between the areas north and south of the solar equator. The variability of magnetic flux especially during sunspot minima is

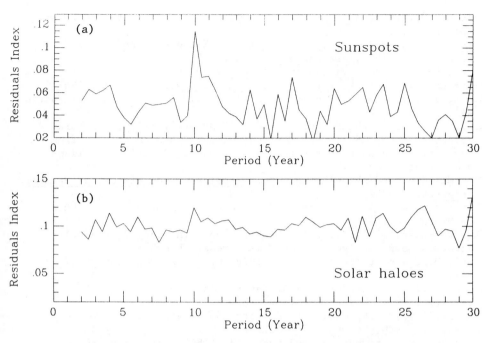

Fig. 13. (a) The spectrogram obtained for the naked-eye sunspot
data using the pseudo-spectral method. (b) The
spectrogram obtained from the solar halo data using the
same method.

inferred. The results obtained here should provide another parameter
for constraining future models on the cyclic variation of the Sun.

The Far Eastern historical records of sunspots have been updated
and revised to include some 71 records from local Chinese histories.
The period search performed on the naked-eye data yielded a mean
period of about 10 years. It is hoped that a few more naked-eye
sunspots will be recovered from other Far Eastern sources like the
biographies of astronomers and diaries of court officials. At the
present, the time for the onset and termination of the Maunder
Minimum is only arbitrarily determined. It is worth pursuing in depth
these two limits if we are to obtain improvements on current models
of protracted periods of low solar activity. Foremost to our understanding of the variations in the sunspot cycles is that the observation
of sunspots should be continued on a systematic basis.

ACKNOWLEDGEMENT

I wish to thank Dr. J.A. Eddy for reading this paper and offering
several helpful comments.

Allen, C.W., 1976, Astrophysical Quantities, 3rd ed., Athlone Press, London University.

Attolini, M.R. and Cecchini, S., 1984, Il Nuovo Cimento, 7C, 245.

Babcock, H.W., 1961, Astrophys. J., 133, 572.

Berry, P.A.M., 1987, Vistas in Astronomy, 30, 97.

Biermann, L., 1941, Vierteljahrsschr. Astr. Gessellsch., 76, 194.

Bloomfield, P., 1976, 'Fourier Analysis of Time Series: An Introduction', Wiley.

Bray, R.J. and Loughhead, R.E., 1964, 'Sunspots', Wiley, New York.

Carrington, R.C., 1863, 'Observations of the Spots on the Sun from November 9, 1853, to March 24, 1861, made at Redhill', Williams and Norgate, London.

Chen Mei-dong and Dai Nian-zu, 1982, Studies in the History of Natural Sciences, 1, 227.

Chen Zun-gui, 1984, 'History of Chinese Astronomy', vol. 3, Renmin Chubanshe, Shanghai.

Chu, Wen-hsin, 1934, 'Li-tai Jih-shih K'ao', (A study of the Solar Eclipses Down the Ages), Commercial Press, Shanghai.

Clark, D.H. and Stephenson, F.R., 1978, Q.J.R. Astron. Soc., 19, 387.

Cohen, T.J. and Lintz, P., 1974, Nature, 250, 398.

Cole, T.W., 1973, Solar Phys., 30, 103.

Cowling, T.G., 1934, Mon. Not. R. astr. Soc., 94, 39.

Cullen, C., 1980, Nature, 283, 427.

De La Rue, W., Stewart, B. and Loewy, B., 1865, Proc. Roy. Soc., 14, 37.

Eddy, J.A., 1976, Science, 192, 1189.

Eddy, J.A., 1983, Solar Phys., 89, 195.

Eddy, J.A., 1988, (this volume).

Evershed, J., 1909, Mon. Not. R. astr. Soc., 69, 454.

Faye, H.A.E.A., 1865, Comptes Rendus 60, 89.

Gilman, P.A. and Howard, R., 1985, Astrophys. J., 295, 233.

Gleissberg, W., 1944, Terr. Magn. Atmos. Electr., 49, 243.

Goldstein, B.R., 1969, Centaurus, 14, 49.

Hale, G.E., 1908, Mount Wilson Solar Observ. Contrib., No. 26.

Herschel, F.W., 1795, Phil. Trans. R. Soc., 85, 61.

Ho Peng Yoke, 1966, 'The Astronomical Chapters of the Chin-shu', Mouton, Paris.

Ho Peng Yoke and Chiu Ling Yeong, 1986, 'Astronomical Records in the Veritable History - Ming Shih-lu', Hong Kong Univ. Press.

Kanda, S., 1933, Proc. Imp. Acad. (Japan), 9, 293.

Keimatsu, M., 1970-1976, Ann. Sci. Kanazawa Univ., Vol. 7-13.

Lockyer, J.N., 1866, Proc. Roy. Soc., 40, 347.

Maunder, E.W., 1890, Mon. Not. R. Astron. Soc., 50, 251.

Maunder, E.W., 1894, Knowledge, 17, 173.

Moore, R. and Rabin, D., 1985, Ann. Rev. Astron. Astrophys., 23, 239.

NAG, 1984, 'FORTRAN Library Manual Mark II', Vol. 6, Chapter G13 Numerical Algorithms Group ltd.

Needham, J., 1959, 'Science and Civilisation in China', Vol. 3, Cambridge University Press, London.

Newton, H.W., 1955, Vistas in Astron., 1, 666.

Otoalo, J.A. and Zenteno, G., 1983, Solar Phys, 89, 209.

Parker, E.N., 1979, Astrophys. J., 230, 905.

RGO, 1954, 'Sunspot and Geomagnetic Storm Data 1874-1954', HMSO, London.

Sarton, G., 1947, Isis, 37, 69.

Schaeberle, J.M., 1890, Mon. Not. R. astr. Soc., 50, 372.

Schove, D.J., 1950, J. Brit. astron. Assoc., 61, 22.

Schove, D.J., 1955, J. Geophys. Res., 60, 127.

Schwabe, H., 1843, Astron. Nachr, 20, 283.

Secchi, A., 1870, 'Le Soleil', Paris.

Sporer, F.W.G., 1889, Bull. Astron., 6, 60.

Stephenson, F.R., 1988, 'Evidence for Solar Variability from Historical
   Records', Il Nuovo Cimento (in press).

Stothers, R., 1979, Astron. Astrophys., 77, 121.

Turner, H.H., 1889, Observatory, 12, 217.

Vyssotsky, A.N., 1949, Medd fran Lunds Astr. Observatorium, Historical
   Papers, 22.

Waldmeier, M., 1961, 'The Sunspot-Activity in the Years 1610-1960',
   Schulthess, Zurich, Switzerland.

Wang, P.K. and Siscoe, G.L., 1980, Solar Phys. 66, 187.

Williams, J., 1873, Mon. Not. R. astron. Soc. 33, 370.

Willis, D.M., Easterbrook, M.G. and Stephenson, F.R., 1980, Nature,
   287, 617.

Wilson, A., 1774, Phil. Trans. R. Soc., 64, 6.

Wittmann, A., 1978, Astron. Astrophys., 66, 93.

Wittmann, A.D. and Xu, Z.T., 1987, Astron. Astrophys. Suppl. Ser.,
   70, 83.

Wolf, R., 1856, Astron. mitt. Zurich, 1, 8.

Xu Zhen-tao and Jiang Yao-tiao, 1979, J. Nanjing Univ. (Nat. Sci.),
   No. 2, 31-38, English translation in Chin. Astron. Astrophys. (1984),
   2, 84.

Yau, K.K.C. and Stephenson, F.R., 1988, 'A Revised Catalogue of Far
   Eastern Observations of Sunspots (165 BC to AD 1918)', Q.J.R. Astron.
   Soc., in press.

Yunnan Observatory, Ancient Sunspot records Research Group, 1976,
   Acta Astron. Sinica, 17, 217.   (English translation in Chinese
   Astron. (1977), 1, 347).

SEASONAL AND SECULAR VARIATIONS OF THE ORIENTAL SUNSPOT SIGHTINGS

D.M. Willis, C.M. Doidge, and M.A. Hapgood
Rutherford Appleton Lab., Chilton, Didcot, Oxon OX11 0QX, UK
K.K.C. Yau and F.R. Stephenson
Department of Physics, University of Durham, South Road,
Durham DH1 3LE, UK

ABSTRACT. An earlier investigation (Willis, Easterbrook and Stephenson, 1980), which showed a clear seasonal variation of the unaided-eye sunspot sightings in the Orient during the pre-telescopic period, is extended to allow for the fact that descriptions of oriental observations recorded in the dynastic histories indicate that particulr sunspots were sometimes seen for several days. Using a revised catalogue of pre-telescopic sunspot observations from the Orient (Yau and Stephenson, unpublished), it is shown that the seasonal variation is even more pronounced when greater weight is given to each entry in the catalogue which indicates clearly that a sunspot was seen on more than just one day. This seasonal variation of the oriental sunspot sightings is attributed to a seasonal variation of suitable atmospheric viewing conditions in the Orient. Furthermore, using statistical theory appropriate to a mixture of Poisson distributions, it is shown that the oriental, unaided-eye, sunspot observations are not purely random events in a continuum of time. In fact, there is clear evidence for a secular variation in the incidence of these sunspot sightings. This secular variation could arise, for example, from actual long-term atmospheric viewing conditions, or variations in the political and social factors influencing the oriental astronomers.

1. INTRODUCTION

Willis et al. (1980) have drawn attention to a clear seasonal variation in the unaided-eye sunspot sightings recorded in the Chinese and Korean dynastic histories during the pre-telescopic period. This previous study was based on a catalogue of pre-telescopic sunspot sightings from the Orient published by Clark and Stephenson (1978). Subsequently, Yau and Stephenson (1988) have produced a revised, more extensive version of the Clark and Stephenson catalogue. For the purposes of the present study, the significant feature of the revised catalogue is that those occasions on which a sunspot was visible with the unaided-eye on more than one day are identified specifically. If greater weight is given to these observations, which indicate clearly that

187

F. R. Stephenson and A. W. Wolfendale (eds.),
Secular Solar and Geomagnetic Variations in the Last 10,000 Years, 187–202.
© *1988 by Kluwer Academic Publishers.*

a particular sunspot was visible for several (consecutive) days, the seasonal variation of the oriental sunspot sightings becomes even more pronounced.

Bray (1974) has claimed that most Chinese observations of sunspots were not made intentionally and that their occurrences were gleaned from casual mention in the diaries of the literati. Likewise, Eddy (1976, 1980, 1983) has argued that pre-telescopic oriental sunspot observations were made not routinely, but randomly, in the nature of an occasional and possibly accidental watch on the Sun. Clark and Stephenson (1978) have contested this circumspect interpretation of the oriental sunspot observations for the following reasons. During almost the whole of Chinese history since Han times, the acknowledged systematic and careful approach to observational astronomy and its compendious documentation, the abundance of records of other day-time phenomena, as well as the deeply held belief that celestial phenomena were precursors of terrestrial events, make it unlikely that only sunspots were omitted from regular patrol. All authors acknowledge, however, that a regular search might have been expected to yield more unaided-eye sunspot observations if sunspot activity prior to the seventeenth century had been at present-day levels. This conclusion follows from the fact that the average rate of occurrence of the oriental sunspot sightings corresponds to about one per decade during the period 43 BC to AD 1608.

The number of observations in the Yau and Stephenson catalogue (1988) is sufficient to perform a statistical test of the hypothesis that the oriental sunspot sightings are absolutely random. By using statistical theory appropriate to a mixture of Poisson distributions, it is proved conclusively that the oriental sunspot sightings are not purely random events in a continuum of time.

2. THE REVISED CATALOGUE OF ORIENTAL SUNSPOT SIGHTINGS

Yau and Stephenson have revised the Clark and Stephenson (1978) catalogue of pre-telescopic sunspot sightings from the Orient. This revised catalogue will be published separately and it is only necessary to mention briefly the significant revisions to the original catalogue. The revised catalogue is based exclusively on historical sunspot records found in the dynastic histories. Table I lists the places and dates of additional sightings that were not recorded in the original catalogue published by Clark and Stephenson (1978). To avoid any possibility of confusion in subsequent papers based on the revised catalogue, a provisional Roman reference number is assigned to each additional sunspot sighting included in Table I. Otherwise, the original Clark and Stephenson reference numbers are used in this paper to identify sunspot sightings that appear in either the original or the revised catalogues. Following the convention adopted in the original catalogue, all dates up to 4 October 1582 are on the Julian calendar while subsequent dates are on the Gregorian calendar.

TABLE I

CATALOGUE OF ADDITIONAL SUNSPOT SIGHTINGS FROM THE ORIENT

Ref. No.	Place	Julian Date					Remarks
(I)	C	43 BC	May	5	- Jun	3	Only month given
(II)	C	32 BC	Mar	8	- Apr	5	
(III)	C	AD 15	Mar	10	- Apr	7	
(IV)	C	AD 505	Jan	4			
(V)	C	AD 947	Nov	26			
(VI)	C	AD 1370	Jan	28	- Feb	3	Visible for 7 days
(VII)	K	AD 1556	Apr	17			
(VIII)	V	AD 1593	Jan	3			
(IX)	V	AD 1603	Apr	4	- May	10	Only month given
(X)	K	AD 1608	May	10			

Place references : C = China, K = Korea, V = Vietnam

Three sunspot sightings that were included in the original catalogue are omitted from the revised catalogue, namely those with reference numbers (5), (52) and (89). Numbers (5) and (52) are excluded because the record is from a secondary source and not a dynastic history. In the revised catalogue, number (89) is combined with (88) to form a single entry that records a sunspot which was visible for five days (AD 1186 May 23 - May 27). Therefore the revised catalogue of oriental sunspot sightings recorded in the dynastic histories contains 146 entries compared with 139 in the original catalogue.

The revised catalogue also incorporates the following corrections and refinements to the Julian dates (which are not given elsewhere in this paper) : (4) 188 Feb 15 - Mar 15, only month given; (9) 302 Dec 6 - 303 Jan 4, only month given;(10) 304 Dec 14 - 305 Jan 11, only month given; (37) 513 April 17; (40) 577 Dec 30? (possible error in month); (43) 826 May 24? (query inserted, possible error in month); (47) 851 Dec 2; (48) 865 Jan 31 - Feb 28, only month given; (53) 974 Mar 3 (query deleted); (59) 1105 Dec 6; (61) 1118 Dec 17; (70) 1138 Mar 16 or Mar 17; (74) 1145 Jul 23 (query deleted) - July 24; (116) 1371 Oct/Nov? (precise date uncertain); (127) 1375 Mar 22. In addition, the dates of sunspot sightings with reference numbers (6), (33), (39), (45) and (139) are refined in Table II. Likewise, the dates of sunspot sightings with reference numbers (72) and (114) are refined in Table III.

3. SUNSPOTS VISIBLE ON CONSECUTIVE DAYS

As noted in the Introduction, in recalculating the seasonal variation of the oriental sunspot sightings, allowance is made for the fact that sunspots were sometimes seen on contiguous days or over an interval of several days. Table II lists each entry in the revised catalogue for which the description in the dynastic history suggests that the sunspot was seen more than once over an interval extending between two and ten days, whereas Table III lists those entries for which the sunspot was apparently seen repeatedly over an interval longer than ten days. The reason for separating the data into these two classes relates to the synodic solar-rotation period of the Sun, as explained in the next section.

TABLE II

SUNSPOTS VISIBLE REPEATEDLY OVER A PERIOD OF UP TO 10 DAYS

Ref. No.	Place		Julian Date	Duration (days)
(6)	C	299	Feb 17 - Mar 18	Several
(14)	C	342	Mar 7 - Mar 11	5
(15)	C	352	Only year given	5
(33)	C	502	Feb 8, 11, 12	3
(39)	C	567	Dec 10 - Dec 15	6
(41)	C	579	Apr 3 - Apr 6	4
(45)	C	837	Dec 22 - Dec 24	3
(58)	C	1079	Mar 20 - Mar 29	10
(65)	C	1131	Mar 12 - Mar 14/15	3/4
(66)	C	1136	Nov 23 - Nov 27	5
(68)	C	1137	Mar 1 - Mar 10	10
(74)	C	1145	Jul 23 - Jul 24	2
(75)	K	1151	Mar 21, 31, Apr 1	3
(76)	K	1160	Feb 28 - Mar 1	3
(81)	K	1183	Dec 4 - Dec 5	2
(86)	K	1185	Apr 18 - Apr 19	2
(88/89)	C	1186	May 23 - May 27	5
(90)	C	1193	Dec 3 - Dec 12	10
(92)	C	1200	Sep 21 - Sep 26	6
(97)	K	1204	Feb 3 - Feb 5	3
(101)	K	1258	Sep 15 - Sep 16	2
(104)	K	1356	Apr 4 - Apr 5	2
(VI)	C	1370	Jan 28 - Feb 3	7
(122)	K	1373	Apr 26 - Apr 27	2
(125)	C	1374	Mar 27 - Mar 31	5
(126)	K	1375	Mar 20 - Mar 21	2
(130)	C	1381	Mar 22 - Mar 25	4
(132)	K	1382	Mar 9 - Mar 11	3
(139)	K	1604	Oct 24 - Oct 25	2

Plate references : C = China, K = Korea

TABLE III

SUNSPOTS VISIBLE REPEATEDLY OVER A PERIOD LONGER THAN 10 DAYS

Ref. No.	Place	Julian Date	Duration (days)
(4)	C	188 Feb 15 onwards	>30
(54)	C	1077 Mar 7 - Mar 21	15
(56)	C	1078 Mar 11 - Mar 29	19
(57)	C	1079 Jan 11 - Jan 22	12
(64)	C	1129 Mar 22 - Apr 14	24
(69)	C	1137 May 8 onwards	>15
(72)	C	1139 Mar 3 onwards	>29
(84)	C	1185 Feb 15 - Feb 27	13
(93)	C	1201 Jan 9 - Jan 29	21
(96)	C	1202 Dec 19 - Dec 31	13
(114)	C	1371 Jun 13 - Jul 12	30

Place reference : C = China

4. CRITERIA FOR DETERMINING THE SEASONAL AND SECULAR VARIATIONS

4.1 Rejection of Sunspot Sightings

Sunspot sightings with reference numbers (67), (116) and (131) are
excluded from both the seasonal and secular statistical studies because
each is believed to provide essentially the same information as the
immediately preceding entry in the catalogue. Similarly, (IX) in
Table I is considered to be the same as (138) in the Clark and
Stephenson catalogue and is therefore also omitted. Finally, the
sunspot sighting with reference number (112) is rejected because the
record merely indicates that there were frequent sunspot sightings
over a period of about one year. Since there were at least five other
independent sunspot sightings in the same year, which may duplicate
the information in (112), it seems sensible to reject this record.

4.2 Criterion for Independent Sunspot Sightings

The time that an individual sunspot actually takes to traverse the
solar disk is half the synodic solar-rotation period at the latitude
of the sunspot, namely about 13.5 days. However, 13.5 days is clearly
an upper bound to the time for which a sunspot is visible with the
unaided eye because of the foreshortening effect when the sunspot
is near the limbs of the Sun. The assumption made in this paper is
that the oriental astronomers were capable of observing the same
sunspot over an interval of 10 days.

Therefore, if two individual sunspot sightings were observed
on distinct days (with precise dates) separated by an interval

extending up to a maximum of 10 days, it is assumed that the <u>same</u> sunspot was observed on the two occasions. By exactly the same principle, if the description recorded in a dynastic history implies that a sunspot was seen repeatedly over an interval of time extending up to a maximum of 10 days, it is assumed that the <u>same</u> sunspot was seen throughout this interval. With this assumption, Table II lists all repeated observations described in the revised catalogue which are believed to refer to the same sunspot.

Conversely, if two individual sunspot sightings were observed on distinct days separated by an interval longer than 10 days, it is assumed that <u>different</u> sunspots were observed on the two occasions. Likewise, if the description recorded in a dynastic history suggests that sunspots were seen repeatedly over an interval longer than 10 days, it is assumed that different sunspots were actually seen at the beginning and end of the interval. With the above assumption, Table III lists all repeated observations described in the revised catalogue which are believed to refer to more than one sunspot.

The class separation criterion of 10 days, which is used to separate the entries in Tables II and III, probably overestimates the optical acuity of the oriental astronomers. The foreshortening factor at the beginning and end of a ten-day interval equally spaced about central meridian is approximately 0.4. However, the intention in this paper is to select the longest plausible time interval for which repeated observations of the same sunspot could have been made by the oriental astronomers. This approach gives the most stringent condition for the identification of statistically independent sunspot sightings. Moreover, with the above criterion, observations of the same sunspot in successive solar rotations are regarded as being statistically independent, which is clearly correct physically in the sense that such repeated observations are a true measure of real sunspot activity.

4.3 Criterion for Repeated Sunspot Observations

It is not known with absolute certainty that sunspots were actually observed on <u>every</u> day in the intervals (durations) quoted in Tables II and III. Although there is evidence for inclusive counting by the oriental astronomers, the conservative assumption made in this paper is that a sunspot must have been visible on at least the first and last day of the interval in order for the oriental astronomers to have cited a duration of visibility in the dynastic history. Therefore, for each entry in Tables II and III, it is assumed that a sunspot was seen only twice. There are two exceptions to this last statement; it is assumed that sunspots were seen on all three days quoted for the records with reference numbers (33) and (75) in Table II. In both these cases, the record is interpreted as a sighting on one day followed several days later by sightings on two successive days.

4.4 Criteria for the Seasonal variation

For the study of the seasonal variation, it is necessary to know the
number of sunspot sightings in each calendar month. All dates are
corrected to the Gregorian calendar to allow for the gradual change
in the dates of the seasons in the Julian calendar. Individual sunspot
sightings on a particular day correspond to a unique date in the
Gregorian calendar. Dates in the revised catalogue that are qualified
by a question mark (?), to indicate some uncertainty in either the
day or month, are rejected. There are only 8 such dates not otherwise
excluded in the revised catalogue, out of a total of 146 entries.
Thus the statistical accuracy of the results, which is set by sample
size, is not significantly reduced by the elimination of these dates.

If it is known only that the sunspot sighting(s) occurred some
time within a month (e.g. (I) in Table I), or that more than one
sighting was recorded within a well-defined range of dates (cf.
Tables II and III), the actual observation is distributed uniformly
over the specified interval with a daily density W/N, where N denotes
the number of days in the interval and W is a weight that has the
value 1 or 2. W = 1 if the catalogue entry suggests that a single
observation was made during the interval; W = 2 if the entry indicates
that more than one observation was made during the interval, which
is essentially equivalent to the conservative assumption that a sunspot
must have been visible on at least the first and last day of an interval of repeated sunspot observations.

The three sightings with reference numbers (15), (49) and (50)
are excluded from the study of the seasonal variation because only
the year of observation is known; however, these sightings are
retained in the study of the secular variation.

4.5 Criteria for the Secular Variation

For the study of the secular variation, it is necessary to know the
number of independent sunspot sightings in each calendar year. Once
again, all dates are corrected to the Gregorian calendar and it is
assumed that the year is then known exactly, even for dates in the
revised catalogue that are qualified by a question mark(?).

As noted in Section 4.2, two individual sunspot sightings
separated by an interval longer than 10 days are assumed to be
distinct and hence independent. Likewise, if the description recorded
in a dynastic history suggests that sunspots were seen repeatedly
over an interval longer than 10 days, it is assumed that two
independent sunspots were seen, one at the beginning of the interval
and the other at the end. Therefore intervals of repeated sunspot
observations, which are longer than 10 days, are assumed to be
essentially equivalent to independent individual sunspot sightings
at either end of the interval. Any two sunspot sightings that are
separated by an interval longer than 10 days are both included in
the analysis of the secular variation because the number of independent sunspot sightings per year is a manifest measure of solar

activity.

Any individual sunspot sighting that is not separated from the preceding individual sunspot sighting by a period longer than 10 days is eliminated from the statistical study of the secular variation of the oriental sunspot sightings. Similar statements can be made in relation to intervals of repeated sunspot observations if such intervals are defined strictly in terms of individual sunspots at the beginning and end of the interval. Any two sunspot sightings that are not separated by an interval longer than 10 days are probably observations of the same sunspot, as discussed in Section 4.2. There-fore, only one of the pair can be considered to contribute to the number of independent sunspot sightings per year.

5. SEASONAL VARIATION OF THE ORIENTAL SUNSPOT SIGHTINGS

Using the criteria and procedures outlined in the previous section, it is possible to determine the number of oriental sunspot sightings in each calendar month. The histogram presented in the lower part of Figure 1 shows the seasonal variation based on the revised catalogue of oriental sunspot sightings. This figure is similar to Figure 1 in the paper by Willis et al. (1980), with the important improvement that proper allowance is made for those descriptions which indicate that the same sunspot was sometimes seen more than once.

The probability of obtaining the histogram in Figure 1 by chance is less than 1 in 10^8, according to the χ^2 test (Kendall and Stuart, 1977) with the theoretically expected frequencies assumed to be prop-ortional to the number of days in the calendar month. Therefore the seasonal variation of the oriental sunspot sightings is highly significant statistically and should not be affected appreciably by the sparseness of the sightings. Furthermore, as noted by Willis et al. (1980), similar seasonal variations exist in the data before (65 records) and after (65 records) AD 1150, which is a convenient date that divides the acceptable data into two subsets containing exactly the same number of records. The seasonal variation also exists separately in the Chinese (96 records) and Korean (32 records) observations included in this statistical study.

An explanation for the seasonal variation of the oriental sunspot sightings, which has been presented previously by Willis et al. (1980), is illustrated schematically in the upper part of Figure 1. The marked spring maximum in the number of oriental sunspot sightings shown in the lower part of Figure 1 is ascribed to the fact that dust storms, or sand storms ("yellow winds"), over central China occur most frequently during late winter and early spring, when aridity due to low winter rainfall and increased surface winds associated with cold fronts contribute to their occurrence (Gherzi, 1951; Watts, 1969; Ing, 1972; Boucher, 1975; Middleton et al., 1986). The source of atmospheric dust is the vast barren land comprising: the Takla Makan Desert in the Tarim Basin; the Kansu Corridor; the loess plateau of Inner Mongolia; and the Gobi Desert, crossing the Chinese border into southern Mongolia (Ing, 1972; Goudie, 1978, 1983; Middleton et

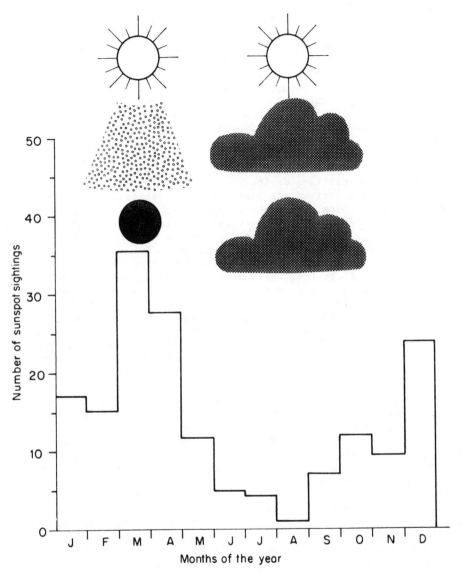

Fig. 1. The lower part of the figure shows the number of unaided-eye sunspot sightings recorded in each month of the year according to a revised catalogue of pre-telescopic sunspot sightings from the Orient. The upper part provides a schematic explanation for the marked seasonal variation of the oriental sunspot sightings. The spring maximum is ascribed to dust haze, which reduces the glare of the Sun and enables sunspots to be seen with the unaided-eye, whereas the summer minimum is attributed to greater obscuration of the Sun by clouds during the rainy season.

al., 1986). Dust carried into the lower troposphere by strong winds can be transported down-wind several thousand kilometres (Ho Peng Yoke, 1966; Ing 1972; Boucher, 1975, Goudie, 1983): the resultant dust haze often obscures the horizon and causes deep red sunsets throughout China at least as far south as Hong Kong (Bell et al., 1970). With this explanation of the seasonal variation of the pre-telescopic oriental sunspot sightings, airborne dust over China, Japan, Korea and northern Vietnam would have acted as a natural filter that reduced the glare of the Sun, whereby enabling the oriental astronomers to perceive sunspots on the solar surface with the unaided eye.

The broad summer minimum in the number of oriental sunspot sightings shown in the lower part of Figure 1 can be attributed to greater cloudiness and precipitation during the rainy season (Watts, 1969; Boucher, 1975). This part of the explanation of the seasonal variation relies on optically thick clouds obscuring the Sun and hence reducing the likelihood of sunspots being detected, particularly at sunrise or sunset, although optically thin clouds or fog may occasionally favour the detection of sunspots. In addition, heavy summer rainfall washes dust out of the lower atmosphere (Bell et al., 1970), which results in intense glare from the Sun on clear days. However, this "clear sky" situation occurs on relatively few days per month because of the greater cloudiness in summer (Chu Ping-hai, 1967). Therefore the small number of sunspot sightings in August (see Figure 1) can be regarded as a summer "base level" determined by a combination of greater obscuration of the Sun on cloudy days and greater glare from the Sun on clear days, with the cloudy days outnumbering the clear days.

With the above interpretation of the seasonal variation of the oriental sunspot sightings, the marked spring maximum shown in Figure 1 is explained in terms of airborne dust reducing the glare of the Sun, which enables fine detail to be seen on the solar surface. It then follows that observations which mention more than one sunspot on the solar disk, which clearly depend on fine optical resolution, would be expected to have occurred preferentially during the spring. Table IV lists all records in the revised catalogue which refer explicitly to observations of two or more sunspots on the solar disk at the same time (see also Stephenson and Clark, 1978). This table confirms that there is a marked tendency for these simultaneously resolvable sunspots to have occurred during either the spring or winter months, when there was more likely to have been a high concentration of airborne dust in the lower troposphere.

6. SECULAR VARIATION OF THE ORIENTAL SUNSPOT SIGHTINGS

If the oriental sunspot sightings are purely random events in a continuum of time, as suggested by Bray (1974) and Eddy (1976, 1980, 1983), the dates of the sightings should comply with the statistics of a Poisson distribution. For this distribution, the chances (probabilities) of recording 0, 1, 2, 3, ... sunspot sightings in a calendar year are, respectively

$$\exp{(-z)} \ [1, \ z, \ z^2/2!, \ z^3/3!, \ \ldots\ldots], \tag{1}$$

where z, the mean (μ) of the Poisson distribution, represents the average number of sunspot sightings per year (Kendall and Stuart, 1977).

It is shown below that the unaided-eye sunspot sightings in the Orient during the pre-telescopic period (43 BC - AD 1608) do not conform to the simple Poisson distribution. Discrepancies should be expected, however, if the likelihood of observing a sunspot varies with time, as would certainly be true, for example, if there were an actual secular variation in solar activity. Greenwood and Yule (1920) have generalized the statistical theory to cover such a possibility; their extension is based on a mixture of Poisson distributions. Following the approach outlined by Kendall and Stuart (1977), assume that the "population" is a mixture of years with different degrees of sunspot detectability, represented by different values of z in a Poisson distribution. Assume further that the distribution of z in the population is of the gamma form, so that the element of frequency, dF, is defined by

$$dF \ = \ [c^p/\Gamma(p)] \ \exp{(-cz)}z^{p-1} \ dz, \ o \leq z \leq \infty; \ p > o, \ c > o,$$

where Γ denotes the gamma function. Hence the frequency of j successes is

$$[c^p/\Gamma(p)]\int_o^\infty \exp{(-cz)} \ z^{p-1} \ \exp{(-z)} \ (z^j/j!) \ dz.$$

Following Kendall and Stuart (1977), the chances (probabilities) of recording 0, 1, 2, 3, ... sunspot sightings in a calendar year then become

TABLE IV

OBSERVATIONS OF TWO OR MORE SUNSPOTS ON THE SOLAR DISK

Ref. No.	Place	Julian Date	No. of Spots
(17)	C	355 Apr 4	2
(23)	C	374 Apr 6	2
(25)	C	388 Apr 2	2
(31)	C	500 Jan 30	3
(33)	C	502 Feb 11, 12	3?
(35)	C	510 Mar 17	2
(36)	C	511 Dec 16	2
(39)	C	567 Dec 13, 14, 15	2
(53)	C	974 Mar 3	2
(60)	C	1112 May 2	2/3
(67)	C	1136 Nov 27	≥ 2
(102)	C	1276 Feb 17	≥ 2
(137)	K	1520 Mar 9	≥ 2
(VIII)	V	1593 Jan 3	5
(IX)	V	1603 Apr 4 – May 10	3
(138)	K	1603 Apr 16	3

Place references : C = China, K = Korea, V = Vietnam

$$\frac{c}{c+1}^{p} \left[1, \; \frac{p}{(c+1)}, \; \frac{p(p+1)}{2!(c+1)^{2}}, \; \frac{p(p+1)(p+2)}{3!(c+1)^{3}}, \; \ldots\ldots \right] \tag{2}$$

The mean (μ) and variance (σ^2) of this distribution are given by the two equations

$$\mu = p/c \tag{3}$$

and

$$\sigma^2 = (p+p/c)/c. \tag{4}$$

Using the criteria and procedures outlined in Section 4, it is possible to determine the number of independent sunspot sightings in each calendar year. In the following statistical analysis it is assumed that the effective time interval of the pre-telescopic oriental sunspot sightings is from 43 BC Jan 1 to AD 1608 Dec 31, that is 1651 complete years. The second column in Table V presents the observed frequencies of the number of independent sunspot sightings per year, namely 0, 1, 2, 3, 4, 5 and ≥ 6, listed in the first column. The numerical values of the mean (μ) and variance (σ^2) of the observed distribution are $\mu = 0.089$ and $\sigma^2 = 0.170$. Substituting these values into (3) and (4), and solving the resulting equations, yields $c = 1.106$ and $p = 0.098$. The pure and generalized Poisson frequencies given in the third and fourth columns of Table V, respectively, are calculated by substituting the numerical values of z $(= \mu)$, c and p into equations (1) and (2) and multiplying the resulting theoretical probabilities by 1651.

TABLE V

OBSERVED AND THEORETICAL FREQUENCIES OF THE SUNSPOT SIGHTINGS

Number of Oriental Sunspot Sightings	Observed Frequencies from the Revised Catalogue	Pure Poisson Frequencies	Generalized Poisson Frequencies
0	1547	1510.35	1549.54
1	79	134.48	72.46
2	15	5.99	18.90
3	6	0.18	6.28
4	0	0.00	2.31
5	4	0.00	0.90
≥ 6	0	0.00	0.61

It is clear from casual inspection of Table V that the generalized Poisson frequencies are in quite good agreement with the observed frequencies, whereas the pure Poisson frequencies are in relatively

poor agreement. According to the χ^2 test, there is a 7% probability
of obtaining the observed frequencies by chance from a population
characterized by the generalized Poisson frequencies. Therefore the
difference between the generalized Poisson frequencies and the observed
frequencies is not statistically significant. Conversely, there is
less than 0.01% probability of obtaining the observed frequencies by
chance from a population characterized by the pure Poisson frequencies.
Thus the difference between the pure Poisson frequencies and the
observed frequencies is statistically significant. Therefore, the
generalized theory based on a mixture of Poisson distributions, which
allows for a temporal variation in the degree of sunspot detectability,
provides good agreement with the oriental sunspot observations. This,
together with the rejection of the pure Poisson distribution, proves
conclusively that there is a true secular variation in the occurrence
of oriental sunspot sightings during the period 43 BC to AD 1608.

Although the statistical analysis presented above provides clear
evidence for a secular variation in the occurrence of the oriental
sunspot sightings, it obviously does not provide any information on
the cause(s) of this secular variation. The secular variation may
be a meaningful manifestation of an actual long-term variation in the
level of solar activity, as discussed by Eddy (1976, 1980, 1983). Other
proxy indicators of solar activity such as the radioactive isotope
^{14}C, which is preserved in the annual growth rings of trees as a
consequence of the bombardment of the upper atmosphere by cosmic rays,
provide clear evidence for a real long-term variation in solar activity
(Eddy, 1976, 1977, 1978). Therefore, a secular variation in the
occurrence of the oriental sunspot sightings is to be expected on
physical grounds, but the sparseness of the sightings recorded in the
dynastic histories makes it difficult to identify a secular variation
that can be attributed unambiguously to solar activity. Moreover,
in view of the compelling evidence presented in Section 5 for a seasonal
variation of the oriental sunspot sightings, the secular variation
of these sightings may be due to a long-term variation in atmospheric
viewing conditions in the Orient, as discussed by Willis et al. (1980).
This latter explanation relies on climatological variations in the
Far East producing a secular variation in either the atmospheric concentration of airborne dust or the level of cloud cover. Finally,
the secular variation of the oriental sunspot sightings may be spurious
in the sense that it arises from political and social pressures on
the oriental astronomers, which clearly influenced the recording of
phenomena of astrological interest. This last possibility has been
reviewed by Clark and Stephenson (1978), who commented in detail on
the completeness of the historical sunspot records. It is quite likely
that the secular variation of the oriental sunspot sightings depends
on all three factors mentioned above, which makes further progress
difficult. Indeed, there may even be other causative factors which
have yet to be discovered.

7. CONCLUSIONS

The seasonal variation in the occurrence of the unaided-eye sunspot
sightings recorded in the official oriental dynastic histories, which
was first reported by Willis et al. (1980), is re-examined using a
revised catalogue of pre-telescopic sunspot sightings from the Orient
(Yau and Stephenson, 1988). An important improvement in the amended
analysis is that greater weight is given to those records in the revised
catalogue for which the descriptions of sunspot sightings indicate
clearly that the same sunspot was seen on several days rather than
on just one day. If proper allowance is made for such repeated observations of the same sunspot, the seasonal variation shown in the
lower panel of Figure 1 is even more pronounced than that reported
previously. In fact, the probability of obtaining the observed seasonal
variation by chance is now less than 1 in 10^8 rather than 4 in 10^5.

The seasonal variation of the oriental sunspot sightings is believed
to result from a seasonal variation in suitable atmospheric conditions
for viewing sunspots with the unaided eye in the Orient, as discussed
by Willis et al. (1980). In particular, the spring maximum in the
number of oriental sunspot sightings is ascribed to dust in the lower
troposphere, as illustrated schematically in the upper part of Figure
1. It is known that airborne dust tends to occur most frequently in
the Orient during late winter and early spring, as a result of low
winter rainfall and increased surface winds from the arid land in
northern China and southern Mongolia. With this explanation of the
seasonal variation, atmospheric dust would have reduced the glare from
the Sun, thus facilitating the observation of sunspots with the unaided
eye. Conversely, the summer minimum is thought to be due to greater
cloudiness during the rainy season. This seasonal increase in optically
thick cloud cover would have reduced the likelihood of observing sunspots
during the summer months. The above explanation of the seasonal variation
of the oriental sunspspot sightings implies that the detection of sunspots with the unaided eye by the oriental astronomers depended sensitively on local atmospheric viewing conditions involving several
meteorological variables.

An entirely new result derived in this paper is that the oriental
sunspot sightings are not purely random events in a continuum of time,
contrary to the conclusions of Bray (1974) and Eddy (1976, 1980, 1983).
By using statistical theory appropriate to a mixture of Poisson processes, it is shown that there is a significant secular variation in
the occurrence of the oriental sunspot sightings. Although the statistical theory provides clear evidence for such a secular variation,
it obviously cannot distinguish between competing causal mechanisms.
For example, possible explanations of the secular variation in the
oriental sunspot sightings may depend on any one, or combination, of
the following factors: (i) long-term variations in the level of solar
activity; (ii) secular variations in the atmospheric viewing conditions in the Orient; and (iii) temporal variations in the political
and social influences on the oriental astronomers.

Further progress would appear to depend on the availability of additional data or information. For example, the sample size could be increased by including sunspot sightings recorded in local gazettes and encyclopaedia. Such sunspot sightings have been excluded from the present study because it is not known if they are as reliable as those recorded in the official dynastic histories. However, the majority of these provincial sunspot sightings occurred in the post-telescopic period, with observations extending into the early part of the twentieth century. Therefore, it should be possible to test the reliability of the later provincial sunspot sightings in the Orient by comparing them with telescopic sunspot observations in Europe. In addition, it may still be possible to glean some further information from the actual discussion of the seasonal variation of the oriental sunspot sightings, it would be necessary to allow for possible regional differences in atmospheric viewing conditions at the various locations where the oriental observations were made, as noted by Willis et al. (1980).

REFERENCES

Bell, G.J., Peterson, P. and Chin, P.C., 1970,'Meteorological Aspects of Atmospheric Pollution in Hong Kong', pp2-3, Royal Observatory, Hong Kong, Technical Note No. 29.

Boucher, K., 1975,'Global Climate', The English Universities Press Ltd., London.

Bray, J.R., 1974, In 'Scientific, Historical and Political Essays in Honour of Dirk J. Struik', pp143-146, eds. Cohen, R.S., Stachel, J.J. and Wartofsky, M.W., (Boston Studies in the Philosophy of Science, Vol. XV), D. Reidel Publ. Co., Dordrecht, Holland.

Chu, Ping-hai, 1967.'Climate of China', Joint Publications Research Service, 41374, US Department of Commerce, Washington.

Clark, D.H. and Stephenson, F.R., 1978, Q. Jl. R. astr. Soc., 19, 387.

Eddy, J.A., 1976, Science, 192, 1189.

Eddy, J.A., 1977, Climatic Change, 1, 173.

Eddy, J.A., 1978, In 'The New Solar Physics', pp. 11-33, ed. Eddy, J.A., (AAAS Selected Symposium 17), Westview Press, Boulder, Colorado.

Eddy, J.A., 1980, In 'The Ancient Sun', (Fossil record in the Earth, Moon and Meteorites), pp. 119-134, eds. Pepin, R.O., Eddy, J.A. and Merril, R.B., Pergamon Press, New York.

Eddy, J.A., 1983, Solar Phys., 89, 195.

Gherzi, E., 1951,'The Meteorology of China', Vols. I and II, Imprensa Nacional de Macau.

Goudie, A.S., 1978, J. Arid Environments, 1, 291.

Goudie, A.S., 1983, Progress in Physical Geography, 7, 503.

Greenwood, M. and Yule, G.U., 1920, Jl. R. statist. Soc., 83, 255.

Ho Peng Yoke, 1966,'The Astronomical Chapters of the Chin Shu', Mouton, Paris.

Ing, G.K.T., 1972, Weather, 27, 136.

Kendall, M.G. and Stuart, A., 1977, 'The Advanced Theory of Statistics', Volume 1: Distribution Theory, Fourth Edition, Charles Griffin and Company Limited, London and High Wycombe.

Middleton, N.J., Goudie, A.S. and Wells, G.L., 1986, In 'Aeolian Geomorphology', pp 237, ed. Nickling, W.G., Allen and Unwin, Boston.

Stephenson, F.R. and Clark, D.H., 1978,'Applications of Early Astronomical Records', (Monographs on Astronomical Subjects: 4), Adam Hilger Ltd., Bristol.

Watts, I.E.M., 1969, In 'Climates of Northern and Eastern Asia', pp. 1-117 ed. Arakawa, H., (World Survey of Climatology, Vol. 8), Elsevier Publ. Co., Amsterdam.

Willis, D.M., Easterbrook, M.G. and Stephenson, F.R., 1980, Nature, 287, 617.

Yau, K.K.C. and Stephenson, F.R., 1988,'A Revised Catalogue of Far Eastern Observation of Sunspots (165BC to AD 1918)', Q. Jl.R. astr. Soc. in press.

THE SOLAR DIAMETER SINCE 1715

[1]John H. Parkinson, [2]F.R. Stephenson, [3] L.V. Morrison
[1]Mullard Space Science Lab., University College, London,
 Holmbury St. Mary, Dorking, Surrey, RH5 6NT, U.K.
[2]Department of Physics, University of Durham, Durham DH1 3LE
 U.K.
[3]Royal Greenwich Observatory, Herstmonceux Castle, Hailsham,
 E. Sussex, BN27 1RP, U.K.

> The question whether the Sun's apparent
> diameter is liable to periodic changes
> has frequently occupied the attention
> of astronomers.
> - Royal Astronomical Society Council
>   Report 1875.

INTRODUCTION

The debate, on whether there is any valid evidence for deducing
significant variations in the diameter of the Sun, still continues over
a century later. Claims for secular decreases, increases and
periodicities on a wide range of time scales have all been proposed
and, at some later time, they have all been disputed. If significant
changes in the size of the Sun were established they would lead to
variations in the gravitational potential energy on time scales
shorter than evolutionary ones and substantial modifications would
need to be made to our current understanding of the solar interior.
     The attention of the present authors was first attracted to this
debate by Jack Eddy (1979) who re-examined the Greenwich meridian
circle measurements and felt that there could be evidence for a larger
Sun over a century ago. We proposed two inherently more accurate
methods (Parkinson et al. 1980) which we discuss in later sections.
We conclude that there is no convincing evidence for any secular change
but there is marginal evidence for a periodic variation.

MERIDIAN CIRCLE MEASUREMENTS

During the 19th century, several meridian circle instruments were
built for the measurement of longitude and, in addition to observing
transits of stars, the right ascension and declination of the Sun

F. R. Stephenson and A. W. Wolfendale (eds.),
Secular Solar and Geomagnetic Variations in the Last 10,000 Years, 203–207.
© *1988 by Kluwer Academic Publishers.*

were also measured. Many of these instruments operated for only a
few years at a time and probably the longest continuous series of
observations with the same instrument were made at Greenwich with
the Airy transit circle commencing in 1851. This instrument introduced a new level of precision to transit measurements and within
a few months it was claimed that "no other instrument ... can be
compared with this for steadiness and accuracy of adjustments or for
accuracy of results".

At Greenwich, the basic method of observing the Sun involved
noting the times at which the leading and following limbs crossed
the meridian wire and measuring the angles from the zenith to the
upper and lower limbs. By averaging these pairs of measurements,
the Sun's position could be determined, whereas subtraction leads
to the apparent horizontal and vertical diameters. Over the years,
various degrees of automation were introduced, but the measurements
always depended on an observer's judgement of when a solar limb was
tangential to a wire in the eyepiece. The influence of observer bias
on the measurements has long been recognised (Thackeray 1885) and
studied by a succession of authors who often reached contradictory
conclusions from the same data set (Gething 1955). We have pointed
out previously that individual discrepancies in the Greenwich
observations can exceed 3 arc seconds in the semi-diameter.

Several other meridian circles have been operated during the
past two centuries, often for only a decade or two. Intercomparisons
are difficult, not just because of the sporadic nature of the
observations but also because each instrument appears to give its
own particular value for the solar diameter. When combined with the
perennial problems of poor and variable seeing, it is clear that
meridian circle measurements are unsuitable for investigating changes
in the solar diameter. Better methods would need to rely less on
skilled judgement and instrumental sophistication.

TOTAL SOLAR ECLIPSES

It is a fortunate coincidence that, when observed from the surface
of the Earth at the present epoch, the Sun and Moon have nearly equal
diameters. In this way total eclipses of the Sun are produced,
typically lasting a few minutes, allowing the Sun's diameter to be
measured using the Moon as a template of known mean diameter and limb
profile.

By simply timing the duration of totality, from when the last
bead of light from the photosphere disappears until the first bead
reappears, the solar diameter can be deduced. With the eclipse shadow
travelling across the Earth at a speed of a few $\times 10^3$ km hr^{-1} these
dramatic events are easily recognised without the aid of a telescope.
During the 1981 total eclipse in Eastern Siberia, we were able to
demonstrate that direct observations led to error bars of less than
0.1 arc seconds in the solar semi-diameter (Parkinson 1983).

The eclipse duration is an elliptical function of position across
the track, making a small error in an observer's position more important

the further he is away from the centre line. By using timings made in the central half of the eclipse track, positional uncertainties are not limiting factors. By contrast, observations made very close to the limits of totality, at well determined positions, allow the size of the shadow to be determined and hence the solar diameter.

Two other uncertainties can affect the accuracy of total eclipse measurements: knowledge of the Moon's limb profile and the Earth's rate of rotation in the past. Charts of the lunar profile as a function of libration have been compiled (Duncombe 1973) from photographic surveys (Watts 1963) and corrected from stellar occultation observations (Morrison and Appleby 1981). The absolute accuracy of any feature on the profile is now thought to be better than ± 0.2. arc seconds.

TRANSITS OF MERCURY

Observations of the time taken for the planet Mercury to cross the face of the Sun also provide a method for measuring the diameter. By timing the internal contacts to a precision of 1 sec, the diameter can be measured to 0.1 arc seconds.

Mercury transits occur much less frequently than total eclipses, on average there are 14 per century. This comparative rarity has led to a great deal of interest over the years with some 30 well observed transits in the past two and a half centuries. Knowledge of the observer's position is less stringent and the lunar limb profile is not required. The main observational difficulty, however, is the "black drop" effect. With Mercury just inside the solar limb there sometimes appears, due to imperfect seeing, to be a connection back to the limb which breaks quite suddenly.

RESULTS

After careful consideration, we have found that the total solar eclipse and Mercury transit methods have comparable accuracies and are far more reliable than meridian circle measurements. We have used some 2150 timings of 30 Mercury transits going back to 1736 (Morrison and Ward 1975a,b) but total eclipse timings are more sparse. Following the pioneering work of Halley for the 1715 eclipse (Halley 1715) interest waned, expedition reports attesting that the predicted times were approximately correct. Accordingly we have used 110 observations from 9 eclipses and all these measurements are shown in Figure 1.

In order to assess the significance of trends in the solar diameter, regression analyses have been performed for each type of observation separately. For each transit or eclipse, the mean semi-diameter has been calculated and weighted with the standard deviation of each set of measurements in order to include the actual spread of the data. The secular trends, δR, are then given below in units of arc seconds cy^{-1} relative to a standard semi-diameter of 959.63 arc

206

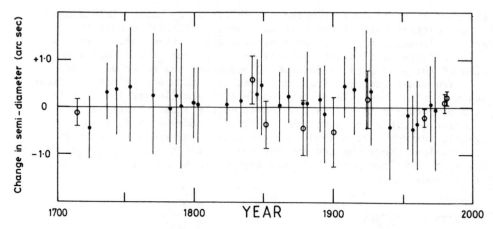

Fig. 1. Apparent changes in the semi-diameter of the Sun deduced
from transits of Mercury (●) and total eclipses (o). The
zero value corresponds to 959.63 arc seconds and the
error bars are standard deviations.

seconds, the error bars are ± 1 standard deviation and T is the time
from 1700 measured in centuries.

Mercury transits
$\delta R = 0.172(\pm0.334) - 0.057(\pm0.202)$ T

Total eclipses
$\delta R = -0.155(\pm0.278) + 0.077(\pm0.111)$ T

These functions give semi-diameter corrections which are less than
0.1 arc seconds for at least the past 150 years indicating that the
two sets of data can be combined.

Mercury transits and total eclipses
$\delta R = -0.004(\pm0.193) + 0.023(\pm0.086)$ T

Since 1700 any secular change is therefore consistent with zero at
the 80% level of confidence.
    In addition, the observations of transits of Mercury have also
been studied for possible periodicities. A least-squares analysis
shows some evidence for an 80 year periodicity given by

$\delta R = 0.24(\pm0.08) \sin 2\pi[T/0.8+0.43(\pm0.04)]$

Recently the question of solar diameter changes was reopened by Ribes
et al. (1987) who analysed observations made at the Paris Observatory
by Picard and La Hire during the late 17th and early 18th centuries.
From these observations they claimed that the Sun was at least 4 arc
seconds larger than at present. However, our analysis of observations
of the 1715 total eclipse (both durations and NS limits) rule out any

possibility of this claim being substantiated (Morrison et al. 1988).
Indeed, for the eclipse, durations would need to be typically 9 sec
shorter and several observers near the edges of totality would need
to be mistaken in their belief that they observed a total eclipse.
In addition, O'Dell and Van Helden (1987) have pointed out that Ribes
et al. have underestimated the effects of optical imperfections·in the
Paris instrument which would increase the image size.

CONCLUSION

An examination of the various methods used for measuring the size of
the Sun leads to our rejection of meridian circle and similar measurements. Measurements of the durations of Mercury transits and total
eclipses mutually lead to a high degree of confidence in there being
no significant secular change in the solar diameter over the past 270
years although there is a hint of a periodicity at approximately 80
years.

REFERENCES

Duncombe, J.S., (1973), US Naval Obs. Circ. No. 141.
Eddy, J.A. and Boornazian, A.A., (1979), Bull. Am. Astr. Soc., 11, 437.
Gething, T.J.D., (1955), Mon. Not. Roy. Astr. Soc., 115, 558.
Halley, E., (1715), Phil. Trans. Roy. Soc. 29, 245 and 314.
Morrison, L.V. and Ward, C.G., (1975), Roy. Gr. Obs. Bull. No. 181.
Morrison, L.V. and Ward, C.G., (1975), Mon. Not. Roy. Astr. Soc., 173,
   183.
Morrison, L.V. and Appleby, G.M., (1981), Mon. Not. Roy. Astr. Soc.,
   196, 1013.
Morrison, L.V., Stephenson, F.R. and Parkinson, J.H., (1988), Nature,
   331,421.
O'Dell, C.R. and Van Helden, A., (1987), Nature, 330, 629.
Parkinson, J.H., Morrison, L.V. and Stephenson, F.R., (1980), Nature
   288, 548.
Parkinson, J.H., (1983), Nature, 304, 518.
Ribes, E., Ribes, J.C. and Barthalot, R., (1987), Nature, 326, 52.
Thackeray, W.G., (1885), Mon. Not. Roy. Astr. Soc., 45, 389, 464 and
   467.
Watts, C.B., (1963), Astron. Papers of Am. Eph. and Naut. Al. XVII.

THE CLIMATE OF THE PAST 10,000 YEARS AND THE ROLE OF THE SUN

T.M.L. Wigley
Climatic Research Unit
University of East Anglia
Norwich NR4 7TJ, U.K.

ABSTRACT. Solar effects on climate are considered for the past 300 years and the past 10,000 years. A box-upwelling-diffusion climate model is used to evaluate the effects of irradiance variations on global-mean temperature. The model shows how periodic solar variations have differing effects on climate according to the period of the forcing oscillations. More rapid fluctuations show greater damping and smaller lags between forcing and response: for a 10-year cyclic forcing the response is only 13-23% of the equilibrium response, while the corresponding fraction for an 80-year cycle is 39-59%. A solar diameter-irradiance relationship in which irradiance is related to the time derivative of diameter is considered and shown to provide a better fit to observations than the conventional theory. The potential climatic effects of cyclic, radius-related irradiance changes are evaluated giving a range of global-mean temperature fluctuations of 0.2-0.3°C over the past 300 years. For the past 10,000 years, the recent glacial chronologies of Röthlisberger (1986) are compared with the ^{14}C anomaly curve of Stuiver et al. (1986). The agreement between times of major ^{14}C anomaly and times of globally-advanced glaciers (i.e. cool summers) is shown to be statistically significant. The implied reduction of solar irradiance during times of maximum century-time-scale ^{14}C anomaly such as the Maunder Minimum is shown to be around 6 Wm^{-2}, equivalent to a net radiative forcing change of about 1 Wm^{-2} at the top of the troposphere. If another major ^{14}C anomaly began early in the 21st century, the associated solar perturbation would be of considerable importance, but still insufficient to fully offset the projected warming due to future greenhouse gas concentration increases.

INTRODUCTION

The relationship between the Sun and climate has always been a con-troversial one, an area ripe for speculation largely because of the dearth of data with which to disprove hypotheses. Most claimed sunspot-climate links have been discredited (Pittock, 1978, 1983;

F. R. Stephenson and A. W. Wolfendale (eds.),
Secular Solar and Geomagnetic Variations in the Last 10,000 Years, 209–224.
© 1988 by Kluwer Academic Publishers.

Shapiro, 1979), usually on statistical grounds, but frequently also because of the lack of any realistic mechanism for relating sunspot variations to the climate of the lower atmosphere. Until recently the most convincing demonstration of a link was Mitchell et al.'s 20-to-22-year midwest U.S. drought cycle (Mitchell et al., 1979; Stockton et al., 1983), but even here the mechanism is obscure. More convincing (and intriguing) is the recent work of Labitzke (1987) who has shown that 30-mb Arctic temperatures are highly correlated with sunspot numbers, provided the data are first stratified according to the phase of the Quasi-Biennial Oscillation. The mechanism for this link is not yet known, nor is its significance for near-surface climate, although more recent analyses suggest that it is more than just an Arctic stratospheric phenomenon (H. van Loon, personal communication).

The basic mystery in sunspot-climate links is that no one expected that solar irradiance could vary sufficiently over a sunspot cycle to produce a noticeable climatic effect. Thus, the mechanism(s) involved must include some indirect amplification. Just how (or even why) a weak solar signal might be amplified to produce a significant climatic effect is unknown, although there have been many speculative suggestions.

In fact, recent satellite data have shown that solar irradiance does vary with sunspot number (Willson et al., 1981; Hoyt and Eddy, 1982) on a daily to monthly time scale. If one extrapolates the Hoyt-Eddy relationship to the 11-year sunspot cycle, then the range of irradiance variation between sunspot minimum and maximum amounts to between 0.05% and 0.12%. The equivalent equilibrium global-mean surface air temperature response could be as much as 0.3° (more likely, around 0.1°C), but the actual response must be much less than this (viz. around 0.03°C) because of the damping effect of oceanic thermal inertia. This is an undetectable effect. A more recent relationship developed by Foukal and Lean (reported by Kerr, 1987) gives irradiance changes over the 11-year sunspot cycle of similar magnitude to the Hoyt-Eddy model, but in the opposite direction. The direct climatic effects of these fluctuations would, necessarily, also be undetectable. So, the mystery still remains - how can a significant sunspot-climate link occur?

As satellite data accumulate, however, the true character of the Sun as a variable star is beginning to emerge. In addition to the direct short time scale effects of sunspot blocking and facular enhancement mentioned above, the Sun's irradiance also varies markedly on the annual to decadal time scale. Balloon data from the 1960s and 1970s, re-analysed by Frölich and Eddy (1984) show an increase in irradiance of 0.3% (but with large confidence limits) between 1970 and 1979, and satellite data show a subsequent decrease of approximately 0.1% (1979-84; irradiance has shown little change since late 1984 through March 1986, see Nimbus-7 ERB/Cloud Newsletter 2(1), pp. 2-3). Such changes may well have noticeable and important surface climate manifestations, as we will discuss further below.

What of longer time scales? Because there are no data, we simply do not know whether the Sun shows longer time-scale irradiance variability, beyond what must be supposed to exist on the basis of recent data and the principle of uniformitarianism. There are, however, two indirect indicators of additional variations, solar diameter changes and atmospheric ^{14}C changes. Although the data are still somewhat controversial, there is evidence of near-periodic solar diameter changes with an amplitude of approximately 0.02% and period of about 76 years (Gilliland, 1982a). There is also evidence that diameter and irradiance changes are related, so the 76-year diameter cycle could be parallelled by a 76-year irradiance cycle. We will review this issue in more detail in a later section. Atmospheric ^{14}C data, inferred from tree-ring ^{14}C measurements, show convincing evidence of solar variability. On decadal to century time scales, the production rate of ^{14}C in the upper atmosphere is controlled by the solar wind. Thus, decadal to century time scale changes in atmospheric ^{14}C concentration, which reflect a balance between production and loss through oceanic sinks and radioactive decay, must tell us a lot about changes in the solar wind (Stuiver and Quay, 1980). Large changes have occurred during the last 10,000 years; but whether these have been associated with changes in solar irradiance is a moot point, one which I will elaborate on below.

In the remainder of this paper, therefore, I will consider the following. First, with the aid of a simple climate model, I will examine the sensitivity of the Earth's climate to solar irradiance changes on various time scales.

The model used differentiates land and ocean in each hemisphere and parameterizes ocean mixing as an upwelling-diffusion process. This is the standard type of model currently used to study the transient climatic reponse to imposed forcing; such models have been recently reviewed by Hoffert and Flannery (1985). The key model parameters which determine its response are: the equilibrium climate sensitivity, i.e., the equilibrium temperature change (°C) per unit of radiative forcing at the top of the troposphere (Wm^{-2}); and the vertical thermal diffusivity of the ocean.

I will then use the model to estimate the magnitude of global-mean temperature changes that should occur in response to those irradiance changes associated with sunspot blocking and facular enhancement. Next, I consider diameter-irradiance links and their climatic consequences, introducing a new form of the diameter-irradiance relationship and calibrating this using recent observations of both variables. I will then examine the ^{14}C-climate link using new paleoclimate data and new ^{14}C data. These data show a sufficiently strong relationship that they justify an estimation of the magnitude of solar irradiance changes associated with the solar wind changes which produce atmospheric ^{14}C anomalies. Finally, I will speculate on the implications of these last results for a future climate in which there may be competing influences from solar variability and anthropogenic greenhouse-gas concentration increases.

CLIMATE SENSITIVITY TO SOLAR IRRADIANCE VARIATIONS

For solar irradiance variations operating on time scales of centuries or less, it is vitally important to distinguish the actual response of the climate from the equilibrium (or potential) response. The actual response may be only a small fraction of the equilibrium response, and the response may lag noticeably behind the forcing. This is a consequence of the ocean's very large thermal inertia. To illustrate this, I will consider two idealized cases, a step-function forcing and sinusoidal forcing. I will calculate the effects of the second type of forcing on global-mean temperature using the simple climate model mentioned above.

In the first case, if the irradiance were suddenly to increase by, say, 1%, then the global-mean temperature would not increase instantaneously, but would rise relatively slowly, approaching a new equilibrium state roughly exponentially on a time scale of decades to centuries. The equilibrium warming would be between about $0.9°C$ and $2.7°C$ (this corresponds to an equilibrium climate sensitivity of about $0.4-1.1°C/Wm^{-2}$, the commonly accepted range of uncertainty).

In the second case, if the irradiance were to vary sinusoidally, the new equilibrium state would never be reached, and the actual response may be very much less than the equilibrium response. Just how much less depends mainly on three factors: the period of the oscillations; the rapidity of ocean mixing (i.e., the vertical diffusivity which determines the effective thermal inertia of the ocean); and the equilibrium climate sensitivity. Results for a diffusivity of $1 \ cm^2sec^{-1}$ are shown in Table 1. For cyclic forcing with a period of 10 years, the response lags the forcing by 1.3-1.4 years and is damped by a factor of 0.23 (for a climate sensitivity of $0.5°C/Wm^{-2}$) to 0.13 (for a sensitivity of $1.0°C/Wm^{-2}$). If the period is longer, the lag becomes more sensitive to the climate sensitivity. For a period of 160 years and a sensitivity of $0.5°C/Wm^{-2}$, the lag is 11.2 years and the damping factor is 0.71. For the same period with double the sensitivity, the lag is 16.0 years and the damping factor is 0.52. The lag results are only weakly dependent on the assumed value of the diffusivity. The damping factor, however, decreases noticeably with increasing diffusivity; for $K = 2 \ cm^2sec^{-1}$ the damping factor is roughly 0.85 times the numbers quoted above.

We can now estimate the response of global-mean temperature to irradiance changes associated with the 11-year sunspot cycle. Hoyt and Eddy (1982) have developed a statistical model relating solar irradiance to sunspot number and faculae area based on recent satellite data. During a strong sunspot maximum (e.g. 1958, $n \cong 205$) the implied reduction in irradiance is about 0.12%; during a weak maximum (e.g., 1905, $n \cong 60$) the reduction is 0.05%. Using a climate sensitivity of $0.5°C/Wm^{-1}$ (a value which accords best with recent global temperature changes; see Wigley, 1987), the implied equilibrium temperature changes are coolings of $0.14°C$ and $0.06°C$, respectively. The actual coolings, however (based on Table 1), are much less, $0.034°C$ and $0.014°C$. (If the climate sensitivity were $1.0°C/Wm^{-2}$, the cooling amounts would be $0.038°C$ and $0.016°C$. The effect of climate

sensitivity on these results is relatively small because of the
increased damping effect at higher sensitivities). It can thus be
seen that the direct effect of sunspot blocking on global-mean temperature is very small, almost certainly below the limits of detection,
given the natural variability of global-mean temperature. Qualitatively similar results have been obtained by North et al. (1983)
using a model with more complex horizontal structure, but with a simpler
method for dealing with oceanic thermal inertia.

The Hoyt-Eddy model predicts an <u>increase</u> in irradiance as sunspot
numbers decline. Over the period 1981-84, as sunspot numbers declined
from the 1980-82 maximum, the increase in irradiance should have been
around 0.06%. In fact, over this four-year period, satellite data
show a <u>decrease</u> in irradiance of about 0.07% (Kyle et al., 1985;
Willson et al., 1986). An alternative model has been developed by
Foukal and Lean (reported by Kerr, 1987) which predicts just such
a decrease. The Foukal-Lean model apparently involves the fact that,
on time scales of months upward, the integrated facular and globally-
distributed magnetic network contribution to irradiance changes
dominates over the effect of sunspot blocking. The model therefore
predicts irradiance changes over the 11-year sunspot cycle that are
of the opposite sign to the Hoyt-Eddy model, but of a similar magnitude.
The climatic response, therefore, would have the same magnitude as
described above, and would, likewise, be almost certainly below the
limits of detection.

Table 1

The response of global-mean temperature to cyclic solar irradiance
changes for a vertical diffusivity of 1 cm^2 sec^{-1} and two climate
sensitivities. The results shown are the lag in years, with the lag
expressed as a ratio of the forcing period given in brackets, followed
by the damping factor (i.e., the ratio of the amplitudes of the actual
temperature response to the equilibrium response). The actual response
in °C can be determined by multiplying any assumed forcing (in Wm^{-2})
by the climate sensitivity and by the damping factor. For example,
for a cyclic perturbation with period 80 yr and forcing amplitude
at the top of the troposphere of 1 Wm^{-2}, if the sensitivity were
$0.5°C/Wm^{-2}$, then the amplitude of the global-mean temperature response
would be 1 x 0.5 x 0.59 ≅ 0.3°C.

Period (yr) sensitivity	5	10	20	40
$0.5°C/Wm^{-2}$	0.7(0.13),0.15	1.3(0.13),0.23	2.5(0.12),0.33	4.3(0.11),0.46
$1.0°C/Wm^{-2}$	0.7(0.13),0.08	1.4(0.14),0.13	2.8(0.14),0.19	5.3(0.13),0.28

	80	160
$0.5°C/Wm^{-2}$	7.1(0.09),0.59	11.2(0.07),0.71
$1.0°C/Wm^{-2}$	9.4(0.12),0.39	16.0(0.10),0.52

Whether or not the recent satellite observations of a 1.5 Wm^{-2} (0.018% per year) decline in solar irradiance over the period 1980-85 can be explained by the Foukal-Lean model is a debatable point. Prior to 1980, there is some evidence of an irradiance increase between 1970 and 1979 at almost 0.03% per year (Frölich and Eddy, 1984; Frölich, 1987). The increase in irradiance subsequent to the sunspot minimum in 1976 accords qualitatively with the Foukal-Lean model, but the earlier increase, if the balloon-based observations can be trusted, is at odds with this model.

An alternative (but not necessarily incompatible) explanation of the irradiance changes post-1970 is that they are related to solar diameter variations (Fröhlich and Eddy, 1984). This explanation is of considerable interest because there are solar diameter data extending back about 300 years (Parkinson et al., 1980; Gilliland, 1981), perhaps into the 17th century (Ribes et al., 1987). For recent decades, Frölich and Eddy (1984) show diameter data which increase by 0.02 ± 0.01% from 1968 to 1976 and decrease subsequently (to 1982); although these data are quite noisy and subject to considerable uncertainty. Earlier diameter estimates are summarized by Gilliland (1981), who suggests that a cyclic variation with a period of about 76 ± 8 years exists with amplitude (i.e., half the range of variations) of 0.024%.

Conventionally, these diameter changes have been related to irradiance changes using the relationship $W = \Delta \ln R / \Delta \ln L$ = constant, where R is solar radius and L is irradiance (luminosity). The value of W is highly uncertain (Gilliland, 1982a). Empirical evidence, based on a climate modelling study, suggests that $W = 0.078$ (Gilliland, 1982b). The irradiance and radius data given by Frölich and Eddy (1984) give a very similar value. This match, however is entirely fortuitous. Gilliland's analysis underestimates W because it was based only on Northern Hemisphere temperature data. If one uses global data from Jones et al. (1986), the implied W value is about 0.2 and the match with more recent data breaks down. Furthermore, Gilliland's analysis suggests that irradiance changes lag behind radius changes by approximately a quarter period, a result which is clearly incompatible with the W theory. (The recent observations used by Frölich and Eddy (1984) also show a lag of irradiance behind radius).

An alternative theory can explain these observations better. Sweigart (1981) has developed a model which leads to the relationship

$$H = \frac{\Delta \ln L}{\frac{d}{dt}(\Delta \ln R)} = \text{constant.}$$

While Sweigart's model is unlikely to be correct since it is subject to the same criticisms of the W theory given by Gilliland (1982a), the possibility that irradiance changes are related to changes in the rate-of-change of radius is worth exploring. It has two major factors in its favour. It implies lags between irradiance and radius changes that are in accord both with recent observations and with the lag inferred by Gilliland (1982b). Secondly, in empirically estimating H, both the recent observations cited by Frölich and Eddy (1984) and the climate modelling approach of Gilliland (1982b) (using global-mean temperatures) give roughly the same value (H ≅ 80 years). The 76-year radius cycle would then have an irradiance signature with amplitude of about 0.16%. The recent changes in radius and irradiance used by Fröhlich and Eddy (1984) represent shorter time scale fluctuations superimposed on the 76-year cycle.

From this discussion of possible solar irradiance changes associated with variations in solar diameter, it can be concluded that irradiance variations with a range of 0.3-0.4% may have occurred over the past 300 years. Although reliable observations of irradiance span only a few years (since 1979) and these show a gross change of only about 0.1%, the possibility of changes up to 0.4% on longer time scales is a realistic one. The corresponding range of global-mean temperature fluctuations (based on the model used earlier and allowing for ocean damping effects and uncertainties in the climate sensitivity) is 0.2-0.3°C.

CLIMATE CARBON-14 RELATIONSHIPS

For times earlier than a few centuries ago, links between solar output and global climate become much more difficult to establish, simply because the only data available are indirect. Although we have no direct observations of the Sun's irradiance or diameter, we do, however, know that some characteristics of the Sun have varied throughout the Holocene. Historical records show that sunspots exhibited similar quasi-periodic behaviour prior to the beginning of the more recent continuous sunspot record, and that this behaviour was punctuated rather irregularly by longer intervals during which there were relatively few sunspots. The prime example of these is the Maunder Minimum, 1645-1715 (Eddy, 1976). A proxy indicator of these longer time scale solar fluctuations is the atmospheric carbon-14 anomaly record (Stuiver and Quay, 1980).

When ^{14}C dates are measured on accurately-dated tree-ring material, the ^{14}C and calendar dates differ. These differences reflect fluctuations (or "anomalies") in atmospheric ^{14}C concentration. These in turn indicate changes in the output of energetic particles from the Sun (the solar wind) which modulate the production rate of ^{14}C in the upper atmosphere. Thus, the ^{14}C anomaly record tells us something about solar variability. This record shows a very long time scale trend (which is thought to have a geomagnetic origin), upon which are superimposed solar-induced decadal to century time scale changes. The dominant features of the latter are intervals of strong positive ^{14}C anomaly which typically span about a hundred years. Documentary historical records show that there is a close correspondence between these intervals of positive ^{14}C anomaly and sunspot minima such as the Maunder Minimum, Spörer minimum, etc.

The solar events which are defined by these high ^{14}C anomaly intervals are not necessarily times when solar irradiance was noticeably perturbed, but one might expect this to be a reasonable possibility. The fact that the main features of the Little Ice Age period (which roughly spans the interval 1450-1850) coincide with the Maunder and Spörer Minima and their associated ^{14}C anomaly intervals, is at least circumstantial evidence in favour of there being a link between the ^{14}C anomaly record and solar irradiance.

The issue of a ^{14}C-climate link has been a controversial one for many years. Eddy (1977) revived the idea. His careful analysis of the available solar evidence showed some evidence of a link extending back for some six millenia. The climate data he used were, however, rather suspect. Subsequent anlayses by Stuiver (1980) and Williams et al. (1980) tended to show that the ^{14}C-climate correspondence was not so good. All studies of the subject, however, have been plagued by data quality problems.

In the last year or so, the ^{14}C anomaly record has been greatly improved (Stuiver et al., 1986). In addition, there has recently been published a uniquely comprehensive study of glacial fluctuations through the Holocene which provides us with a better climate time series than available previously (Röthlisberger, 1986). It is therefore possible to re-examine the issue of a ^{14}C-climate link using these new data.

Röthlisberger's glacial data come from twelve regions in both the Southern and Northern Hemispheres. He has combined these into six larger-region time series for the following locations; southeastern Alaska, tropical western South America (around 10°S), southern South America (35-55°S), the European Alps, the Himalayas, and southern New Zealand. In addition, he has included data for northern Scandinavia obtained from Wibjorn Karlen. I have combined these data with rough area weights in order to produce hemispheric- and global-mean time series. Since all time series are highly correlated (an interesting result in itself) the mean series do not depend much on the actual values used for the weights. The series are shown in Figure 1.

The global time series of glacial advances and retreats is clearly not a perfect paleoclimatic record since glaciers respond to climatic fluctuations in quite complex ways and their movements are affected by both temperature and precipitation (at least). However, it is likely that the dominant signal recorded by the glaciers is one of summer temperature. There are, of course, other proxy climate data that could be used to supplement this glacial record, but to critically synthesize all available data would be a massive task. The reader can get some idea of the problems involved from the paper by Williams and Wigley (1983), a work that considers only Northern Hemisphere data and spans only the past 2000 years.

The calender dates and durations of major positive ^{14}C anomalies have been obtained from Stuiver et al. (1986). Over the past 7000 years, fourteen major anomaly intervals can be identified (compare this with the nine intervals identified by Eddy, 1977). In order to make a comparison between the continuous climate record of Figure 1

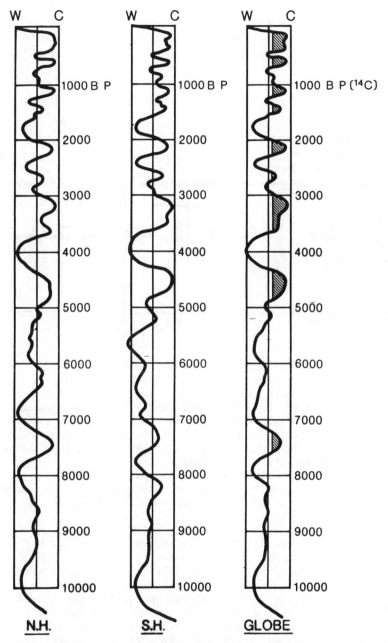

Fig. 1. Hemispheric- and global-mean temperature fluctuations during the Holocene based on the glacial chronologies of Röthlisberger (1986). Note the remarkable correspondence between the hemispheres.

and the discrete ^{14}C anomaly record, it is necessary to convert the former to a similar discrete record (of major cold periods). The two records are shown in Figure 2, together with the corresponding results given by Eddy (1977).

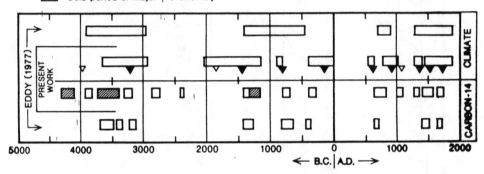

Fig. 2. A comparison of major climatic and ^{14}C anomalies. The upper and lower series are from Eddy (1977). The major ^{14}C anomalies have been extracted from the ^{14}C anomaly data of Stuiver et al. (1986). Hatched intervals represent periods of notable anomaly that are relatively less well defined, containing marked decadal time scale ^{14}C variability. The climate record (second row from top) comes directly from the global curve in Figure 1, with radiocarbon dates converted to calendar dates using Stuiver et al. (1986). Major and minor temperature minima are identified.

In Eddy's case, there was a clear visual correspondence between major ^{14}C anomalies and cold periods, apparent in Figure 2. However, it is equally clear from Figure 2 that there are substantial errors in Eddy's climate and ^{14}C records. Nevertheless, when the more recent data are compared, there is still a visual correspondence between the two records. Most of the nine major temperature minima line up with a major ^{14}C anomaly to within a century. The leading question, however, is ... "is this correspondence significantly better than would be expected to occur by chance?"

To answer this question, we must first recognize some uncertainties in the data and in the expected degree of correspondence. First, the climate data have been dated using ^{14}C. Röthlisberger (1986) gives his evidence in radiocarbon years, which I have converted to

calendar years using the Stuiver et al. (1986) anomaly data. There are
thus two types of uncertainty in the dating of the climate record;
those due to basic ^{14}C dating uncertainties inherent in dating glacial
records (direct ^{14}C counting uncertainties are obviously a problem
– other uncertainties in dating glacial deposits are described by,
e.g. Porter, 1981), and those due to uncertainties in the conversion
to calendar dates (which arise from the short-time-scale variability
of the ^{14}C anomaly record). It is difficult to quantify these uncertainties precisely. Dating errors could be reduced by the following
facts: that many dates are involved, that the records must be stratigraphically consistent, and that the global record is a composite
of a number of independently dated records. In spite of this, it
is doubtful that the dates of the major temperature minima shown
in Figure 2 are better than ± 100 years (two-sigma limits).

The second (and lesser) source of uncertainty lies in the expected
correspondence itself. Just how close would one expect this correspondence to be? The primary source of both the cold periods and
the major ^{14}C anomalies is hypothesized to be the Sun. However,
an event on the Sun will not be immediately reflected either in the
terrestrial ^{14}C or climate records. In both cases, there will be
lags effected by the damping effect of the ocean's chemical or thermal
inertia. It is likely that these two lags will be of similar magnitude;
but there is an additional lag in the glacial record due to the
response time of glaciers to a change of climate. The net effect
of these factors is probably an overall uncertainty of the order
of decades.

One final source of uncertainty lies in the ^{14}C record. I have
defined the major ^{14}C anomaly intervals subjectively. This will
not affect the general location of the major anomalies, but it could
affect their durations – i.e. their precise beginning and end dates.
The magnitude of this uncertainty could be up to a century, but it
is likely to be of order decades in most cases.

One can now consider the statistical problem of determining
the significance of the correspondence between ^{14}C anomalies and
climate in the following way. I have identified nine temperature
minima which have estimated 95% confidence limits in their dates
of ± 100 years. These represent 200-year-wide "bullets" which, as
a null hypothesis, can be considered to have been fired at random
at the fourteen ^{14}C anomaly "targets". How many "hits" would we
expect to occur?

As a reasonable approximation, this can be considered as a binomial
problem. The total target area, allowing for the ± 100 year uncertainty in the climate dates, is (from Figure 2) approximately
4300 years, so the probability of a "hit" is 4300/7000 = 0.6143.
Inspection of Figure 2 shows that eight of the nine major temperature
minima hit a ^{14}C anomaly interval to within 100 years. The probability of this occurring by chance is therefore given approximately
by

$$P = 9(0.6143)^2 (1-0.6143) = 0.0704.$$

A slightly different result obtains if one eliminates the less well-defined ^{14}C anomaly intervals (shaded in Figure 2). In this case the target area reduces to about 3400 years. There are still eight hits, and the probability that this could occur by chance is

$$P = 9(0.4857)^8 (1-0.4857) = 0.0143.$$

The first of these two results is not significant at the 5% level, while the second result is. In any event, it is clear that the probability of the observed results occurring by chance is quite small. Even in the one case where there is a "miss" (viz. the temperature minimum at about 2150 calendar years BP), the temperature minimum _interval_ does overlap a major ^{14}C anomaly. It seems reasonable, therefore, to conclude that the correspondence between climate and major ^{14}C anomalies is a manifestation of some real physical process.

There is, however, one nagging question. Why are there some ^{14}C anomaly intervals that appear to have no corresponding climate anomaly? For the earliest of these, a minor temperature minimum does match the ^{14}C anomaly (within 100 years) - see Figure 2. However, the two ^{14}C anomalies between 4000 and 5000 calendar years BP have no obvious climate anomaly analogues. They both occur within the timespan of the warmest period in the climate record, around 4000 BP (radiocarbon years) in Figure 1.

Although I have described this analysis as a test of the hypothesis that there are major solar irradiance perturbations that occur in parallel with the solar events which cause major ^{14}C anomalies, there is one other possible explanation. The ^{14}C anomalies could, themselves, be the result of the climate perturbations. This is unlikely, since the ^{14}C anomaly link with the Sun has been convincingly demonstrated (Stuiver and Quay, 1980), at least for the last 1000 years. However, if major cold periods were associated with a large change in the rate of oceanic bottom water formation (as has been hypothesized for the late-glacial), then parallel perturbations in the atmospheric ^{14}C content would certainly occur. This is a topic which deserves further attention.

IRRADIANCE CHANGES FOR MAUNDER-MINIMUM-TYPE EVENTS

I will now assume that the ^{14}C-climate link is real and estimate the change in solar irradiance that is likely to be associated with ^{14}C-anomaly-producing solar events like the Maunder Minimum. To do this, it is first necessary to estimate the global-mean temperature deviation associated with the major glacial advances documented in Figure 1. This can be done (albeit with considerable uncertainty) by reference to the most recent period of the record. Over the last 100-150 years, the globe has warmed by approximately 0.5°C (Jones et al., 1986). This interval corresponds to that part of the most recent warming trend in Figure 1 which lies above the "zero" reference line. From this, it can be deduced that the glacial advance maxima correspond roughly to an annual, global-mean cooling of 0.4°C below

the reference line. (One can also deduce that the global-mean
temperature at the height of the Little Ice Age was roughly 0.9°C
cooler than today.)

This 0.4°C cooling represents only a transient response effect.
To estimate the solar forcing, we need to know the corresponding
equilibrium temperature change. This depends on the time scale of
the forcing. The average duration of the major ^{14}C anomaly episodes
is around 100 years; but the ratio of transient to equilibrium temperature change due to a forcing episode lasting 100 years depends
not only on the duration, but also on the time history of the forcing.
For example, if the forcing were comprised of two step-like changes,
the precise ratio would differ from that which would occur if the
forcing had a more sinusoidal character. Nevertheless, an estimate
can be obtained from Table 1, using the damping ratio for cyclic
forcing with a 200-year period (i.e. assuming the forcing perturbation
to be a half-cycle).

If the climate sensitivity were 0.5°C/Wm^{-2}, then the ratio is
0.74 and the equilibrium temperature change would be about 0.4/0.74
= 0.54°C. The corresponding forcing at the top of the troposphere
would therefore be about 1.1 Wm^{-2} (corresponding to a solar output
change of 6.3 Wm^{-2}). If the sensitivity were 1.0°C, the equilibrium
temperature change would be 0.4/0.57 = 0.70°C, corresponding to a
forcing of 0.7 Wm^{-2} (i.e., a solar perturbation of 4.0 Wm^{-2}). Since,
as noted before, recent empirical evidence favours the lower value
for the climate sensitivity, the best estimate of the decline in
solar forcing at the top of the troposphere associated with major
^{14}C anomalies like the Maunder Minimum is about 1 Wm^{-2} (i.e., a solar
irradiance perturbation of 5.7 Wm^{-2}).

These estimates can be checked by using the model to estimate
the transient response to isolated solar irradiance perturbations
with different time histories: for example, a step-like reduction
followed by a step-like increase 100 years later; or a linear decline
for 50 years followed by a linear increase over the next 50 years
to the original level. For a climate sensitivity of 0.5°C/Wm^{-2} the
corresponding damping ratios are 0.91 and 0.71 (compared with 0.74
assumed above); while for a sensitivity of 1.0°C/Wm^{-2} the ratios
are 0.80 and 0.54 (cf. 0.57 assumed above). The case of step-like
solar changes minimizes the irradiance perturbation required to explain
the observed cooling, reducing it by a factor of approximately 0.7-0.8
compared with the results based on a sinusoidal forcing assumption.

The only other estimate of this quantity is 0.14% (i.e., 1.9 Wm^{-2})
attributed by Kerr (1987) to Foukal and Lean and based solely on
their solar irradiance model. In the Foukal-Lean model, the irradiance
perturbation associated with Maunder-Minimum-type events would be
similar to a pair of step-like changes in a continuum of decadal
time scale fluctuations associated with the usual 11-year sunspot
cycle. For an isolated event with magnitude as estimated by Foukal
and Lean, the maximum cooling estimated by the model is 0.15°C for
a climate sensitivity of 0.5°C/Wm^{-2} and 0.26°C for a sensitivity
of 1.0°C/Wm^{-2}. The magnitude of the cooling is somewhat reduced
if the effect of the normal sunspot continuum is included. It appears,

therefore, that the observed coolings are two to four times larger
than can be explained by the Foukal-Lean model.

The estimate of a solar irradiance reduction of around 6 Wm^{-2}
associated with Maunder-Minimum-type events is clearly sensitive
to the assumption that these events cause a global-mean cooling of
0.4°C. I have based this on the temperature depression below the
central reference line in Figure 2, a line which represents the base
level for the past 1500 years only. Prior to that, the glacial
advances appear to be deeper and more long-lasting perturbations.
The past 1500 years have five ^{14}C events more closely spaced than
in any earlier period, and this would probably have a long-term effect
on the base level. One might, therefore, reasonably assume that
the amount of cooling was somewhat larger than 0.4°C, with a correspondingly larger irradiance change. However, it would be foolish
to suppose that all of the details of the Holocene climate fluctuations
shown in Figure 1 can be explained by solar Maunder-Minimum events,
or indeed that the record in Figure 1 is totally reliable. Over the
7000-year period, very large changes in the seasonal character of
solar insolation have occurred due to Milankovitch effects, and these
must have had a marked influence on the long-term climate record.
The implications I have drawn from this record, albeit based on better
data than previous studies and supported by a physically realistic
model, already stretch credibility to the limit. Additional, more
complex interpretations or speculations are unwarranted.

CONCLUSIONS AND IMPLICATIONS FOR THE FUTURE

In this paper I have tried to evaluate the climatic record of the
Holocene using a quantitative, model-based framework. Both direct
and indirect observations of solar irradiance changes over the past
300 years were used to estimate the range of possible solar effects
on global-mean temperature. The estimated range was 0.1-0.3°C; i.e.
fluctuations of roughly ± 0.1°C about the hypothetical steady-state
which would prevail in the absence of external forcing.

For times earlier than 300 years ago there is indisputable
evidence of the Sun's variability, but not direct evidence of solar
irradiance changes. By re-examining the possible link between major
atmospheric ^{14}C anomalies and climate I concluded that this link
was real. This implies that significant irradiance variations occur
in parallel with the solar fluctuations responsible for the ^{14}C
anomalies. Reference to recent global-mean temperature changes, and
to the earlier-described model results, allowed two estimates to
be made; that the global-mean temperature perturbation associated
with the major Holocene glacial advances was 0.4°C and that this
represented a decline in solar irradiance of about 6 Wm^{-2} (i.e.,
a net perturbation of 1 Wm^{-2} at the top of the troposphere).

These results have implications for the future. Today and over
the past century the dominant climate forcing mechanism has probably
been Man's input of greenhouse gases into the atmosphere. The combined radiative forcing due to changes in the atmospheric con-

centrations of carbon dioxide, methane, nitrous oxide and the chlorofluorocarbons is, between the late 18th century and 1985, approximately 2.2 Wm^{-2} (Wigley, 1987). Between now and 2030, the expected additional forcing at the top of the troposphere is about 3 Wm^{-2}. Should another Maunder Minimum event begin within the next few decades (as has been suggested by some authors, e.g. Landscheidt, 1983; Fairbridge and Shirley, 1987), then this would clearly offset future greenhouse-gas-induced climatic change to some considerable degree. However, the magnitude of Man's impact is such that it would still be the dominant factor in future climatic change.

REFERENCES

Eddy, J.A., 1976, Science, 192, 1189.
Eddy, J.A., 1977, Climatic Change 1, 173.
Fairbridge, R.W., and Shirley, J.H., 1987, Solar Physics (in press).
Fröhlich, C., 1987, J. Geophys. Res., 92, 796.
Fröhlich, C., and Eddy, J.A., 1984, Adv. Space Res., 4, 121.
Gilliland, R.L., 1981, Astrophys. J., 248, 1144.
Gilliland, R.L., 1982a, Astrophys. J., 253, 399.
Gilliland, R.L., 1982b, Climatic Change, 4, 111.
Hoffert, M.I., and Flannery, B.P., 1985, In, 'Projecting the Climatic Effects of Increasing Carbon Dioxide', (eds. M.C. MacCracken and F.M. Luther), U.S. Dept. of Energy, Carbon Dioxide Research Division, 149.
Hoyt, D.V., and Eddy, J.A., 1982, NCAR Tech. Note TN-194 + STR, Nat. Ctr. for Atmos. Res., Boulder, Colo.
Jones, P.D., Wigley, T.M.L., and Wright, P.B., 1986, Nature, 322, 430.
Kerr, R.A., 1987, Science, 236, 1624.
Kyle, H.L., Ardanuy, P.E., and Hurley, E.J., 1985, Bull. Amer. Met. Soc., 66, 1378.
Labitzke, K., 1987, Geophys. Res. Letts., 14, 535.
Landscheidt, T., 1983, In, 'Weather and Climate Responses to Solar Variations', (ed. B.M. McCormac), Colo. Assoc. Univ. Press, 293.
Mitchell, J.M., Jr., Stockton, C.W., and Meko, D.M., 1979, In, 'Solar-Terrestrial Influences on Weather and Climate', (eds. B.M. McCormac and T.A. Seliga), D. Reidel, Dordrecht, 125.
North, G.R., Mengel, J.G., and Short, D.A., 1983, In, 'Weather and Climate Responses to Solar Variations', (ed. B.M. McCormac), Colo. Assoc. Univ. Press, 243.
Parkinson, J.H., Morrison, L.V., and Stephenson, F.R., 1980, Nature, 288, 548.
Pittock, A.B., 1978, Rev. Geophys. Space Phys., 16, 400.
Pittock, A.B., 1983, Quart. J. Roy. Met. Soc., 109, 23.
Porter, S.C., 1981, In, 'Climate and History', (eds. T.M.L. Wigley, M.J. Ingram and G. Farmer), Cambridge Univ. Press, 82.
Ribes, E., Ribes, J.C., and Barthalot, R., 1987, Nature, 326, 52.
Röthlisberger, F., 1986, '10000 Jahre Gletschergeschichte der Erde', Verlag Sauerländer, Aarau.

Shapiro, R., 1979, J. Atmos. Sci., 36, 1105.

Stockton, C.W., Mitchell, J.M. Jr., and Meko, D.M., 1983, In, 'Weather and Climate Responses to Solar Variations', (ed. B.M. McCormac), Colo. Assoc. Univ. Press, 507.

Stuiver, M., 1980, Nature, 286, 868.

Stuiver, M., and Quay, P.D., 1980, Science, 207, 11.

Stuiver, M., Pearson, G.W., and Braziunas, T., 1986, Radiocarbon, 28, 980.

Sweigart, A.V., 1981, In, 'Variations of the Solar Constant', NASA Conf. Publ. 2191, 143.

Wigley, T.M.L., 1987, Climate Monitor, 16 (in press).

Williams, L.D., and Wigley, T.M.L., 1983, Quarternary Research, 20, 286.

Williams, L.D., Wigley, T.M.L., and Kelly, P.M., 1980, In, 'Sun and Climate', CNES, Toulouse, France, 11.

Willson, R.C., Gulkis, S., Janssen, M., Hudson, H.S., and Chapman, G.A., 1981, Science 211, 700.

Willson, R.C., Hudson, H.S., Fröhlich, C., and Brusa, R.W., 1986, Science, 234, 1114.

THE CLIMATIC IMPACT OF SECULAR VARIATIONS IN SOLAR IRRADIANCE

George C. Reid and Kenneth S. Gage
Aeronomy Laboratory
National Oceanic and Atmospheric Administration
Boulder, Colorado, U.S.A.

ABSTRACT. Variations in solar luminosity are expected to give rise
to variations in average sea-surface temperatures (SST). We explore
this possibility using both indirect evidence on the variability of
tropical SST based on measurements of tropopause height, and a recent
compilation of direct SST measurements by ships over the past 130
years. Both data sets show modulation that could be construed as
arising from solar variability. A simple advective-diffusive model
of the response of the ocean to thermal forcing is used to estimate
the amplitude of the solar variations needed to explain the observed
variability. We conclude that a modulation of the sun's luminosity
with a period of about 80 years and an amplitude of about 0.5% is
consistent with the globally averaged direct SST measurements, but
that the consistency of the results is weakened if the calculated
response to the increase in atmospheric CO_2 is included in the model.
Any 11-year component is unlikely to have had a large enough
amplitude to cause a significant effect, except possibly during the
great solar cycle that peaked in the late 1950s.

1. INTRODUCTION

The search for some kind of relationship between solar activity and
the earth's climate has a long history, going back almost to the days
of Galileo. The subject has acquired a faint aura of pseudo-science,
largely because it has relied heavily on correlations, usually between
sunspot number and some climate parameter, with a shortage of convincing
physical mechanisms to explain the correlations. The current state
of the subject has been well reviewed in recent publications (Herman
and Goldberg, 1978; National Academy of Sciences, 1982; McCormac,
1983).
   The search for mechanisms has taken two distinct routes. One
has concentrated on the more exotic part of the solar output, in the
form of energetic particles, solar-wind plasma, and radiation in the
near and far ultraviolet, all of which are known or suspected to vary
with solar activity. Because the total energy flux represented in

225

F. R. Stephenson and A. W. Wolfendale (eds.),
Secular Solar and Geomagnetic Variations in the Last 10,000 Years, 225–243.
© *1988 by Kluwer Academic Publishers.*

this output is small, however, and its direct effects are confined to the upper and middle atmosphere, some kind of trigger mechanism is needed to produce any significant effect in the troposphere, and the search for a convincing mechanism has so far proven fruitless. The other approach, which is more hopeful, postulates the existence of a small variation in the sun's luminosity (the solar "constant") associated with solar activity, and utilizes the generally accepted model result that the earth's climate is sensitive to small changes in solar heating. This approach also has a long history, but not until direct measurements of the solar constant above the atmosphere from spacecraft became a reality (Hickey et al., 1980; Willson et al., 1981) did any quantitative assessment become possible. We now know that the sun's luminosity does vary on time scales ranging from days to years (Willson et al., 1986). The short-term variations are associated with the appearance of sunspots, and thus negatively correlated with solar activity; the relationship between the longer-term variations and solar activity is not yet clear.

The choice of a climate parameter to use in studying the effects of solar variability is also critical. Many of the past studies have used regional parameters that are likely to be much more dependent on changes in local conditions than on small changes in solar luminosity. It seems essential to use parameters that are representative of as wide an area as possible, and indeed to use global parameters if they are available.

In this paper we present the evidence for solar luminosity variations on the time scales associated with solar activity, using sea-surface temperatures as the basic climate parameter. Given that 70% of our planet's surface is ocean, sea-surface temperature seems to be an appropriate parameter to use, provided that the area sampled is large enough to average out local effects. Even so, however, the climate system is exceedingly complex, and one could well expect that variations produced by changes in the sun's luminosity would be masked by variations that are internally generated. In the present state of our understanding it is difficult to separate these, and the evidence can only be circumstantial in nature. Such circumstantial evidence is useful, however, since it provides us with some of the clues needed to construct a theory of the solar-terrestrial interaction that governs terrestrial climate.

2. TROPICAL TROPOPAUSE VARIATIONS

An apparent relationship between the height of the tropical tropopause and the sunspot number was first suggested by Stranz (1959), using data from a single African station (Leopoldville) for a limited time period (June 1953 to December 1958). The relationship was confirmed by Rasool (1961), who also reported a similar but weaker variation in the height of the tropopause at a few extratropical locations, and by Cole (1975), who examined data from a number of tropical Pacific and Caribbean stations over a period of about 20 years and concluded that a statistically significant correlation existed for several of the stations.

None of these authors suggested a quantitative mechanism that might explain the relationship, although they alluded to the possibility that a solar-cycle-related variation in stratospheric ozone might be involved. The amplitude of any solar-cycle variation in ozone concentration in the tropical lower stratosphere is predicted to be small (Garcia et al., 1984), however, and the height of the tropopause is in any case insensitive even to large changes in ozone concentration (e.g. Reck, 1976), being controlled rather by the intensity of the convective mixing that takes place in the troposphere (e.g. Reid and Gage, 1981). This implies that the cause of the solar-cycle-related variation in tropopause height, if real, probably lies in some corresponding variation in tropical convective activity.

In the course of a study of the climatology of the tropical tropopause, the relationship between tropopause heights and sunspot number was re-examined (Gage and Reid, 1981; Reid and Gage, 1985) for stations in the tropical western Pacific with a high-quality 30-year data base, and the result is shown in Figure 1. The top curve shows the 12-month running mean sunspot number, and the middle curve shows the heavily smoothed tropopause height variation at Koror, which is typical of the tropical western Pacific island stations. The smoothing was done to remove the pronounced annual cycle and the weaker quasi-biennial cycle in tropical tropopause heights. The bottom curve

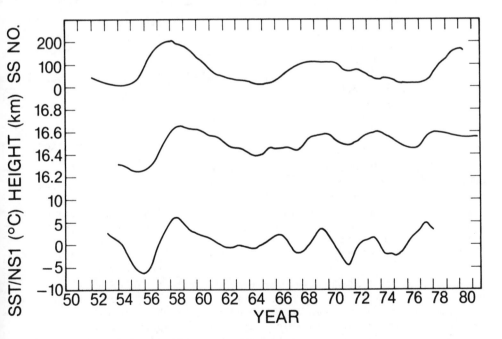

Fig. 1   12-month running mean sunspot number (top), smoothed tropopause height at Koror (middle), and principal non-seasonal component of variation in Pacific SST (bottom).

is a measure of Pacific sea-surface temperature (SST) anomalies (Weare et al., 1976), and will be discussed later.

The overall correspondence between the upper two time series is clear, though it is more obvious in the earlier part of the data than in the later part. We suggested (Gage and Reid, 1981) that the relationship could be explained by the existence of a small variation in solar irradiance that was in phase with the 11-year solar cycle. The mechanism is illustrated in Figure 2 (Reid and Gage, 1981). Small changes in surface insolation give rise to corresponding small changes in average SST (neglecting any changes in the average level of other factors that might affect SST, such as wind-driven upwelling), and hence to changes in the absolute humidity of the troposphere, assuming that relative humidities remain nearly constant. Since latent heat of condensation provides the main tropospheric energy source in the tropics, the convective mixing rate is also changed, leading to changes in the depth of the troposphere, i.e. in the height of the tropopause. The direction of the change is such that an increase in insolation leads to a rise in the tropopause. An increase in tropical convective activity also intensifies the vertical Hadley circulation in the lower stratosphere, causing adiabatic cooling, which also tends to raise the tropopause. The

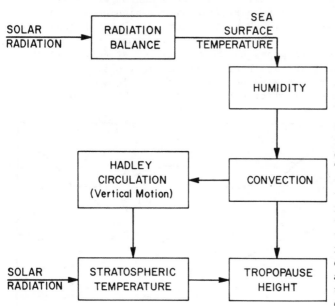

SOLAR RADIATION AND THE HEIGHT OF THE TROPICAL TROPOPAUSE

Fig. 2. Schematic illustration of the mechanism by which changes in solar irradiance can directly affect the tropical tropopause.

height (and also, incidentally, the temperature) of the tropical tropopause is thus a rather sensitive indicator of the intensity of convective activity averaged over a wide area of the tropics.

Using a simple model in which the ocean was approximated as a uniform layer of water, and air parcels were assumed to conserve their equivalent potential temperature while moving vertically upward in the protected cores of deep convective clouds, we estimated that the observed change in tropopause height between sunspot minimum and sunspot maximum could be explained by a change in the solar constant of about 0.5%, in phase with the solar cycle. The phasing is the same as that proposed by Eddy (1977) to explain the apparent relationship between solar activity and climate on a century-long time scale, but opposite to that required by sunspot blocking effects (Eddy et al., 1982).

Figure 3 shows the time series of annual mean tropopause heights averaged over 5 tropical western Pacific stations, together with 6-month means for the northern winter and summer half years. There is an obvious long-term trend in all three time series, amounting to a rise of about 9 meters per year in the annual means, implying a long-term warming of the troposphere (and/or cooling of the stratosphere) whose origin will not be discussed here. It was removed by fitting a least-squares straight line and subtracting it from the data points. The same procedure was carried out for the annual mean sunspot numbers, and the cross-correlation coefficient between the two detrended time series was determined. The result is shown in Figure 4, in which the maximum correlation coefficient is +0.62 for a lag of 1 year, in the sense of sunspot numbers leading tropopause heights. The correlation is significant at about the 95% level, taking into account autocorrelation in the individual time series. When the 32-year period was broken up into two consecutive 16-year periods and the correlation analysis carried out for each separately, however, Figure 5 shows the result. As one might suspect from Figure 1, the maximum correlation coefficient for the earlier period is very high (about +0.87, again for a lag of 1 year), but is not statistically significant for the later period. Because of the small number of degrees of freedom in the shortened time series, even the high correlation coefficient is significant at only about the 95% level.

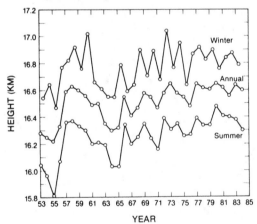

Fig. 3.   Interannual variation of topopause height averaged over five tropical western Pacific island stations. The middle time series consists of 12-month means of monthly averages, and the other two are 6-month means for the northern hemisphere winter and summer half years.

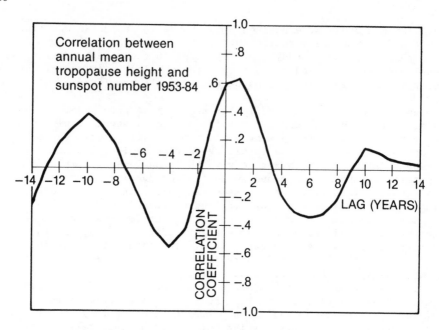

Fig. 4. Cross-correlation coefficient between detrended annual
mean sunspot number and tropopause height for 1953–84. Lag
is positive for sunspot number leading.

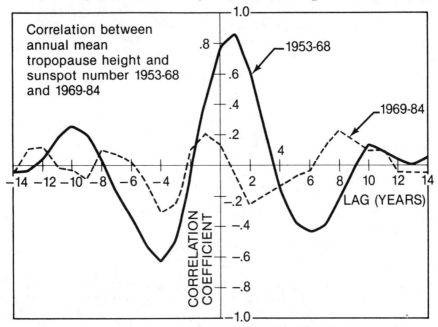

Fig. 5. Same as Figure 4 for 1953–68 and 1969–84 separately.

Possibly the real reason for the apparent relationship is shown by the bottom curve in Figure 1. This shows the time series of the amplitude of the strongest non-seasonal component in an empirical orthogonal function analysis of Pacific SSTs by Weare et al. (1976), and can be regarded as a measure of the surface warmth of the Pacific Ocean. The positive peaks generally correspond to years of strong El Niño events when warm water spreads eastward across the equatorial Pacific in unison with a weakening of the trade winds. The event of 1957-58 (Bjerknes, 1966) and the deep minimum that preceded it in 1955-56 are especially noteworthy. The rise in tropopause heights between these points can perhaps be regarded as simply a response to the corresponding large swing in SST that was probably unrelated to changes in solar irradiance. If there is any causal relationship between the intensity and phase of El Niño events and the solar cycle, it would have to involve some relatively subtle change in the general circulation of the atmosphere. No such mechanism has yet been proposed.

The overall result of the study is thus rather inconclusive. Clearly no case can be made for a general relationship between solar activity and tropical tropopause heights, and at least part of the apparent relationship seen in the 1953-68 period can be attributed to the large-amplitude swing in Pacific SST during the 1953-60 period. One could probably dismiss the entire relationship as coincidental were it not for the facts that the solar-activity cycle during which it was seen was the most intense in recorded solar history, when any 11-year variation in solar luminosity might be expected to have its greatest amplitude, and that the correlation peak occurred for a lag of about 1 year between sunspot numbers and tropopause heights, which is about the lag that would be predicted for a forcing period of 11 years in solar luminosity as a result of the thermal inertia of the ocean (Hoffert et al., 1980). The study also served to point out a potentially important mechanism: if the solar constant does vary, any resultant changes in average sea-surface temperatures are likely to be amplified through the exponential dependence of saturation vapor pressure on temperature, and the dependence of the vigour of tropical convection on the humidity of the troposphere. Evaporational cooling is likely to impose a negative feedback on this interaction, however, and may lead to a practical upper limit on tropical sea-surface temperatures (Newell et al., 1978).

3. SOLAR LUMINOSITY AND SEA-SURFACE TEMPERATURES

Two recently published data sets have revived the interest in a possible connection between solar luminosity variations and sea-surface temperatures. The first is the series of measurements of the solar constant by the Active Cavity Radiometer Irradiance Monitor on the Solar Maximum spacecraft (Willson et al., 1981) and by the Nimbus-7 Earth Radiation Budget experiment (Hickey et al., 1980). These measurements have revealed a long-term downward trend in the solar constant, amounting to about 0.02% per year during the period 1980-85 (Willson et al., 1986), superimposed on more sporadic fluctuations

caused by sunspot blocking. Since the 1980-85 period was one of decreasing solar activity, the trend is at least consistent with the
existence of a cyclical variation in solar luminosity in phase with
the 11-year cycle. At the rate observed, however, the peak-to-peak
variation between solar maximum and solar minimum would amount to
only about 0.1%, barely enough to be climatically significant. Fröhlich
(1987) has recently combined these measurements with earlier balloon,
rocket, and satellite data obtained during the 1967-71 period, and
has inferred that the solar constant increased at a rate of about
0.03% per year between 1967 and 1980. He suggests that the combined
measurement set is consistent with a 22-year cyclical variation in
luminosity with a peak-to-peak range of about 0.4%, which would probably
be enough to cause a significant effect on global climate.

The second important compilation is the Meteorological Office
Historical Sea Surface Temperature data Set (Folland and Kates, 1984;
Folland et al., 1984), which is derived from 50 million individual
measurements of SST made by ships since 1854. These data have been
carefully treated to take account as far as possible of changes in
measuring techniques over the 130-year period, and they have been
combined to produce hemispheric and global averages. Since the measurements are biased toward the principal shipping lanes, and are
particularly sparse at high southern latitudes, these averages have
to be treated with caution. The facts that the two hemispheres
individually show similar long-term variations, at least prior to
1960, and that the data for the four seasons taken separately all
show essentially identical long-term variations (Folland and Kates,
1984) lends confidence to the reality of the variation, which is similar
to that deduced by Paltridge and Woodruff (1981) using a more
restricted data base.

Figure 6 shows the most recent version of the global average
SST (C.K. Folland, private communication) smoothed by taking an 11-year
running mean. (The original SST values used were 3-month seasonal
departures from an arbitrary "normal" value). Also shown is the sunspot number smoothed by the same numerical filter to produce a quantity
proportional to the envelope of the solar cycle. While the two curves
are far from identical, they have enough points of similarity to lead
one to speculate once again on the existence of a causal relationship.
Both curves have minima in the 1901-10 decade and maxima in the 1951-60
decade. While the structure in the sunspot curve prior to 1890 is
not reproduced in the SST curve, the rise between 1910 and 1950 is
similar in both, and even the dip between 1960 and 1980 appears in
both. Folland et al. (1984) report that maximum entropy power spectral
analysis of the SST data yielded a major peak at a period of 83 years,
which is close to that of the Gleissberg cycle of solar activity.

If one wishes to draw an inference from this, it is that the
solar luminosity might vary roughly in phase with the envelope of
the solar activity cycle, rather than with the solar activity cycle
itself, i.e., with a quasi-period of roughly 80 years. Interestingly
enough, a study by Gilliland (1981) of solar radius variations based
on measurements made over a 265-year period revealed a modulation
in radius with a period of 76 years, antiphased with the Gleissberg

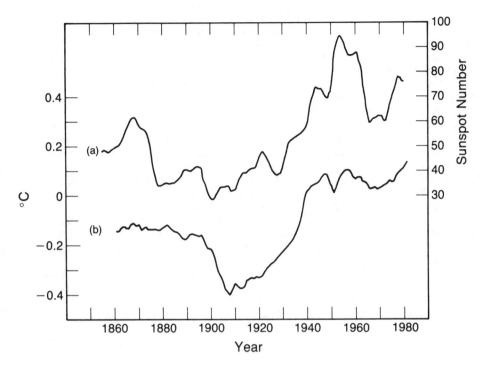

Fig. 6. 11-year running means of (a) sunspot number, and (b)
global average SST anomalies.

cycle (i.e. maximum radius occurring at Gleissberg-cycle minima).
If there is indeed a luminosity variation that is in phase with the
Gleissberg cycle, it would imply an antiphase relationship between
solar radius and luminosity. We shall return to this point later.

   If the gross features of the global SST variation are indeed
due to a cyclical variation in solar luminosity with an 80-year period,
the intriguing possibility arises that the decrease in the solar
constant seen during the 1980-85 period may be part of such a variation.
If so, the peak-to-peak range in luminosity would be of the order
of 1%. Is a variation of such a magnitude capable of explaining the
amplitude of the SST variation shown in Figure 6? When combined with
the amplitude of the long-term variation in solar radius reported
by Gilliland (1981), what does it imply for the solar radius-luminosity
relationship? We shall address these questions in the next section.

   Meantime, however, we should note that past attempts to explain
long-term variations in climate have invoked other forcing functions
in addition to solar luminosity variations. In particular, varying
combinations of volcanic dust veils and atmospheric carbon dioxide
concentrations, as well as solar variations, have been included in
a number of model studies (Schneider and Mass, 1975; Miles and

Gildersleeves, 1977; Gilliland and Schneider, 1984; Harvey and
Schneider, 1985). By assuming that the overall shape of the SST
variation is determined by the solar luminosity, we are implicitly
assuming also that time variations in other forcing factors are less
important, at least over most of the 130-year period. This point
will also be discussed later, with particular reference to the influence
of increasing CO_2 concentrations.

The solar-constant measurements in the late 1960s reported by
Fröhlich (1987), which suggested that the luminosity increased during
the 1967-80 period, might be taken as evidence against a variation
with the Gleissberg cycle, which has been declining since its maximum
in the late 1950s, and will presumably reach its next minimum around
the end of the century. The Gleissberg cycle, however, is a cycle
only in the crudest sense, and the envelope of the 11-year cycle shown
in Figure 6 actually shows a substantial, but probably transient,
increase during the 1967-80 period. In fact, the running mean sunspot
number increased by 14 (from 62 to 76) between 1967 and 1980, while
the solar constant increased by about 0.35% according to the data
presented by Fröhlich (1987). Applying this same rate of change to
the entire range of sunspot number shown in Figure 6 (about 60) would
lead to a total range in solar luminosity of about 1.4%, which is
reasonably consistent with the 1% estimate above, based on the observed
decay in the 1980-85 period. Thus the data presented by Fröhlich
(1987) are at least as consistent with a luminosity variation proportional to the envelope of the 11-year cycle as they are with a 22-year
cycle.

4. MODEL CALCULATIONS

The influence of a varying solar constant on global temperatures has
been modeled by a number of authors, as mentioned in the last section,
but most of these calculations have been directed toward explaining
the historical variability of surface air temperatures rather than
sea-surface temperatures, and have included factors other than solar
variations. While some impressive matches have been obtained, the
number of adjustable parameters is often large enough to raise doubts
about their uniqueness, while surface air temperature measurements
are usually biassed toward land areas, comprising only 29% of the
earth's surface, and their historical record shows significant
differences from that of sea-surface temperatures (Paltridge and
Woodruff, 1981; Folland et al., 1984).

In order to estimate the effect of variations in solar luminosity
alone on global average SST, we have used the one-dimensional ocean
thermal model of Hoffert et al. (1980), which will be described only
briefly here. The interested reader should consult the reference
for full details. The model ocean consists of three separate components:
an upper well mixed layer of uniform temperature, a deep ocean whose
temperature varies with depth, and a polar sea where the sinking of
dense surface water drives a thermohaline circulation through the
rest of the global ocean.

The global radiative balance at the surface can be expressed as

$$(S/4)(1-\alpha) - I = 0 \tag{1}$$

where S is the solar constant, α is the planetary albedo, and I is the outgoing infrared radiative flux. A convenient parameterization for I was introduced by Budyko (1969) in the form

$$I = A + BT \tag{2}$$

where T is the surface temperature in degrees Celsius. The global equilibrium temperature can then be defined as

$$BT_{eq} = (S/4)(1-\alpha) - A \tag{3}$$

Hoffert et al. (1980) show that the temperature T_m of the oceanic mixed layer is given approximately by

$$dT_m/dt = \{T_{eq}(t) - T_m\}/\tau - h_m^{-1}(\partial Q/\partial t)$$

$$Q = \int_0^{h_d} T(z,t)dz \tag{4}$$

where h_m and h_d are the thickness of the mixed layer and the depth of the main ocean respectively, z is the vertical coordinate measured downward from the base of the mixed layer, and τ is an effective radiative time constant for the mixed layer. The first term on the right-hand side represents radiative heating (or cooling) at the surface, while the second term represents transfer of heat to (or from) the deep ocean, whose temperature $T(z,t)$ is governed by

$$\partial T/\partial t = K(\partial^2 T/\partial z^2) + w(\partial T/\partial z) \tag{5}$$

where K is an eddy diffusivity and w is a vertical upwelling speed representing the slow thermohaline circulation of the ocean driven by the formation of deep water in the high-latitude oceans of both hemispheres. The temperature is thus determined by a balance between eddy heat conduction from above and vertical advection of cold water from below.

The procedure is to solve the coupled equations (4) and (5), where T_{eq} in (4) is determined from the time-varying solar constant via (3), and (5) is subject to the boundary conditions

$$T = T_m \qquad (z = 0) \tag{6}$$

$$K(\partial T/\partial z) + wT = wT_p \qquad (z = h_d) \tag{7}$$

Here T_p is the (potential) temperature of the deep water formed in the polar oceans.

The values of the parameters used in the calculations were mostly identical to those used by Hoffert et al. (1980). The planetary albedo α was taken as 0.3, the mixed-layer radiative time constant τ as 3.9 years, the mixed-layer and deep-ocean depths as 100m and 4000m respectively, the eddy diffusivity as 2000 m^2yr^{-1}, and the global average upwelling velocity w as 4m yr^{-1}. The infrared cooling parameter B was taken as 2.2 W $m^{-2}K^{-1}$, and A was fixed by the requirement that the reference condition of S = 1360 W m^{-2} and T_{eq} = 14.8°C (Hoffert et al., 1980) be met. This condition gave A = 205.4 W m^{-2}.

The equations were solved numerically, dividing the deep ocean into 40 layers of 100m thickness and taking time steps of 0.5 years. For most of the calculations, the polar ocean temperature T_p was fixed at 1.2°C, but several runs were made in which the upwelling speed w was determined by the mixed-layer temperature T_m in order to test the effect of feedback through deep water formation.

Figure 7 shows the predicted response of the mixed layer and the upper two layers of the deep ocean to a sinusoidal variation in the solar constant with an amplitude of 1% and a period of 83 years. The amplitude of the temperature oscillation in the mixed layer is 0.8°C, lagging behind the forcing function by 6 years. The lags at depths of 100m and 200m below the mixed layer are 11 years and 18 years respectively, illustrating the rapid increase in thermal inertia with increasing depth.

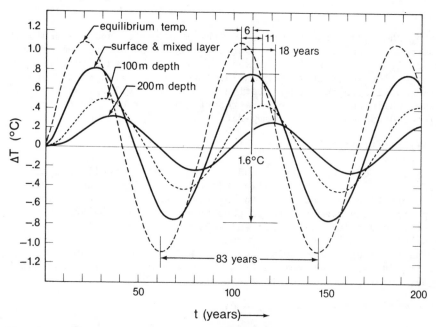

Fig. 7. Response of model ocean temperatures to variations in the solar constant with a period of 83 years and an amplitude of 1%. The mixed layer is 100m thick and the depths are measured from its base.

In order to test the assumed relationship between the envelope of the 11-year sunspot cycle and the solar luminosity against the observed global average SST, we must adopt a functional form for the relationship. We have assumed a simple linear form, i.e.

$$S(t) = S_o[1 + \beta\{N(t) - N_o\}] \tag{8}$$

where $S(t)$ is the time-dependent solar constant, $N(t)$ is the 11-year running mean sunspot number, and S_o and N_o are reference values, chosen to give reasonable values for modern SSTs in the model calculation. The values finally adopted were $S_o = 1362$ W m^{-2} and $N_o = 18.8$, the latter value being applicable to the so-called Dalton minimum of solar activity in the early 19th century. These values led to average SSTs of about 19.5°C during the mid-20th century, in good agreement with current estimates (Hoffert et al., 1980).

The model calculations were carried out for several values of the slope β of the luminosity versus sunspot-number envelope relation, and Figure 8 shows a comparison between the model SST and the annual values of global average SST (C.K. Folland, private communication) for $\beta = 1.08 \times 10^{-4}$, which gave a reasonable match to the two amplitudes. Since the sunspot-number envelope ranged from a minimum of about 33 near 1910 to a maximum of about 90 near 1960, the corresponding fractional range in solar luminosity is $\Delta S/S_o = \beta\Delta N \cong 0.6\%$, which is not an unreasonable variation to expect on a decadal time scale.

As in Figure 6, the agreement between the two curves in Figure 8 is not perfect, but it is close enough to suggest that the relationship is real. As a further illustration, Figure 9 shows a scatter plot of decadal averages of the data and the model results, the latter being expressed as departures from an arbitrary zero value chosen so as to make the averages for the data and the model agree over the

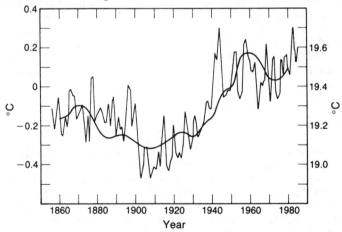

Fig. 8.   Response of model surface temperature (RH scale) to solar luminosity variations proportional to the solar-cycle envelope (heavy curve), and annual variation in global mean anomaly of observed SST (LH scale).

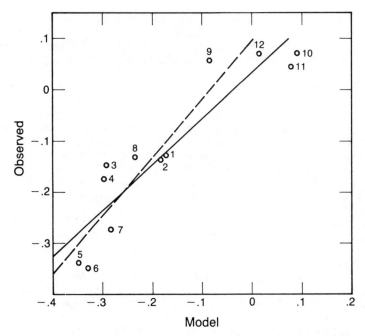

Fig. 9.   Decadal averages of anomalies in observed and model SST.
Points are numbered to show consecutive decades from
1 = 1861-70 to 12 = 1971-80.

three decades 1951-80.   The solid straight line is a least-squares
fit to the entire 1861-1980 period, while the broken line is a similar
fit to the data prior to 1941, i.e. excluding the four high points.
The two lines are reasonably similar, with slopes of 0.90 and 1.15
respectively, indicating that the assumed relationship has shown a
certain amount of stability over the period of 120 years.   The linear
correlation coefficient between the two time series is 0.92, but the
number of degrees of freedom is too small to give much confidence
in its significance.

5.   THE CO_2 CONTRIBUTION

The chief conclusion at this point is that the simple assumption of
a linear variation of solar luminosity with the envelope of the solar-
activity cycle over the past century or so leads to a calculated
variation of globally averaged SST that has a considerable amount
of similarity to the observed variation.   The required range in solar
luminosity of about 0.6%, of course, implies that there have been
no other factors affecting globally averaged SST during that time.
In fact, however, we know that the atmospheric CO_2 oncentration has
been increasing, and it seems reasonable to assume that the greenhouse
warming due to the additional CO_2 should have contributed to the

temperature change.

The model of Hoffert et al. (1980) assumes a logarithmic dependence of equilibrium temperature on CO_2 concentration, so that equation (3) becomes

$$BT_{eq} = (S/4)(1-\alpha) - A + C \ln \{c(t)/c_o\} \qquad (9)$$

where $c(t)$ is the CO_2 mixing ratio and c_o is a reference pre-industrial value. The value of the constant C was chosen to give an increase in T_{eq} of 2.5°C for a doubling of CO_2 concentration, in general accord with the results of model calculations (National Academy of Sciences, 1983) and allowing for uncertainties associated with the growing concentrations of other greenhouse gases (Ramanathan et al., 1985).

Figure 10 shows the effect of using equation (9) instead of equation (3) in the model, with the solar constant held fixed at 1370 W m^{-2}, and the time variation in atmospheric CO_2 taken from Hoffert et al.'s (1980) parameterization of the data of Keeling and Bacastow (1977). The "full CO_2" case refers to the value of C given above, i.e. corresponding to an increase of 2.5°C in T_{eq} for a doubling of CO_2, while the "half CO_2" case corresponds to an increase of 1.25°C. The predicted temperature rise between 1860 and 1980 is 0.45°C for the full CO_2 case, and 0.25°C for the half CO_2 case, both amounting to a substantial fraction of the observed temperature change. Clearly the shape of the observed temperature variation cannot be explained by the CO_2 effect unless some unidentified and highly non-linear effects are involved, but the rise in CO_2 concentration may help to account for the observed temperature rise between 1910 and 1960, thus reducing the amplitude of the solar-luminosity variation needed. The characteristic shape of the model temperature curve, which is apparently so successful in following the shape of the observed curve, has its maxima and minima damped out, however, as the CO_2 contribution is emphasized relative to the solar contribution.

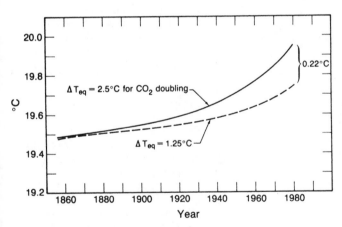

Fig. 10. Response of model surface temperature to increasing atmospheric CO_2 with a fixed solar constant.

Figures 11 and 12 show the model results for the full CO_2 case, again using (8) to describe the variation of solar constant with solar-cycle

240

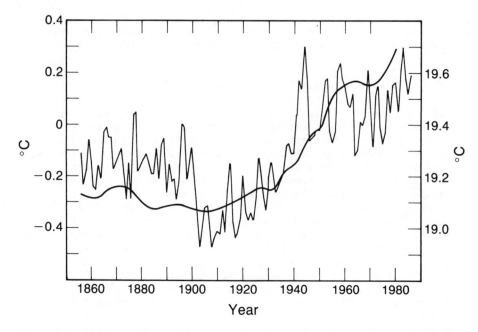

Fig. 11. Same as Figure 8, but with the effect of increasing
          atmospheric CO_2 included.

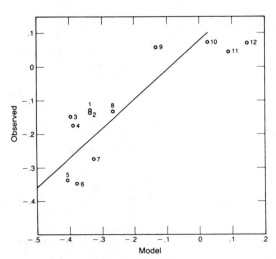

Fig. 12. Same as Figure 9, but
with the effect of increasing
atmospheric CO_2 included.

envelope. The best fit
in this case was obtained
with $\beta = 0.65 \times 10^{-4}$,
corresponding to a range
in solar luminosity of
0.4% over the 1860-1980
period instead of 0.6%.
Figure 12 shows that the
fit is not as good as in
the case with no CO_2 effect,
the correlation coefficient
between decadal averages
dropping from 0.92 to 0.85.
In general, the fit is
better the smaller the
CO_2 contribution, and is
reasonably good for the
half CO_2 case, with a rather
larger value for β. The
CO_2 warming by 1980 in
this case is closer to
that predicted by ocean
models other than the
advective-diffusive model of Hoffert et al. (1980) (e.g. Siegenthaler

and Oeschger, 1984).

&. DISCUSSION

Summarizing the results, we have shown that an apparent relationship between the height of the tropical tropopause and the sunspot number that had been pointed out by several authors did not hold good for the period following 1969, implying that any variation in solar luminosity accompanying the 11-year solar cycle had an amplitude considerably smaller than that needed to produce significant changes in tropical sea-surface temperatures, i.e. less than several tenths of a percent. During the 1953-68 period, however, which included the greatest peak in solar activity in recorded history, the 11-year component in solar luminosity may have exceeded that threshold.

At longer periods, the historical record of globally averaged SST since about 1855 was found to show a remarkable similarity to the envelope of the 11-year sunspot cycle, both having a principal period of 75-85 years. Calculations with a simple advective-diffusive ocean thermal model showed that the observed variation in SST could be produced by a variation in solar luminosity proportional to the envelope of the 11-year cycle and approximately in phase with it, with a range of less than 1%.

A periodicity of 70-100 years is a pervasive feature of solar and solar-terrestrial physics. It appears in the record of recent solar activity as the Gleissberg cycle (e.g. Sonett, 1982), and the existence of a pronounced 88-year period in medieval auroral records was reported by Feynman and Fougere (1984). A period of about 100 years was found in the Precambrian varves of the Elatina Formation in southern Australia (Williams and Sonett, 1985), which have been related to ancient solar activity, and an indication of a similar period has been reported in the thickness of recent varves in a glacial lake in Alaska (Sonett and Williams, 1985). Johnsen et al. (1970) found major periods of about 80 and 180 years in the oxygen isotope ratio in the Camp Century, Greenland, ice core, and as mentioned above Gilliland (1981) reported a 76-year cycle in the solar radius, based on transits of Mercury, solar eclipse reports, and meridian transit measurements over a period of 258 years. This latter result is of particular importance when combined with the SST measurements, since it carries with it the implication of a relationship between solar radius and luminosity, a much-debated issue in solar physics. Gilliland (1981) concluded that the 76-year cycle in solar radius was negatively correlated with the Gleissberg cycle in solar activity, i.e. maxima in radius occurred near minima in the Gleissberg cycle; the last maximum in radius occurred about 1911, close to the minimum in SST shown in Figure 6.

In a similar vein, Ribes et al. (1987) have presented evidence that the radius of the sun was significantly larger than its present value during the Maunder Minimum of solar activity in the late 17th century, and that it decreased when solar activity resumed in the early 18th century. As Eddy (1977) has argued, the coincidence between the Maunder Minimum and the peak of the Little Ice Age, and the rather less clear positive correlation between terrestrial temperatures and earlier solar activity as revealed by the C^{14} record, form a body of circumstantial evidence in support of a positive relationship between solar activity and solar luminosity on long time scales, to which the evidence presented here lends additional support. The solar radius measurements, which show an anticorrelation between radius and sunspot activity, then imply an anticorrelation between radius and luminosity. Much additional theoretical and observational work will be needed, however, to place this circumstantial evidence on a secure footing.

In summary, we have tried to present the point of view that the climate record over the past 12 decades or so does show evidence of variability in the sun's luminosity with a dominant period of about 80 years and an amplitude of a few tenths of a percent. Circumstantial and speculative as the evidence may be, we feel that it should not be ignored, since it carries implications for solar physics, as well as for our understanding of the forces that shape terrestrial climate, and thus ultimately for our ability to predict both natural and artificial climate change.

REFERENCES

Bjerknes, J.A., 1966, Tellus, 18, 820.

Cole, H.P., 1975, J. Atmos. Sci., 32, 998.

Eddy, J.A., 1977, Clim. Change, 1, 173.

Eddy, J.A., Gilliland, R.L. and Hoyt, D.V., 1982, Nature, 300, 689.

Feynman, J. and Fougere, P.F., 1984, J. Geophys. Res., 89, 3023.

Folland, C. and Kates, F., 1984, 'Milankovitch and Climate, Part 2', ed. A.L. Berger et al., D. Reidel Publishing Co., 721.

Folland, C.K., Parker, D.E. and Kates, F.E., 1984, Nature, 310, 670.

Fröhlich, C., 1987, J. Geophys. Res., 92, 796.

Gage, K.S. and Reid, G.C., 1981, Geophys. Res. Lett., 8, 187.

Garcia, R.R., Solomon, S., Roble, R.G. and Rusch, D.W., 1984, Planet. Space Sci., 32, 411.

Gilliland, R.L., 1981, Astrophys. J., 248, 1144.

Gilliland, R.L. and Schneider, S.H., 1984, Nature, 310, 38.

Harvey, L.D.D. and Schneider, S.H., 1985, J. Geophys. Res., 90, 2191.

Herman, J.R. and Goldberg, R.A., 1978, 'Sun, Weather and Climate', NASA SP-426, National Aeronautics and Space Administration.

Hickey, J.R., Stowe, L.L., Jacobowitz, H., Pellegrino, P., Maschhoff, R.H., House, F. and Vonder Haar, T., 1980, Science, 208, 281.

Hoffert, M.I., Callegari, A.J. and Hsieh, C.-T., 1980, J. Geophys. Res., 85, 6667.

Johnsen, S.J., Dansgaard, W., Clausen, H.B. and Langway, C.C., 1970, Nature, 227, 482.

Keeling, C.D. and Bacastow, R.B., 1977, 'Energy and Climate', Geophysical Study Committee, National Academy of Sciences, Washington, 72.

McCormac, B.M., 1983, ed., 'Weather and Climate responses to solar Variations', Colorado Associated University Press, Boulder, Colorado.

Miles, M.K. and Gildersleeves, P.B., 1977, Meteorol. Mag., 106, 314.

National Academy of Sciences, National Research Council,'Solar Variability, Weather, and Climate', Geophysics Study Committee, Washington, 1982.

National Academy of Sciences, National Research Council,'Changing Climate', Report of the Carbon Dioxide Assessment Committee, 1983.

Newell, R.E., Navato, A.R. and Hsiung, J., 1978, Pure Applied Geophys., 116, 351.

Paltridge, G. and Woodruff, S., 1981, Mon. Wea. Rev., 109, 2427.

Ramanathan, V., Cicerone, R.J., Singh, H.B. and Kiehl, J.T., 1985, J. Geophys. Res., 90, 5547.

Rasool, S.I., 1961, Pure Appl. Geophys., 48, 93.

Reck, R.A., 1976, Science, 192, 557.

Reid, G.C. and Gage, K.S., 1981, J. Atmos. Sci., 38, 1928.

Reid, G.C. and Gage, K.S., 1985, J. Geophys. Res., 90, 5629.

Ribes, E., Ribes, J.C. and Barthalot, R., 1987, Nature, 326, 52.

Schneider, S.H. and Mass, C., 1975, Science, 190, 741.

Siegenthaler, U. and Oeschger, H., 1984, Ann. Glaciol., 5, 153.

Sonett, C.P., 1982, Geophys. Res. Lett., 9, 1313.

Sonett, C.P. and Williams, G.E., 1985, J. Geophys. Res., 90, 12019.

Stranz, D., 1959, J. Atmos. Terr. Phys., 16, 180.

Weare, B.C., Navato, A.R. and Newell, R.E., 1976, J. Phys. Oceanogr., 6, 671.

Williams, G.E. and Sonett, C.P., 1985, Nature, 318, 523.

Willson, R.C., Gulkis, S., Janssen, M., Hudson, H.S. and Chapman, G.A., 1981, Science, 211, 700.

Willson, R.C., Hudson, H.S., Fröhlich, C. and Brusa, R.W., 1986, Science, 234, 1114.

THE SOLAR COMPONENT OF THE ATMOSPHERIC ^{14}C RECORD

Minze Stuiver and Thomas F. Braziunas
Department of Geological Sciences and Quaternary
  Research Center
University of Washington
Seattle, WA 98195, U.S.A.

ABSTRACT. Heliomagnetic modulation of the cosmic ray flux causes variations in annual ^{14}C production during the 11 yr solar cycle. Sunspot, auroral, and ^{10}Be records suggest that century type ^{14}C oscillations are also primarily influenced by the sun rather than climate. The history of decadal and bi-decadal ^{14}C production rates during the past 9700 years, as derived from the ^{14}C content in tree-rings using a carbon reservoir model, indicates at least two recurring patterns of solar change. Increases in ^{14}C production rate that peak at 30% and last ca. 280 years ("Spörer types") are repeated 8 times at irregular intervals while similar fluctuations that last only ca. 200 years ("Maunder types") are found 9 times. The decadal and bi-decadal records of atmospheric Δ^{14}C and model-derived ^{14}C production rates, with long-term trend removed, are displayed in 1000 yr intervals.

    The flux of cosmic ray particles arriving in the upper atmosphere is not constant. Yearly global ^{14}C production, as calculated from observed neutron fluxes, varies with the 11 yr solar cycle (O'Brien, 1979; Lingenfelter, 1963). Such modulation of the cosmic ray flux is tied to the magnetic properties of a solar wind that has variable particle distribution and velocity fields. Annual ^{14}C production rate changes of up to 25% are experienced during a strong solar cycle. The solar wind interaction with the cosmic rays is such that high ^{14}C production is encountered during a "quiet" sun (low sunspot number).
    A production rate change from one year to the next will not influence atmospheric ^{14}C content unduly because the decay rate of ^{14}C is a relatively small 1/8260 yr^{-1}. The small decay rate leads to a build up of a large ^{14}C reservoir on earth of 8260 yr of production (only then will decay balance production). The amount stored in the atmosphere is about 1.2% of the total, or a rounded 100 yr of ^{14}C production. A production rate change of 25% from one year to the next can only produce a perturbation in atmospheric ^{14}C level of 1/4 yr in a 100 yr reservoir, or 2.5‰. Even such a small change is an upper limit because exchange with other reservoirs (ocean and biosphere) takes place simultaneously.

F. R. Stephenson and A. W. Wolfendale (eds.),
Secular Solar and Geomagnetic Variations in the Last 10,000 Years, 245–266.
© 1988 by Kluwer Academic Publishers.

Past changes in atmospheric ^{14}C content can be detected in tree-rings. The cellulose carbon in the rings, formed from photosynthetic products, is derived from atmospheric CO_2 taken up during the year of formation. The cellulose $^{14}C/^{12}C$ ratio will not be identical to the air CO_2 $^{14}C/^{12}C$ ratio because the photosynthetic process discriminates against the heavier ^{14}C isotope. However, such tree specific behaviour can be corrected for by normalizing on a fixed $^{13}C/^{12}C$ ratio. The activity determination involves several steps: (a) a high precision measurement of tree-ring cellulose ^{14}C activity, (b) a normalization on constant $^{13}C/^{12}C$ ratio, (c) a correction for radioactive decay since the time of formation (for which the age of the ring has to be known) and (d) a comparison of the age corrected activity with the ^{14}C activity of a standard (NBS oxalic acid). The resulting $\Delta^{14}C$ term (Stuiver and Polach, 1977) represents atmospheric ^{14}C activity during the year of formation of the tree ring, expressed relative to the oxalic acid standard.

Atmospheric $\Delta^{14}C$ values change in response to geomagnetic modulation as well as solar modulation of the cosmic ray flux. A substantial portion of the observed long-term $\Delta^{14}C$ trend (from $\Delta^{14}C$ = + 9% about 9000 yr ago to 0% last century) can be explained by geomagnetic field intensity changes (Sternberg and Damon, 1983). A redistribution of ^{14}C in the carbon reservoirs, possibly induced by glacial-interglacial changes in ocean circulation, also may play a role (Siegenthaler et al., 1980). But whatever the cause, it is unlikely that the long-term trend in ^{14}C is associated with solar modulation.

Multiple evidence exists that the century type ^{14}C oscillations around the main trend are caused by heliomagnetic modulation of the cosmic ray flux. For instance, the maximum in $\Delta^{14}C$ near A.D. 1700 fits extremely well with the lack of sunspots experienced during the A.D. 1654-1714 Maunder Minimum interval (Stuiver and Quay, 1980; Eddy, 1976). Auroral evidence, and naked eye sunspot observations also are quite compatible with the $\Delta^{14}C$ history of the last millennium (Stuiver and Quay, 1980; Stuiver and Grootes, 1980). In addition, there is supporting evidence from the cosmogenically produced ^{10}Be isotope. The timing and magnitude of the ^{10}Be increases in ice cores agree with the features of the $\Delta^{14}C$ record (Beer et al., 1987).

As an alternative to solar modulation, climatic change could possibly play a role in atmospheric $\Delta^{14}C$ change by redistributing ^{14}C between the carbon reservoirs. These effects cannot be calculated because the extent of change in carbon reservoir parameters, induced by climatic changes such as those experienced during the Little Ice Age, is not known. But climate would have to play a devious role in that atmospheric $^{14}CO_2$ (a global indicator) and ^{10}Be in ice cores (a regional indicator) by coincidence would be altered to such an extent that the magnitude of change in both isotopes fully agrees with those calculated from solar modulation (see Beer et al., 1987). Furthermore, the ^{14}C and ^{10}Be excursions of the last 4000 years coincide in time quite well with each other whereas a relationship between climate and ^{14}C is either non-existent (Stuiver, 1980; Williams et al., 1980) or weak when the timing of climate change is accorded large errors (Wigley, 1988). Given the evidence currently available, we consider heliomagnetic modulation of the cosmic ray flux to be the primary causative factor of $\Delta^{14}C$ change on time scales of centuries. Of course, climate change could be responsible for second order effects.

Atmospheric $\Delta^{14}C$ level is dependent not only on the ^{14}C produced in the upper atmosphere, but also on the $^{14}CO_2$ exchange with terrestrial carbon reservoirs. To calculate ^{14}C production rates from the atmospheric record, the fluxes of ^{14}C between the atmosphere, ocean and biosphere must be determined. For this purpose we use the four reservoir, box-diffusion model of Oeschger et al. (1975). A description of our procedures has been given elsewhere (Stuiver et al., 1986b; Stuiver and Quay, 1980, 1981).

For the calculation of ^{14}C production rates we use both a 4500 yr decadal $\Delta^{14}C$ record (Stuiver and Becker, 1986) and a 9700 yr bi-decadal Δ^{14} C record derived from several sources (Kromer et al., 1986; Linick et al., 1985, 1986; Pearson et al., 1986; Pearson and Stuiver, 1986; Stuiver et al., 1986a; Stuiver and Pearson, 1986). This atmospheric $\Delta^{14}C$ record is rather precisely known, with a typical precision of 2 per mil back to 5200 B.C. and 5 per mil for the 5200-7700 B.C. interval. Removing the long-term trend by fitting a cubic spline to the record (the spline curve itself is closely approximated by a 400 yr moving average) yields the atmospheric $\Delta^{14}C$ fluctuations around the main trend. These fluctuations are given in 1000 yr intervals in Figure 1 for the decadal record and in Figure 2 for the bi-decadal composite record.

When calculating the production rate changes in the carbon reservoir model we assume variations in ^{14}C production rate Q to be responsible for the observed $\Delta^{14}C$ changes. All other parameters, such as oceanic eddy diffusion coefficient and ocean-atmosphere exchange rate, are kept constant. Once Q has been calculated, we remove the long-term trend Q_t by fitting a cubic spline to the Q record. The remaining Q variations, expressed as percent $\Delta Q/Q_t$ change, are given in Figures 3 and 4.

A change in any bi-decadal $\Delta^{14}C$ value affects the model-derived $\Delta Q/Q_t$ for a 140 year duration (due to, for example, biospheric feedback) with the strongest effect by far during the first couple of bi-decades. The typical precisions in $\Delta^{14}C$ cited above translate into contemporaneous $\Delta Q/Q_t$ uncertainties of 3% back to 5200 yr B.C., and of 6-8% for the 5200-7700 yr interval. Thus a bi-decadal $\Delta Q/Q_t$ value of 20% read from the graphs would be 20 ± 8% prior to 5200 B.C. and 20 ± 3% for the younger portion. A precision of 2‰ in the decadal $\Delta^{14}C$ record yields an uncertainty of 5-7% in contemporaneous decadal $\Delta Q/Q_t$ with a much reduced influence on the subsequent 100 years of model output.

The different responses of the model to the bi-decadal and decadal changes is interpreted here as reflecting the different data spacing involved. Changing an atmospheric ^{14}C level from what it was 20 years ago requires less adjustment in average production rate than would the same change over a period of 10 years. Thus the decadal model output is more sensitive to any change in the atmospheric input than is the bi-decadal output.

The $\Delta Q/Q_t$ parameter reflects solar change more faithfully than $\Delta^{14}C$ because the $\Delta^{14}C$ response to the Q input is frequency dependent.

Fast Q changes are drastically attenuated; slower ones much less. The Spörer and Maunder type of oscillation, as identified for the current millennium, recur often in the earlier record. In Figures 5 and 6 we have compiled nine oscillations of the Maunder, and eight of the Spörer variety. The only "subjective" choice made for the Figure 5 and 6 compilations was the start (yr) of an oscillation. The vertical scale is identical to the Figure 3 and 4 scale; no vertical adjustment was made.

There is a surprising degree of similarity within each class of oscillations (see Figures 5 and 6) as well as between average Spörer and Maunder prototypes. The average Maunder and Spörer oscillations both increase their production rate by 30 percent in 80 years, and both decline back to "normal" in 80 years. The basic difference is the duration of the maximum production rate which is short (about two decades) for the Maunder but lasts at least 60 years for the Spörer prototype (Figure 7).

The power spectrum of the Q record reveals the previously reported periods near 200 and 150 years (Stuiver, 1980) in addition to a distinct peak at 87 yr. The latter may well be representative of the Gleissberg cycle deduced from historical sunspot observations.

The main purpose of this paper is to give a graphical representation (see figures) of the solar features of the ^{14}C record. Our discussion is kept brief, and a more comprehensive analysis will be given elsewhere.

REFERENCES

Beer, J., Bonani, G., Suter, M., Wölfli, W., Siegenthaler, U., Oeschger, H. and Finkel, R.C., 1987, Nature, in press.
Eddy, J.A., 1976, Science 192, 1189.
Kromer, B., Rhein, M., Bruns, M., Schoch-Fischer, H., Münnich, K.O., Stuiver, M., and Becker, B., 1986, Radiocarbon 28, 954.
Lingenfelter, R.E., 1963, Rev. of Geophysics 1, 35.
Linick, T.W., Long, A., Damon, P.E. and Ferguson, C.W., 1986, Radiocarbon 28, 943.
Linick, T.W., Suess, H.E., and Becker, B., 1985, Radiocarbon 27, 20.
O'Brien, K., 1979, J. Geophys. Res. 84, 423.
Oeschger, H., Siegenthaler, U., Schotterer, U., and Gugelmann, A., 1975, Tellus 27, 168.
Pearson, G.W. and Stuiver, M., 1986, Radiocarbon 28, 839.
Pearson, G.W., Pilcher, J.R., Baillie, M.G.L., Corbett, D.M. and Qua, F., 1986, Radiocarbon 28, 911.
Siegenthaler, U., Heimann, M., and Oeschger, H., 1980, Radiocarbon 22, 177.
Sternberg, R.S. and Damon, P.E., 1983, Radiocarbon 25, 239.
Stuiver, M., 1980, Nature 286, 868.
Stuiver, M., 1982, Radiocarbon 24, 1.
Stuiver, M., and Becker, B., 1986, Radiocarbon 28, 863.
Stuiver, M., and Grootes, P.M., 1980, Proc. Conf. Ancient Sun, 165.

Stuiver, M., Kromer, B., Becker, B., and Ferguson, C.W., 1986a,
  Radiocarbon 28, 969.
Stuiver, M., and Pearson, G.W., 1986, Radiocarbon 28, 805.
Stuiver, M., Pearson, G.W., and Braziunas, T., 1986b, Radiocarbon 28,
  980.
Stuiver, M., and Polach, H.A., 1977, Radiocarbon 19, 355.
Stuiver, M., and Quay, P.D., 1980, Science 207, 11.
Stuiver, M., and Quay, P.D., 1981, Earth and Planetary Science Letters
  53, 349.
Wigley, T.M.L., 1988, This volume.
Williams, L.D., Wigley, T.M.L., and Kelly, P.M., 1980, 'Sun and Climate',
  CNES, France, 11.

The Figures referred to in this paper are to be found on the following
pages.

250

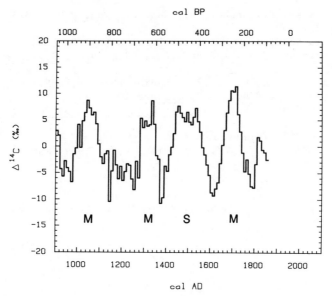

Fig. 1a
Figs. 1 a-e   A 4500 yr record of decadal atmospheric (tree-ring)
Δ^{14}C after removal of the long-term trend as approximated by a
spline similar to a 400 yr moving average. M = Maunder type
fluctuation, S = Spörer type fluctuation (see text).

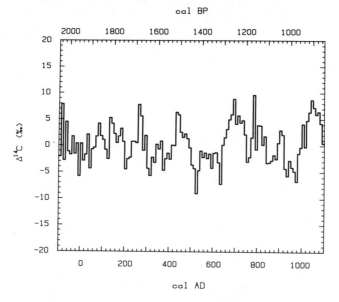

Fig. 1b (see Fig. 1a for caption)

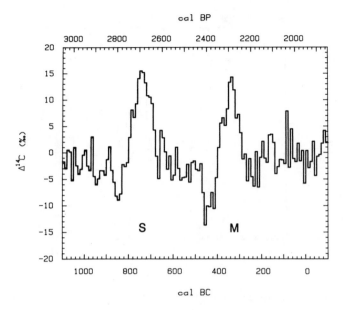

Fig. 1c (see Fig. 1a for caption).

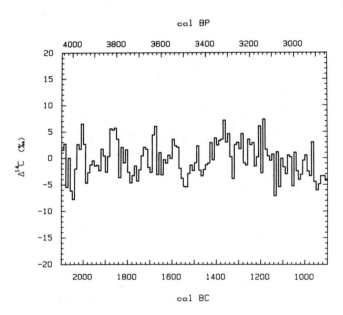

Fig. 1d (see Fig. 1a for caption).

252

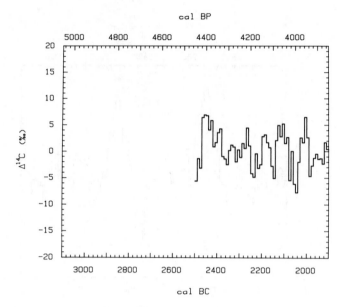

Fig. 1e (see Fig. 1a for caption)

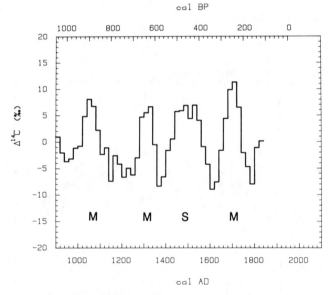

Fig. 2a
Figs. 2 a-j. A 9700 yr record of bi-decadal atmospheric (tree-ring)
$\Delta^{14}C$ after removal of the   long-term trend as in Fig. 1. Symbols
as in Fig. 1.

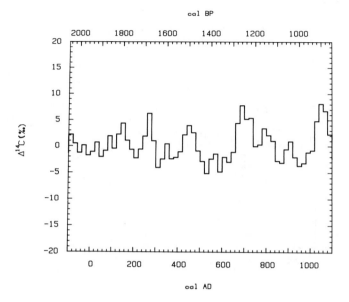

Fig. 2b (see Fig. 2a for caption)

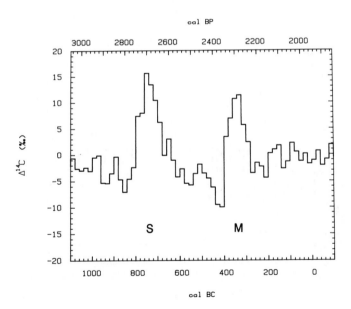

Fig. 2c (see Fig. 2a for caption)

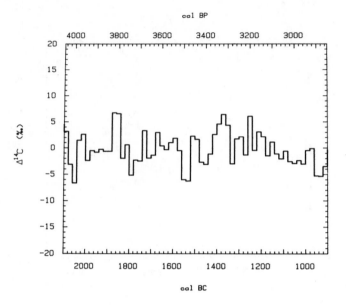

Fig. 2d (see Fig. 2a for caption)

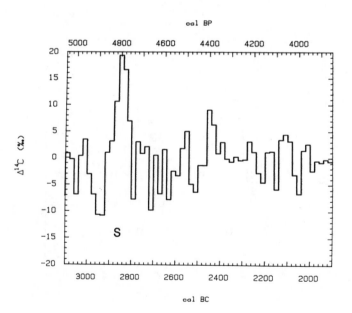

Fig. 2e (see Fig. 2a for caption)

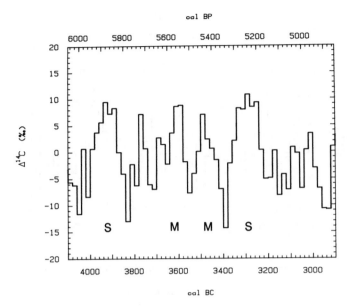

Fig. 2f (see Fig. 2a for caption)

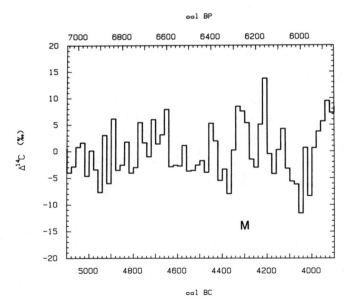

Fig. 2g (see Fig. 2a for caption)

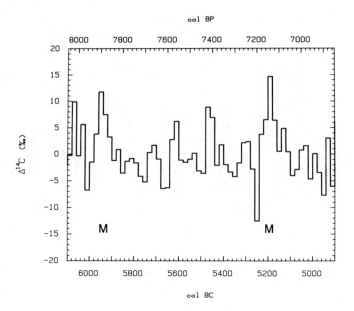

Fig. 2h (see Fig. 2a for caption)

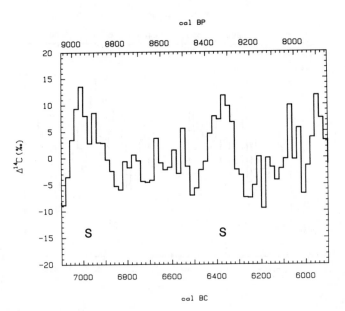

Fig. 2i (see Fig. 2a for caption)

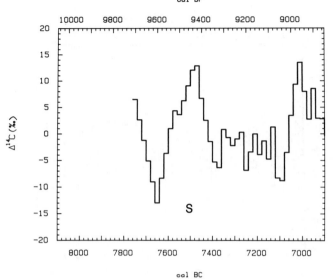

Fig. 2j (see Fig. 2a for caption)

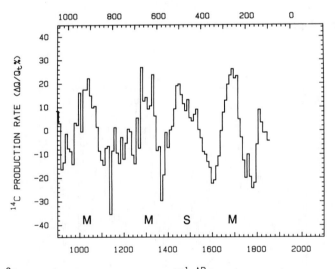

Fig. 3a

Figs. 3 a–e. A 4500 yr record of decadal model-derived ^{14}C production rates Q after removal of the   long-term trend as approximated
by a spline similar to a 400 yr moving average. The ^{14}C production
rate units are percent $\Delta Q/Q_t$ where ΔQ is the difference between the
time-specific ^{14}C production rate Q and the concurrent value Q_t for
the long-term trend in ^{14}C production as represented by the spline.
M = Maunder type fluctuation, S = Spörer type fluctuation (see
text).

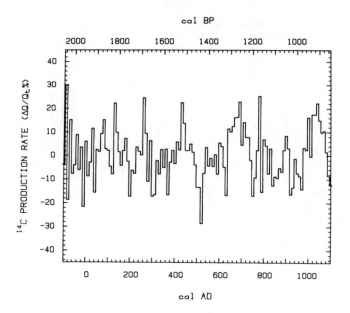

Fig. 3b (see Fig. 3a for caption)

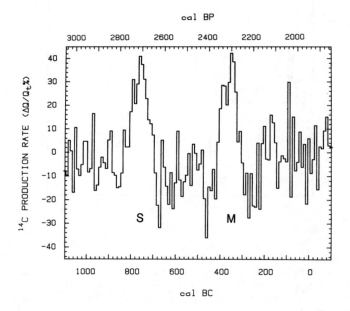

Fig. 3c (see Fig. 3a for caption)

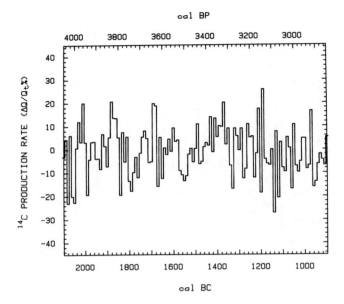

Fig. 3d (see Fig. 3a for caption)

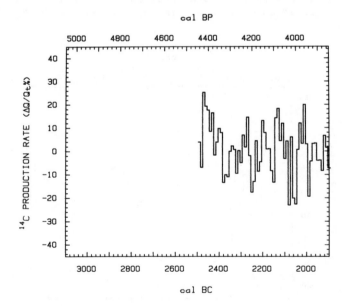

Fig. 3e (see Fig. 3a for caption)

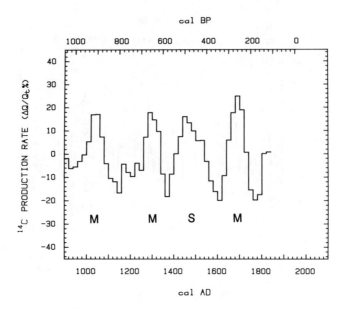

Fig. 4a
Figs. 4 a–j. A 9700 yr record of bi-decadal model-derived ^{14}C production rates as defined in Fig. 3. Symbols as in Fig. 3.

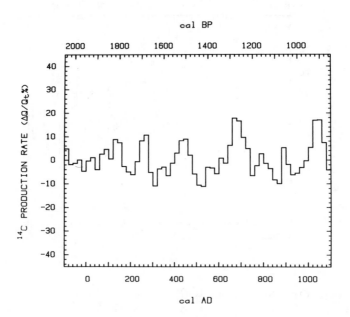

Fig. 4b (see Fig. 4a for caption)

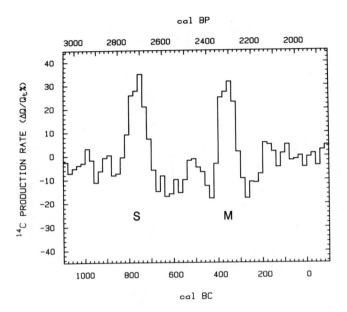

Fig. 4c (see Fig. 4a for caption)

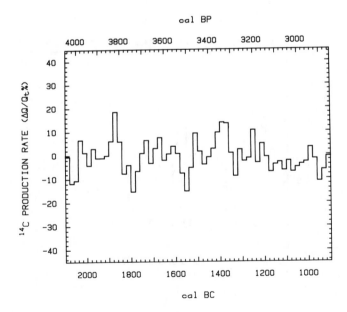

Fig. 4d (see Fig. 4d for caption)

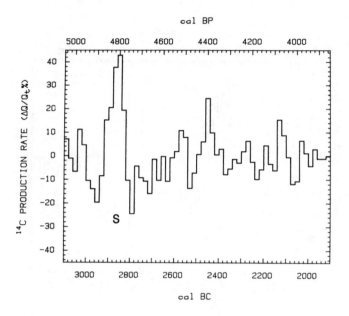

Fig. 4e (see Fig. 4a for caption)

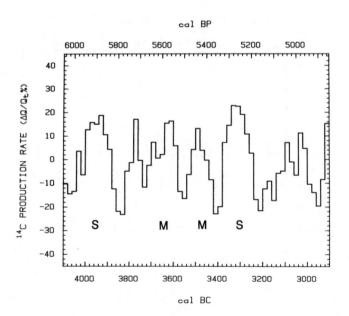

Fig. 4f (see Fig. 4a for caption)

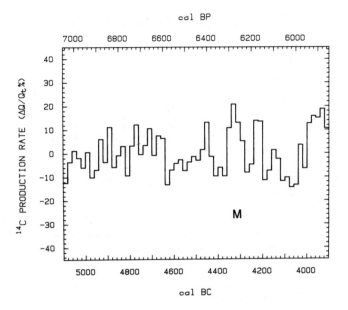

Fig. 4g (see Fig. 4a for caption)

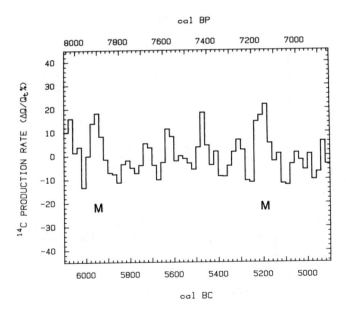

Fig. 4h (see Fig. 4a for caption)

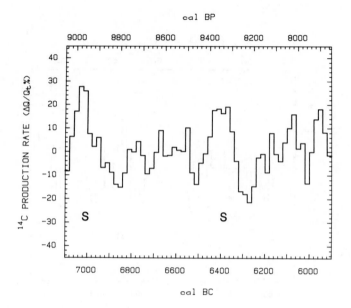

Fig. 4i (see Fig. 4a for caption).

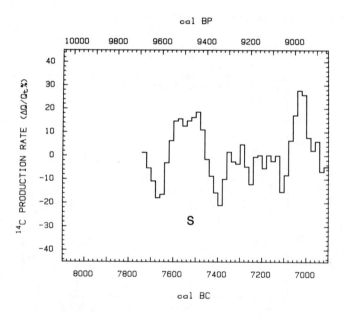

Fig. 4j (see Fig. 4a for caption).

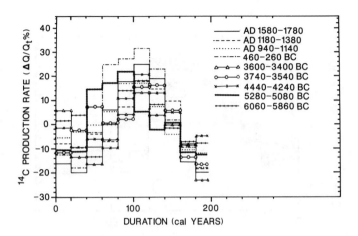

Fig. 5. Maunder type fluctuations in ^{14}C production rate (see Figure 3 legend) as taken directly from Figure 3. The corresponding 200 yr calendar periods for the fluctuations are also listed.

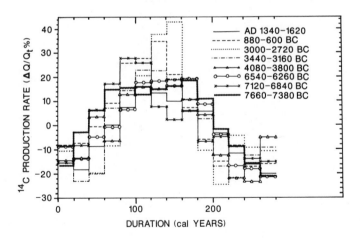

Fig. 6 Spörer type fluctuations in ^{14}C production rate (see Figure 3 legend) as taken directly from Figure 4. The corresponding 280 yr calendar periods for the fluctuations are also listed.

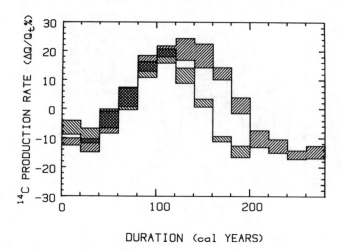

Fig. 7 The average Maunder type and Spörer type oscillation in
^{14}C production rate (see Figure 3 legend) as derived from the
individual fluctuations displayed in Figures 5 and 6. The shaded
areas represent the standard deviations in the means.

PRODUCTION AND DECAY OF RADIOCARBON AND ITS MODULATION BY GEOMAGNETIC
FIELD–SOLAR ACTIVITY CHANGES WITH POSSIBLE IMPLICATIONS FOR GLOBAL
ENVIRONMENT

Paul E. Damon
Laboratory of Isotope Geochemistry
Department of Geosciences
Gould-Simpson Building No. 77
University of Arizona
Tucson, Arizona USA 85721

ABSTRACT. Seven estimates of the global ^{14}C production rate at solar
minimum (1965, S=15) and solar maximum (1969, S=106) were published
between 1970 and 1980. Six of these are is good agreement at
2.47±0.19 (σ, S=15) and 2.02±0.13 (σ, S=106). The seventh estimate is
22% lower. Nevertheless, modelers of ^{14}C fluctuation have used this
lowest production rate because it is consistent with their estimates
of the ^{14}C inventory. Standard models for the global ^{14}C cycle do not
include sedimentary reservoirs because the flux to sediments is considered to be negligibly small. However, ^{14}C accumulates in sediments
over its 8270-year mean life and constitutes a significant part of the
total pre-anthropogenic ^{14}C inventory (24%) and should be included in
^{14}C fluctuation models. Addition of sedimentary sink reservoirs
yields global ^{14}C decay rates that are consistent with the six higher
estimates of global ^{14}C production rates. The carbon flux to sediments is being anthropogenically enhanced and may account for a
significant part of the "missing" carbon in the 20th century global
carbon cycle.
    The atmosphere responds to and integrates changes in the global
rate of ^{14}C production with a lag time of only a few years. Unfortunately other proxy indicators of the geophysical environment, such as
climate and geomagnetism, are strongly biased by regional variations. Considering this limitation, the archaeomagnetic record is
consistent with long-term modulation of ^{14}C production by changes in
the Earth's magnetic field yielding the envelope of the $\Delta^{14}C$ temporal
fluctuation curve. The envelope in turn is modulated by solar activity with a strong 200-year quasi-cyclic "wiggle" component. The most
intense Maunder minimum-type "wiggles" occur every 2100 to 2400
years. These wiggles, unlike the wiggles of lesser intensity, appear
to be independent of the strength of the Earth's dipole moment.
Results of the NASA Solar Maximum Mission indicate a direct linear
relationship between the total solar irradiance and sunspot number
with a decrease in luminosity of 0.1% over the five-year period from

F. R. Stephenson and A. W. Wolfendale (eds.),
Secular Solar and Geomagnetic Variations in the Last 10,000 Years, 267–285.
© *1988 by Kluwer Academic Publishers.*

solar maximum to solar minimum. This is consistent with the previously suggested correlation of climate with $\Delta^{14}C$ and sunspots. However,
confirmation requires a longer record of total solar irradiance and
integration of proxy climate records on a reasonable approximation to
global coverage.

1. INTRODUCTION

Following a generation of research, it is now well known that the
production of radiocarbon and other cosmogenic isotopes have been
measurably modulated by changes in the Earth's geomagnetic field and
solar activity. We have reviewed this subject elsewhere (Damon et
al., 1978) to which the reader is referred for a more extensive
account. Briefly it is a necessary consequence of the interaction of
charged particles with a magnetic field that an increase in the geomagnetic field will reduce entry of cosmic rays into the Earth's
atmosphere where they result in the production of the cosmogenic
isotopes. Radioactive isotopes are produced directly by spallation or
by the neutrons released during spallation. Radiocarbon is produced
by an $^{14}N(n,p)^{14}C$ reaction and its production rate Q is proportional
to $M(t)^{-\alpha}$ where $M(t)$ is the geomagnetic dipole moment and $\alpha=0.52$
(Elsasser et al., 1956).

There is an empirical inverse relationship between the global
neutron flux and solar activity. During the 11-year solar cycle, when
solar activity is a maximum, the neutron flux is a minimum. Because
the neutron flux produces ^{14}C, its production rate is also modulated
by solar activity. The theoretical basis for solar activity modulation of cosmic rays is still not fully understood (Lin, 1977).
However, the inverse relationship between ^{14}C production and solar
activity during the 11-year cycle also holds for the various solar
minima such as the Maunder, Spörer and Wolf minima (Eddy, 1976a,
1976b; Stuiver and Quay, 1980) which are quasi-periodic with a ca.
200-year period (Suess, 1980; see Figure 1, this paper).

Changes in reservoir constants resulting from climate change is
another mechanism that can modulate atmospheric ^{14}C concentration.
This does not seem to be important during most of the Holocene (Damon,
1970). However, there is now evidence from the Greenland and
Antarctic ice caps for a large variation in the concentration of
atmospheric carbon dioxide during the glacial-interglacial transition. The general trend shows low CO_2 concentration during the
glacial period and higher values during the post-glacial. The
increase is about 50% (Berner et al., 1980). This effect alone would
result in a 2.5% increase in $^{14}C/C$ ratio during the glacial period
(Siegenthaler et al., 1980). Expected changes in ocean circulation
could cause a larger effect. Consequently, the effect of climate
change on atmospheric ^{14}C concentrations will compete with changes in
production rate to modify atmospheric ^{14}C concentrations during the
transition from glacial to interglacial. This makes the interpretation of the proxy record of ^{14}C variations contained in the carbon
isotope content of dendrochronologically dated tree rings more complex

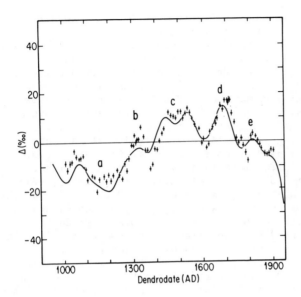

Figure 1. Δ^{14}C fluctuations during the last millennium. The data points are the high precision (±2‰) measurements of Stuiver and Quay (1980). The curve was produced by Fourier analysis of the residuals around the sixth-order logarithmic trend curve based on a compendium of medium precision data from an N.S.F. workshop [Klein et al., 1980]. The minimum of Δ^{14}C data occurs during the Medieval Warm Epoch (a) when solar activity was high. The episodes of high Δ^{14}C occur during periods of low solar activity (b = Wolf minimum; c = Spörer minimum; d = Maunder minimum; e = Dalton minimum).

than a simple record of production rate as the Holocene-Würm boundary is approached.

The intent of this paper is to demonstrate that global production of radiocarbon can be reconciled with its global decay rate; to discuss the implications of this balance in linear system modeling of natural radiocarbon; and to discuss changes in atmospheric ^{14}C resulting from changes in the Earth's magnetic field and solar activity. I will also discuss the possible relationship between atmospheric radiocarbon and global climate.

2. PRODUCTION AND DECAY OF RADIOCARBON

As previously mentioned, ^{14}C is produced by the secondary neutron flux resulting from spallation of atmospheric isotopes. These neutrons are almost quantitively involved in interaction with ^{14}N to produce radiocarbon plus tritium in the reactions $^{14}N(n,p)^{14}C$ and $^{14}N(n,T)^{12}C$. There were seven different calculations of the production rate of ^{14}C during the decade from 1970 to 1980. The estimated production rates for the 1965 solar minimum and 1969 solar maximum are shown in Table 1. Six of the calculations are in good agreement and average 2.47 ± 0.19 ^{14}C atoms/s cm_e^2 for solar minimum and 2.02 ± 0.13 ^{14}C atoms/s cm_e^2 for solar maximum. Surprisingly, the lowest estimate by O'Brien (1979) is commonly used in modeling ^{14}C in the carbon cycle (e.g., Stuiver and Quay, 1980) because it is in agreement with the decay rate as estimated from ^{14}C inventories (e.g., Siegenthaler, 1985). Siegenthaler calculated a decay rate of 1.73 dps cm_e^{-2}. We obtained a higher value, 2.01 dps cm_e^{-2}, and suggested that the global decay rate was compatible with the higher estimates of the production rate (Damon et al., 1982). The difference lies in the amount of ^{14}C contained in the sedimentary reservoir. According to Siegenthaler et al. (1980), "sedimentation is neglected as well as other minor fluxes, because it removes only about 0.5×10^{-5} parts of the total oceanic amount of carbon (or ^{14}C) per year." However ^{14}C accumulates in sediments during its mean life of 8,270 years and it can be shown that sediments contain a significant part of the total global inventory.

Table 1. Comparison of radiocarbon production calculations (atoms/sec cm_e^2).

Year	Solar Max+Min	Merker (1970)	L&R (1970)	Light et al. (1973)	Povinec (1977)	O'Brien (1979)	K&M (1980)	C&L (1980)
1965	15.1	2.15	2.40	2.42	2.6	1.91	2.58	2.66
1969	105.6	1.86	2.18	1.93	2.1	1.60	1.93	2.12

L&R: Lingenfelter and Ramaty
K&M: Korff and Mendell
C&L: Castagnoli and Lal

Past changes in atmospheric ^{14}C concentrations ($^{14}C/C$) are evaluated by measuring the carbon isotopic content of dendrochronologically dated tree rings. The ^{14}C concentration is corrected for radioactive decay and isotopic fractionation and compared with mid-18th century wood. The per mil (‰) difference is referred to as $\Delta^{14}C$. Figures 1 and 2 are plots of $\Delta^{14}C$ for different periods of time. The trend of $\Delta^{14}C$ excluding short-term fluctuations can be approximated by a

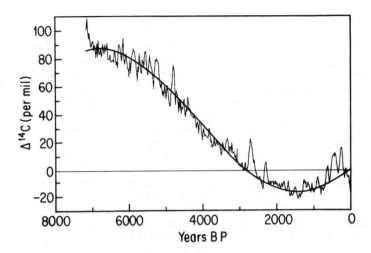

Figure 2. High precision data [±2‰; Stuiver and Pearson, 1986; Pearson and Stuiver, 1986; Pearson et al., 1986] for dendrochronologically dated, 20-year tree-ring intervals. The equation of the curve is: $\Delta^{14}C(‰) = 35 + 51 \sin (5.826 \times 10^{-4} t - 2.401)$ corresponding to a period of 10,780 years with a phase lag of 2.401 radians or 137.6°.

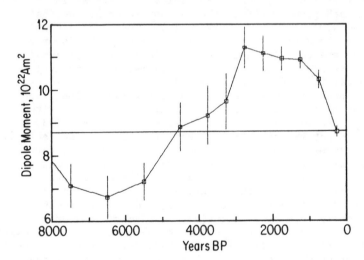

Figure 3. Archaeomagnetic determination of the strength of the Earth's dipole moment. The data are from Table 4.2 of Merrill and McElhinny (1983). The vertical bars are 95% error. The ^{14}C time scale was calibrated by the above authors using an older smoothed calibration scheme to 6500 BP (Clark, 1975) and a linear interpolation to an assumed ^{14}C-true age convergence at 10,000 ^{14}C years.

sinusoidal curve with a period of about 11,000 years back to 6,500 BC (Bruns et al., 1983; Damon and Linick, 1986) but may deviate from a sinusoidal curve prior to that time (Kromer et al., 1986). Since the Twelfth Nobel Symposium held in Uppsala, Sweden in August 1969, this long term trend of $\Delta^{14}C$ has been attributed to secular variation of the strength of the geomagnetic dipole moment resulting in changes in the rate of production of ^{14}C (Olsson, 1970). ^{14}C integrates changes in production rate resulting from changes in the dipole moment on a global basis whereas individual archaeomagnetic measurements are strongly affected by the non-dipole components. Figure 3 shows a plot of a synthesis of archaeomagnetic data by McElhinny and Senanyake (1982). Generally, it appears that the dipole moment field strength was lower than at present prior to the 5th century BC and higher afterwards. Thus, the ^{14}C production rate, $Q(t)$, would have been higher prior to the 5th century BC and lower more recently. The global ^{14}C inventory can be calculated from the following relationship (Sternberg and Damon, 1979; Damon et al., 1982):

$$I = - \eta \int_{\infty}^{0} Q(t) \ e^{-\eta t} \ dt \qquad (1)$$

where I (inventory) is the global decay rate, η is the ^{14}C decay constant, and $Q(t)$ is the production rate as a function of time, t. As mentioned previously, $Q(t)$ is proportional to $M(t)^{-\alpha}$ where $\alpha = 0.52$. The function $M(t)^{-\alpha}$ is asymmetric. A decrease in the dipole moment has a much greater effect than a corresponding increase (Sternberg and Damon, 1979, Figure 12). Thus, fortuitously, the higher earlier production rates offset the more recent lower production rates and the calculated inventory I is not significantly different than the average production rate during recent solar cycles which is approximately 2.2 ^{14}C aps cm_e^{-2} or 132 apm cm_e^{-2}.

There is general agreement concerning the radiocarbon inventory of pre-anthropogenic ambient reservoirs, i.e., reservoirs with transfer times that are much less than the half-life of ^{14}C (5,730 years). The inventory for ambient reservoirs is given in Table 2. Note that the summed inventory is about 20% less than the current average production rate. If the model from which I in equation 1 was estimated is correct, the difference must lie either in the calculation of production rates or in a missing component of the inventory. The missing component is the controversial sedimentary reservoir.

Much attention has been given in recent years to storage of carbon in sediments because of the need to account for the "missing" anthropogenic carbon resulting from the combustion of fossil fuels and agricultural practices such as the clearing of forests. I have taken advantage of this flourishing literature to estimate the ^{14}C inventory of sediments (Table 3). This estimate differs from our earlier estimates in the addition of organic matter in the coastal wetlands which are now considered to be as important as the marine carbonate reservoir. Much of the carbon storage in this reservoir is in the form of macrophyte (large marine plants such as eel grass, kelp and intertidal marsh grasses). A higher fraction of macrophyte than phytoplankton escapes oxidation and is buried in marsh and other

Table 2. Radiocarbon inventory of pre-anthropogenic ambient reservoirs.

Reference	Reservoir	C in Reservoir g/cm_e^2	[14]C Activity of Reservoir dpm/gC	Reservoir Decay Rate dpm/cm_e^2
1	Atmosphere	0.116±0.002	14.1±0.5	1.64±0.06
2	Terrestrial bio-sphere (rapid, vascular, hetero-troph and litter)	0.172±0.020	13.6±0.6	2.34±0.29
2	Soil humus	0.393±0.029	13.2±0.6	5.19±0.45
2	Hydrosphere, mixed layer of ocean (0-75 m)	0.122±0.005	13.4±0.5	1.63±0.09
2	Hydrosphere, inter-mediate and deep ocean	7.420±0.030	11.8±0.5	87.56±3.73
2	Saprosphere (total organic matter in oceans)	0.274±0.078	13.3±0.5	3.64±1.05
3	Fresh surface water	0.0002±0.001	13.0±0.6	0.03±0.02
		$\Sigma = 8.50±0.14$		$\Sigma = 102±4$

1. Oeschger et al. (1985), 2.80±5 ppm; 2. Olson et al. (1985);
3. Kemp (1979)

coastal sediments (Baes et al., 1985) or transported to the shelf area (Walsh, 1984). Carbon which is rapidly cycled is not included in the inventory. The inventory only includes carbon which is stored for a sufficiently long time to allow most of the ^{14}C to decay. Eventually, the bulk of the stored carbon is returned to the surface through diagenesis before accumulating as sedimentary rock.

The combined inventory is given in Table 4. The estimated global decay rate, 135±8 dpm/cm$_e^2$ is close to the estimated production rate, 132±12 apm/cm$_e^2$. The marine hydrosphere accounts for most of the inventory with a third of the inventory contained in all other reservoirs. The sedimentary reservoir which is frequently neglected accounts for 24% of the total inventory. O'Brien's (1979) estimate of production is nearly equal to the ^{14}C decay rate in the ambient reservoir. Hence, as will be shown, it is preferred by modelers who neglect the ^{14}C stored in sediments.

Table 3. Radiocarbon inventory in buried sediment (long-term storage).

Refer-ence	Reservoir	C accumulated during mean life of ^{14}C g/cm_e^2	^{14}C Activity of C flux dpm/gC	Reservoir Decay Rate dpm/cm_e^2
1	Continental shelves (carbonate)	0.033±0.016	13.4±0.6	0.44±0.21
1	Continental slopes and deep ocean (carbonate)	0.853±0.178	13.4±0.6	11.43±2.86
2	Continental shelves (organic)	0.178±0.089	13.4±0.6	2.38±1.20
2	Continental slopes and deep ocean (organic)	0.026±0.013	13.4±0.6	0.35±0.18
3	Coastal wetlands, lagoons, marshes, and deltas	1.135±0.378	13.2±0.6	14.98±5.95
4	Freshwater lakes and wetlands	0.24±0.03	13.4±0.6	3.22±0.42
		Σ=2.46±0.42		Σ=32.86±6.7

1. Hay (1985); 2. Berner (1982); 3. Baes et al. (1985); 4. Kemp (1979)

Table 4. Distribution of radiocarbon in terrestrial nature.

Reservoir	Decay Rate dpm/cm_e^2	Relative Amount %
Hydrosphere (mixed, intermediate, and deep ocean)	89.2	66.2
All other ambient reservoirs	12.8	9.5
Coastal wetlands (lagoons, bays, marshes, and deltas)	15.0	11.1
Carbonate in marine sediments	11.9	8.8
All other sediments	6.0	4.4
		100.0

$$\Sigma = 135\pm8 \ dpm/cm_e^2$$

3. MODELING THE ^{14}C CYCLE

Figure 4 presents various published carbon reservoir models of varying complexity and analogy to the natural, external, solar driven cycle of carbon. These are the three-box, first-order exchange model of Houtermans et al. (1973); the six-box, first-order exchange model of Bacastow and Keeling (1973) which for most purposes can be converted into a five-box model by combining the stratosphere and troposphere into a single atmospheric reservoir; and the box-diffusion model of Oeschger et al. (1975). Most modelers prefer the box-diffusion model as the better analog to nature. Each model is driven by a production function, $Q(t)$. The carbon content of each reservoir and exchange rates between reservoirs must be estimated and a set of differential equations formulated for each model. All the models are linear systems and can be analyzed by Fourier transforms.

I will limit the discussion to the significance of the global production and decay rates in the modeling process. In reservoir modeling, the production rate, $Q(t)$, and the atmospheric ^{14}C content, N_a^*, can be separated into steady state (DC) and variable parts:

$$Q(t) = \overline{Q} + \Delta Q(t) \tag{2}$$

$$N_a^*(t) = \overline{N}_a^* \pm \Delta N_a^*(t) \tag{3}$$

and

$$\overline{N}_a^* = g\overline{Q} \tag{4}$$

where g is the DC gain of the system given by equation 4. For strictly valid modeling of the ^{14}C content of the atmosphere, the variable component should fluctuate around a steady state value during the period of time to be modeled. Thus the DC gain calculated from equation 4 must be equal to the DC gain of the model. If the steady state production rate, $\overline{Q}$, is not equal to the sum of the decay rates for all of the reservoirs, the model will not be balanced and will seek a new steady state for $\overline{N}_a$ and other parameters (Figure 5).

The standard models do not include the sedimentary sink reservoirs that have been added in figure 4. In order to balance the model, either $\overline{Q}$ must be decreased or more ^{14}C added to the model. If my analysis of the ^{14}C inventory is correct, a closer analogy to nature is provided by adding a sedimentary reservoir to achieve a steady state. This has the effect of draining off the excess production rate that occurs without the sedimentary sink in figure 4. Figure 6 shows that the models with sedimentary sinks behave better as analogs to the 18th and 19th century atmospheric ^{14}C fluctuations.

When relative values are used, the existence of an unbalanced model may not be immediately apparent if the resulting $\Delta^{14}C$ values are forced to pass through an arbitrary reference point as in Figure 7 where Backastow and Keeling (1973) have forced the $\Delta^{14}C$ curve to pass through zero at AD 1890. Thus the existence of a floating N_a^* and reference point is obscured by the use of relative values. The authors noted that they found poor correspondence between the measured "wiggles" and the modeled "wiggles." Their model appears to attenuate

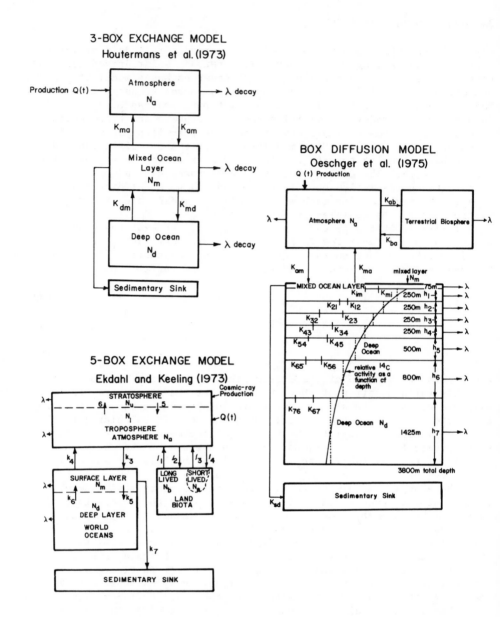

Figure 4. C reservoir models. The original models do not include a sedimentary sink.

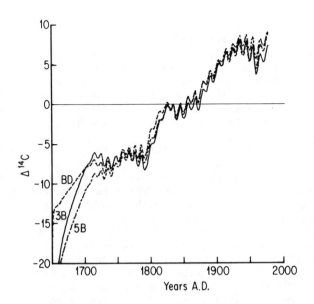

Figure 5. Performance of unbalanced models that do not include a sedimentary sink. Standard reservoir parameters are used with the Lingenfelter and Ramaty (1970) ^{14}C production rate, hence production exceeds decay. The models are not constrained to pass through an AD 1890 reference point.

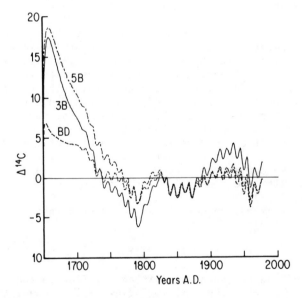

Figure 6. Above models with a sedimentary sink added to balance decay with the Lingenfelter and Ramaty (1970) production rate.

by a factor of five. However, their model, as parameterized, is unbalanced. They used the Lingenfelter and Ramaty production function and standard reservoir parameters. Thus, production of ^{14}C exceeds decay and the model must seek a new steady state. When a sedimentary sink is added to balance production and decay rates, the model behaves more in accord with observation as can be seen by comparing figure 7 with figure 1. The balanced model shows the "wiggle" at AD 1820 that follows the solar activity minimum (Dalton minimum) at AD 1810 with a predictable phase lag.

Figure 8 shows the Fourier transform of the gain for the three models. The DC gain is the plateau value at long periods ($>10^5$ years). The attenuation is obtained by dividing the gain at any frequency by the DC gain. Thus the DC gain acts as a scaling factor and the attenuation is one-third less with a sedimentary sink. This, however, does not alone explain the observed attenuation by a factor of five. The 33% greater attenuation only holds if the models without sedimentary sinks are balanced by using a lower production rate and hence the use of O'Brien's (1979) relatively low production rate by modelers. The factor of five attenuation must result from the fact that the unbalanced model is no longer linear and cannot be separated into steady state and variable parts. The system is not fluctuating around a steady state but must seek a steady state and as a result overattenuate other variable components. In other words, the Fourier transform is not valid for an unbalanced system.

4. THE MISSING SINK FOR ANTHROPOGENIC CARBON

The DC gain contains useful information concerning the carbon cycle. For example, with reference to equation 4, for modeling solar activity fluctuations, we can estimate $\bar{N}_a$ to within a few %. If we have an accurate value for the production rate, we can estimate the observed DC gain. This then provides information concerning the total ^{14}C in the carbon cycle, the ^{14}C stored in sediments, and the total carbon flux to the sedimentary reservoir. We have seen that the production rate of ^{14}C (Table 1) can be reconciled with the global inventory (Table 4). This requires a sedimentary carbon sink of 2.46 ± 0.42 g/cm^2_e during the mean life of ^{14}C and a carbon flux of 1.52 ± 0.26 gty^{-1} (1×10^{15} gy^{-1}) to the sedimentary reservoir. It is of interest to compare this flux with the missing flux of anthropogenic carbon (Broecker et al., 1979).

This missing flux may be calculated for appropriate values from the following equations:

$$F_o + F_a + F_m = F_f + F_b \qquad (5a)$$

That is the sum of the net anthropogenic annual flux to the oceans (F_o), atmosphere (F_a) plus the "missing" flux (F_m) is equal to the sum of the annual increment due to fossil fuel consumption (F_f) and land conversion (F_b). Solving for F_m:

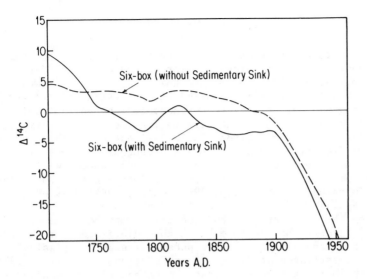

Figure 7. Bacastow and Keeling (1973) six-box model with and without sedimentary sink. The sedimentary sink was added to balance the production rate of Lingenfelter and Ramaty (1970) with the model decay rate. Note that the peak at AD 1820 resulting from the Dalton solar minimum was attenuated by a factor of five in the unbalanced model. The 11-year cycle (not shown here) is also attenuated in the unbalanced model. The comparison six-box model with sedimentary sink is a composite of solar modulation input to the six-box model to AD 1870 with smoothed measured data from AD 1870 to AD 1950. (The industrial or Suess effect was not modeled for the six-box model with sedimentary sink).

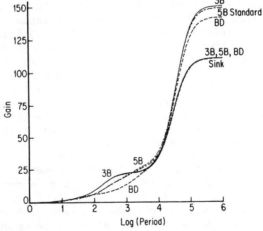

Figure 8. Gain of models with and without a sedimentary sink. The plateau at long periods is the steady state or DC gain. The attenuation is the gain at any frequency divided by the DC gain which acts as a scaling factor. The gain at higher frequencies is model dependent but independent of DC gain.

$$F_m = F_f + F_b - F_o - F_a \qquad (5b)$$

The authors of a recent attempt to evaluate this equation concluded that "we currently need a missing sink (or sinks) of 0.6×10^{15} to 3.4×10^{15} gC/year around 1980 to balance the global cycle" (Trabalka et al., 1985). This is similar in magnitude to the estimated 1.5 gty^{-1} of pre-anthropogenic carbon flux to sedimentary reservoirs (Table 3). This carbon flux has been enhanced during the last century by human activity. Baes et al. (1985, Table 5.2) estimated this enhancement at 0.2 to 0.8 gty^{-1}. Most of the anthropogenic increase is considered to be deposited in the coastal and continental wetlands which have become the subject of much recent interest as a potential sink of the "missing" carbon (Kemp, 1984; Walsh, 1984; Baes et al., 1985; Olsen et al., 1985; Trabalka et al., 1985). Also, part of the carbon lost to the biosphere by land conversion may be compensated by nitrogen fertilization of the forests which would increase tree growth and storage of carbon in soils. This potential sink of "missing" carbon has been estimated at 0.1 gty^{-1} by Peterson and Melillo (1985).

The point to be made here is that the carbon flux to sediments is not negligible. It is needed to balance production and decay of ^{14}C and it may be needed to account for the "missing" carbon in the global cycle.

5. ^{14}C AND SOLAR ACTIVITY

The sinusoidal trend curve in Figure 2 has been subtracted from the $\Delta^{14}C$ values to obtain Figure 9. Five prominent peaks occur in the $\Delta^{14}C$ "wiggle" curve that are equivalent in magnitude to the Maunder and Spörer minima ($\Delta^{14}C$ maxima) peaks. These occur at 250 BP (Maunder minimum), 420 BP (Spörer minimum), 2700 BP, 4800 BP and 7140 BP or, combining the Maunder and Spörer peaks, after lapses of 2365, 2100 and 2340 years. An equivalent periodicity (2100-2400 years) has also been observed in the earlier intermediate precision $\Delta^{14}C$ data sets (Houtermann, 1971; Sonett, 1984; Damon and Linick, 1986). Peaks of lower intensity occur at approximately 200-year intervals as has been emphasized by Suess (1980). It can be seen from figures 2, 3 and 9 that these ca. 200-year periodicity peaks are of greater intensity during the 8th through 6th millennia BP when the Earth's dipole moment was low and subdued in later millennia when the dipole moment was considerably higher. The five intense Maunder minima type peaks seem to be exceptions to the rule in that they appear to be relatively independent of the strength of the Earth's magnetic field.

Stuiver (1961, 1965) was the first to show a convincing relationship between these medium term $\Delta^{14}C$ fluctuations and solar activity (see Damon et al., 1978, for review). Stuiver assumed the same relationship between neutron flux and solar activity as observed during the short-term 11-year solar cycle. The relationship between solar activity and ^{14}C production is inverse. When solar activity is low, the neutron flux is high and consequently more ^{14}C is produced. Eddy (1976a) ascribed the intense maxima during the Maunder and Spörer minima to very low solar activity with a virtual absence of sunspots during much of that time. In a later paper, Stuiver and Quay (1980) have suggested that further modulation may occur even when sunspot activity is absent.

The production of ^{14}C as a function of solar activity is dependent on the strength of the Earth's magnetic field (Lingenfelter and Ramaty, 1970). The effect of solar activity on ^{14}C production rate decreases as the magnetic field increases. The enhancement of all but the largest Maunder-type wiggles in Figure 9, taking into consideration measurement errors, is about 50%, in accord with prediction by Lingenfelter and Ramaty (1970, see their Figure 3). Much greater enhancement was predicted for dipole moments significantly lower than 7×10^{22} Am^2 which tends to confirm the validity of the minimum field strengths in Figure 3 (this paper). The relative apparent independence of the most intense Maunder-type wiggles was not predicted. If it is not mere coincidence, this phenomena may provide further information concerning the modulation process.

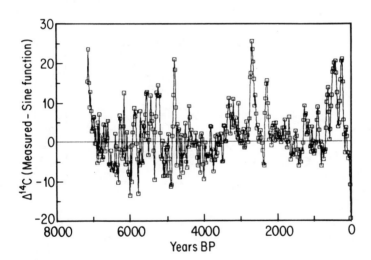

Figure 9. Δ^{14}C after subtraction of sinusoidal trend curve (Figure 2). Note that the ca. 200 yr. peaks are more attenuated when the Earth's dipole moment exceeds 8.75×10^{22} Am^2 which is the strength of the present dipole moment (see Figure 3). Prior to ca. 5000 years BP, when the geomagnetic field was low, the peaks are less attenuated by a factor of 1.6. The five largest peaks equivalent to the Maunder minimum peak appear to be an exception in that they are of nearly the same amplitude.

6. ^{14}C, SOLAR ACTIVITY AND CLIMATE

As pointed out by a number of authors (deVries, 1958, 1959; Damon,
1968, 1970; Suess, 1970; Eddy, 1976b, 1977; see Damon et al., 1978,
for review), there seems to be a relationship between the ^{14}C content
of the atmosphere and northern hemisphere temperature variations. The
Spörer and Maunder minima of solar activity occur during the Little
Ice Age (15th through 18th centuries) and the Medieval Warm Epoch
(12th and 13th centuries) has been associated with high solar activity
(Schove, 1955). Thus low solar activity has been associated with cool
temperatures and high solar activity with warm temperatures in inverse
relationship to the Δ^{14}C fluctuations. The first five years (1980 to
1985) of the National Aeronautics and Space Administration's (NASA)
solar maximum mission have tended to confirm the sense if, perhaps,
not the magnitude of this relationship (Willson et al., 1986). Total
solar irradiance observations by the first Active Cavity Radiometer
Monitor (ACRIM I) have shown a long-term downward trend in total solar
irradiance during that time from solar maximum to solar minimum of
about 0.1%. I have plotted the trend of the average daily mean values
versus annual sunspot number in Figure 10. There is a linear relationship but a longer time series will be needed to confirm the

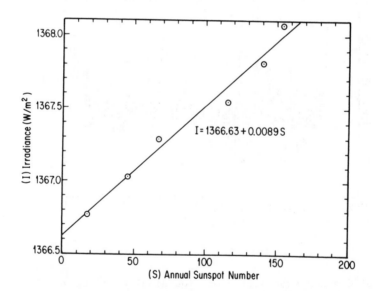

Figure 10. Total solar irradiance (I) from NASA Solar Maximum Mission
(Willson et al., 1986) vs. annual sunspot number (S). The irradiance
is taken from the trend line of the measured values. To a close
approximation there is a linear increase in I with increase in S given
by the equation.

relationship between total solar irradiance and sunspots. There
remains the possibility that solar luminosity is following a longer-
term trend such as the 22-year magnetic cycle. However, Willson et
al. (1986) point out that "past climate variability would seem to
preclude a lengthy continuation of the present downward trend of solar
luminosity" (p. 1117).

Eddy has suggested that "the sunspot number may follow changes in
the solar constant, through a kind of amplitude modulation of an
otherwise more uniform cycle" (Eddy, 1977, p. 186). This suggests a
low-pass phenomenon in which small changes in solar luminosity would
accompany short-term variations such as the 11-year solar cycle and
larger changes in solar luminosity would correlate with the envelope
of solar activity. This relationship would be similar to that low-
pass dependence of $\Delta^{14}C$ on ^{14}C production (Figure 8). Thus, Eddy
suggests that short-term changes in solar luminosity might be imper-
ceptible but longer-term changes of a hundred years or more would be
felt directly on Earth as perceptible climate changes (Eddy, 1976b).
This would seem to be confirmed by the observation of Sonett and Suess
(1984) of a 200-year cycle in the tree-ring widths of the 5290-year
Campito Mountain tree-ring chronology of LaMarche and Harlan (1973).
On the other hand, Stuiver (1980) found no consistent relationship
between ten proxy climate records and $\Delta^{14}C$ as an indicator of solar
variability during the current millennium. The geographical variability of climate change must be emphasized. Radiocarbon equili-
brates within the atmosphere in a few years and consequently
integrates global changes in production rate, whereas the proxy
climate records are for specific localities and strongly affected by
regional variations that may obscure global changes. The difficulty
in defining global climate changes during the current century
illustrates this point very well (Damon and Kunen, 1976). It is
tempting to correlate the changes in ^{14}C production rate associated
with the large Maunder-type wiggles in Figure 9 with global climate
change but the evidence is not compelling. Demonstration of a
correlation must await the integration of proxy climate indicators on
a sufficiently representative global basis.

ACKNOWLEDGEMENTS

I thank James Abbott for drafting the figures and Jo Ann Overs for
typing the manuscript. I am especially indebted to Drs. Song Lin
Cheng and Robert Sternberg for computer programming. I also benefited
from stimulating conversations with Drs. Austin Long, Valmore LaMarche
and Robert Sterberg and participation in the Durham Workshop. This
work was supported by N.S.F. Grants EAR-8314067 and ATM-8607161 and
the State of Arizona.

REFERENCES

Baes, C.F., Jr, Björkstrom, A., and Mulholland, P.J., 1985, in
    Trabalka, J.R., ed, 'Atmospheric carbon dioxide and the global
    carbon cycle', US Dept. of Energy Report DOE/ER-0239, 81.
Bacastow, R., and Keeling, C.D., 1973, in Woodwell, G.M. and Pecan,
    E.V., eds., 'Carbon and the biosphere', Upton, New York, USAEC
    Conf. 720510, 86.
Berner, R.A., 1982, Am. J. Science 282, 451.
Berner, W., Oeschger, H., and Stauffer, B., 1980, Radiocarbon, 22(2),
    227.
Broecker, W.S., Takahashi, T., Simpson, H.J., and Peng, T.H., 1979,
    Science, 206, 409.
Bruns, M., Rhein, M., Linick, T.W., and Suess, H.E., 1983, in Mook,
    W.G. and Waterbolk, H.T., eds., '14C and archaeology', PACT,
    Strasbourg, 8, 511.
Castagnoli, G., and Lal, D., 1980, Radiocarbon, 22(2), 113.
Damon, P.E., 1968, Meteorol Monographs, 8, 106.
Damon, P.E., 1970, in Olsson, I.U., ed., 'Radiocarbon variations and
    absolute chronology', Proc XII Nobel Symp: New York, Wiley, 571.
Damon, P.E., and Kunen, S.M., 1976, Science, 193, 447.
Damon, P.E., and Linick, T.W., 1986, Radiocarbon, 28(2A), 266.
Damon, P.E., Lerman, J.C., and Long, A., 1978, Ann. Rev. Earth Planet
    Sci., 6, 457.
Damon, P.E., Sternberg, R.S., and Radnell, C.J., 1982, Radiocarbon,
    25(2), 249.
de Vries, H., 1958, K. Ned. Akad. Wet., Proc. Ser. B, 61, 94.
de Vries, H., 1959, in Abelson, P.H., ed., 'Researches in geochemistry',
    New York, Wiley, 169.
Eddy, J.A., 1976a, Science, 192, 1189.
Eddy, J.A., 1976b, in Williams , D.J., ed. 'Physics of solar planetary
    environments', 2, Washington, D.C., Amer Geophys Union, 958.
Eddy, J.A., 1977, Climatic Change, 1, 173.
Elsasser, W.E., Ney, E.P., and Winckler, J.R., 1956, Nature 178, 1226.
Hay, W.H., 1985, in Sundquist, E.T. and Broecker, W.S., eds, 'The
    carbon cycle and atmospheric CO_2: Natural variation Archean to
    Present', Amer. Geophys. Union Monograph 32, 573.
Houtermans, J.C., 1971, 'Geophysical interpretations of Bristlecone
    Pine radiocarbon measurements using a method of Fourier analysis for
    unequally-spaced data', Ph.D. Dissertation, Univ. Bern, Switzerland.
Houtermans, J.C., Suess, H.E., and Oeschger, H., 1973, Jour. Geophys.
    Research, 78, 1897.
Kemp, S., 1979, in Bolin, B., Degens, E.T., Kemp, S., and Ketner, P.,
    eds. 'The global carbon cycle, Scope 13', John Wiley and Sons,
    New York, 317.
Klein, J., Lerman, J.C., Damon, P.E., and Linick, T., 1980, Radiocarbon,
    22(3), 950.
Korff, S.A., and Mendell, R.B., 1980, Radiocarbon, 22(2), 159.
Light, E.S., Merker, M., Verschell, H.J., Mendell, R.B., and Korff,
    S.A., 1973, Jour. Geophys. Research., 78(16), 2741.

Lin, R., 1977, in White, O.R., ed., 'The solar output and its variation', Colorado Associated Univ. Press, 39.

Lingenfelter, R.E., and Ramaty, R., 1970, in Olsson, I.U., ed., 'Radiocarbon variations and absolute chronology', Nobel Symposium, 12th Proc., New York, John Wiley and Sons, 513.

McElhinny, M.W., and Senanyake, W.E., 1982, Jour. Geomag. Geoelect., 34, 39.

Merker, M., 1970, 'Solar cycle modulation of fast neutrons in the atmosphere', Ph.D. Dissertation, New York Univ., New York, University Microfilms, Ann Arbor, Michigan, 1971.

Merrill, R.T., and McElhinny, M.W., 1983, 'The Earth's magnetic field, its history, origin and planetary perspective', Academic Press, London.

O'Brien, K.J., 1979, Jour. Geophys. Research, 84, 423.

Oeschger, H., Siegenthaler, U., Schotterer, U., and Gugelmann, A., 1975, Tellus, 27, 168.

Olson, J.S., Garrels, R.M., Berner, R.A., Armentano, T.V., Dyer, M.I., and Yaalon, D.H., 1985, in Trabalka, J.R., ed., 'Atmospheric carbon dioxide and the global carbon cycle', US Dept. of Energy Report DOE/ER-0239.

Olsson, I.U., 1970, ed., 'Radiocarbon variations and absolute chronology', Proc. XII Nobel Symp: New York, Wiley.

Pearson, G.W., and Stuiver, M., 1986, Radiocarbon, (Calibration Issue), 28(28), 839.

Pearson, G.W., Pilcher, J.R., Baillie, M.G.L., Corbett, D.M., and Qua, F., 1986, Radiocarbon, (Calibration Issue), 28(28), 911.

Peterson, B.J., and Mellilo, J.M., 1985, Tellus, 37B, 117.

Povinec, P., 1977, Acta Physica Comeniana, 18, 139.

Siegenthaler, U., 1985, in Fontes, J.C., and Fritz, P., eds. 'Handbook of environmental isotope geochemistry', v.3, Elsevier, Amsterdam (in press).

Siegenthaler, U., Heimann, M., and Oeschger, H., 1980, Radiocarbon, 22(2), 177.

Sonett, C.P., 1984, Rev. Geophysics Space Physics, 22(3), 239.

Sternberg, R.S., and Damon, P.E., 1979, in Berger, R., and Suess, H.E., eds., 'Radiocarbon dating', Berkeley, Univ. California Press, 691.

Stuiver, M., 1961, Jour. Geophys Research, 66, 273.

Stuiver, M., 1965, Science, 149, 533.

Stuiver, M., and Pearson, G.W., 1986, Radiocarbon (Calibration Issue), 28(28), 805.

Stuiver, M., and Quay, P.D., 1980, Science, 207(4426), 11.

Suess, H.E., 1980, Radiocarbon, 22(2), 200.

Trabalka, J.R., Edmonds, J.A., Reilly, J.M., Gardner, R.H., and Voorhees, L.D., 1985, in Trabalka, J.R., ed. 'Atmospheric carbon dioxide and the global carbon cycle', US Dept. of Energy Report DOE/ER-0239.

Walsh, J.J., 1984, Bioscience, 34, 499.

Willson, R.C., Hudson, H.S., Frohlich, C., and Brusa, R.W., 1986, Science, 234, 1114.

^{10}Be AS A PROXY INDICATOR OF VARIATIONS IN SOLAR ACTIVITY AND GEOMAGNETIC FIELD INTENSITY DURING THE LAST 10,000 YEARS

G.M. Raisbeck and F. Yiou
Laboratoire René Bernas du Centre de Spectométrie
   Nucléaire et de Spectrométrie de Masse,
91406 Campus Orsay,
France.

INTRODUCTION

The study of the concentration of long lived cosmogenic (cosmic ray produced) isotopes in various natural reservoirs is a potentially powerful method for investigating time variations in cosmogenic production rates (Raisbeck and Yiou, 1984). In principle at least, these variations can then be interpreted in terms of a temporal record of the parameters which affect cosmogenic production rate, including, in the context of the present meeting, solar activity and the geomagnetic field intensity. The study of ^{14}C in tree rings is the most developed example of such an application, and is treated in detail elsewhere in this volume. The technique of accelerator mass spectrometry (AMS) offers the possibility of extending such studies to a number of other cosmogenic isotopes, and in particular to ^{10}Be (half-life 1.5 My). Indeed the first published measurements of ^{10}Be by AMS emphasized the potential of this type of application (Raisbeck et al. 1978), and one of the first AMS profiles of ^{10}Be in a natural reservoir included a tentative identification of an enhanced production rate associated with reduced solar activity during the Maunder Minimum (Raisbeck et al., 1981).

Since that time, the efforts of our group have been somewhat dispersed, including the pursuit of other applications, and the installation and development of a dedicated AMS facility to permit numerous and accurate ^{10}Be measurements. This capability now exists (Raisbeck et al., in press), thus offering us the possibility of carrying out a much more comprehensive and detailed program of research using ^{10}Be as a "proxy" indicator of cosmogenic production rate changes. In the meantime, several other groups have begun similar studies, in particular a Bern-Zurich collaboration (see Beer et al., this volume, for a review of that work).

On the time scale relevant to the present meeting (last 10,000 years) the potentially most appropriate reservoirs for ^{10}Be measurements are probably lacustrine sediments and polar ice cores. While deep sea marine sediment cores can go back much further in time (and are thus very important for investigating possible changes in primary cosmic

F. R. Stephenson and A. W. Wolfendale (eds.),
Secular Solar and Geomagnetic Variations in the Last 10,000 Years, 287–296.
© 1988 by Kluwer Academic Publishers.

ray intensity (Raisbeck, 1985)), their low sedimentation rates, frequently combined with mixing by bioturbation, limit the time resolution available with them. Coastal marine sediments, while often having relatively high sedimentation rates, are expected to be severely influenced by climatic conditions on nearby continents.

While we have measured [10]Be profiles in a number of lake cores, most of the results remain unpublished. The reason is that, although all of these profiles show significant concentration (and deposition flux) variations, these are not readily interpretable in terms of production rate changes. For one thing, the measured variations are not coherent from one lake to another. Secondly, with one possible exception (Raisbeck et al., 1986), the observed [10]Be deposition fluxes in the sediments are considerably larger (sometimes by an order of magnitude) than can be accounted for by the estimated production rate. We interpret this as arising from [10]Be input from the (time variable?) surrounding watershed and/or, especially in the case of crater lakes with small and fixed watersheds, [10]Be associated with aeolian transported dust particles. Thus, while we continue to look carefully at potential lacustrine reservoirs, we believe it is presently necessary to be very prudent in interpreting [10]Be profiles in such sediments in terms of production rate changes.

For the reasons mentioned above, we concentrate in this paper on the [10]Be record in Antarctic ice. We have earlier discussed the advantages of polar ice as a [10]Be reservoir, particularly for obtaining a proxy record of solar activity (Raisbeck and Yiou, 1980a). One of the potential difficulties is that climate related effects (atmospheric circulation, precipitation rate changes) can also give rise to [10]Be concentration variations in polar ice. We feel however, that there is some evidence to suggest that the influence of such parameters during the last 10,000 years may have been minor, or at least resolvable. Thus, by combining [10]Be records from several ice cores, preferably both from the Arctic and Antarctic, we believe it should be possible to extract a physically significant production rate signal from the climatically induced "noise". As one contribution toward such a record, we report here a nearly continuous [10]Be profile from a 180m ice core drilled at Dome C, Antarctica.

[10]Be IN ICE AT DOME C

In Figure 1 we show the [10]Be concentrations measured in a 180m ice core drilled at Dome C, Antarctica, (Gilet and Rado, 1979) a few meters from the 906m core described by Lorius et al. (1979). The [10]Be was extracted from samples typically $\sim$ 80 cm in length, and 1 kg weight, using procedures described elsewhere (Raisbeck et al., 1981). The AMS measurements were carried out on the Gif-sur-Yvette Tandetron. For the majority of samples, the number of [10]Be ions counted was $\sim$ 1000 leading to a one standard deviation statistical error of $\sim$ 3%. This, combined with a conservatively assumed instrumental uncertainty of 5% (Raisbeck et al., in press), leads to a ± 6% uncertainty in the final concentrations. This is small compared to the observed variations

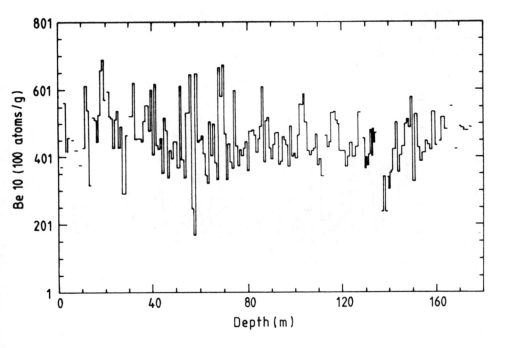

Fig. 1.   Concentration of [10]Be as a function of depth in Dome C
ice core.

in Figure 1. Approximately half the samples from the upper 50 m had
been prepared earlier for measurement at the Grenoble cyclotron
(Raisbeck et al., 1981). Although these samples were remeasured with
the Tandetron, the six times larger quantity of [9]Be carrier used led
to fewer [10]Be ions being counted, and resulting uncertainties of up
to ± 10%.
        Because the annual precipitation rate at Dome C is very low, it
is not possible to reliably extract annual signals of stable isotopes,
dust or ionic impurities. Thus the establishment of a time scale must
rely on a glaciological model. Fortunately Dome C, as its name indicates,
is located at a local maximum in the ice sheet, and flow properties
are expected to be particularly simple, especially in the upper portion
of the ice sheet considered here.
        The largest uncertainty in applying a glaciological model in the
present case involves the choice of an appropriate precipitation rate.
In earlier work, the average precipitation rate at Dome C has been
estimated as $2.7-3.7 g/cm^2$ yr. (Lorius et al., 1979, Petit et al., 1982).
More recently, Legrand & Delmas (1987) have estimated an average precipitation
rate of 3.06 g/yr during the past $\sim$ 200 years, based on an acid signal
believed to have come from the Tambora volcanic eruption of 1815. We
have used this latter estimate in constructing the time scale of Figure 2,

although we clearly recognize the uncertainty in extrapolating this value by a factor of 20 in time. Indeed the difference in the precipitation estimates cited above gives some idea of the time variability to be expected in this parameter at Dome C.

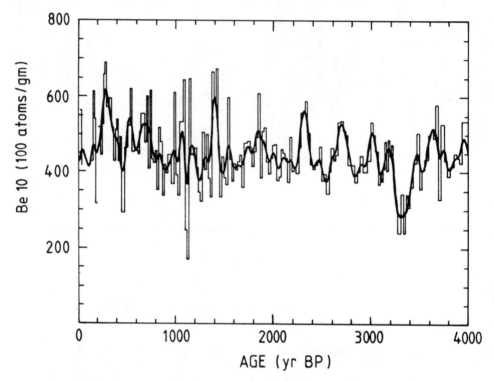

Fig. 2. (Histogram) Concentration of ^{10}Be as a function of estimated age of ice at Dome C. Ages are calculated using a precipitation rate of 3.0g/cm^2 yr.
(Heavy curve) Concentration smoothed with spline function.

SHORT TERM ^{10}Be VARIATIONS

From Figure 2 one sees that our ^{10}Be record covers the past ∿ 4000 years, and that the most prominent variations are on time scales of the order of 100 years. This is quite reminiscent of variations on a similar time scale observed in ^{14}C tree ring records (Stuiver, this volume). In order to more readily compare these two records, we have smoothed the ^{10}Be profile with a spline function (heavy curve of Figure 2). This is appropriate for two reasons. First, it may help smooth out some high frequency meteorological noise in the ^{10}Be record. Secondly, the ^{14}C record is itself highly damped on this timescale due to exchange between ^{14}C in the atmosphere, biosphere and ocean surface water (which is the main reason the amplitude of its variations are much smaller than those of the ^{10}Be).

The above smoothing procedure is much less sophisticated than propagating the ^{10}Be signal through a carbon cycle model, as done by Siegenthaler and Beer (this volume). However, it is not clear to us that the latter procedure is justified, since the ^{10}Be record in polar ice may be preferentially recording high latitude production rate changes, while the ^{14}C signal represents a global response. Assuming the carbon cycle is modelled correctly, and that the cycle has itself remained constant over the time period of interest, the relative amplitudes of the ^{14}C responses predicted by the procedure of Siegenthaler and Beer should be correct. However, until a quantitative relationship can be established between ^{14}C and ^{10}Be responses to the same solar activity variations, it is probably not justified to quantitatively interpret one isotope record with respect to the other. We will return to this question below in connection with geomagnetic intensity variations. The same argument applies to the statement of Lal (1987) that observed ^{10}Be variations in polar ice cannot be due to production changes, because they are larger than (global) theoretical predictions.

Fortunately as discussed earlier (Raisbeck and Yiou, 1980a) it may be possible to establish a direct "calibration" between ^{10}Be in polar ice and solar activity itself, by measuring detailed profiles of ^{10}Be during the past few solar cycles, for which there are fairly accurate measurements of various indicators of solar activity.

Keeping in mind the uncertainties mentioned above regarding a quantitative relationship, it is nevertheless useful to qualitatively compare the Dome C ^{10}Be profile with that of the ^{14}C tree ring record over the same time period (Figure 3). Leaving aside the long term trend (discussed below), there are both similarities and apparent differences between the two records, as suggested by the tentative correlation indicated in Figure 3. The most striking similarities are peaks at about 300, 2300 and 2700 B.P. The first of these coincides with the Maunder Minimum, probably the best documented period of reduced solar activity, as discussed by Eddy in this volume. The peaks at $\sim$ 2300 and 2700 B.P. are also very prominent in the ^{14}C record, and we feel that their presence in the ^{10}Be records of both Dome C, and Camp Century ice cores (Beer et al., this volume) is strong support for a production rate (probably solar modulation) origin.

Two other prominent peaks in the ^{14}C record, labelled by Eddy (1976) as the Sporer ($\sim$ 500 B.P.) and Wolf ($\sim$ 700 B.P.) minima appear to be poorly reproduced in the Dome C record, both in amplitude, width and age. Some of these differences might be explained by variations in precipitation rate at Dome C. It is interesting to note however, that the corresponding ^{10}Be peaks in Greenland ice also appear to be much less prominent than for the Maunder Minimum (Beer et al., this volume). Unfortunately, this is also the part of the Dome C ^{10}Be profile where our results have the greatest uncertainties, so we prefer not to make any definitive conclusions for the moment.

There are two relatively brief, but large, peaks in the Dome C record at $\sim$ 1050 and 1400 B.P. (the former is more readily seen in the raw data than in the smoothed curve). Allowing for uncertainties in

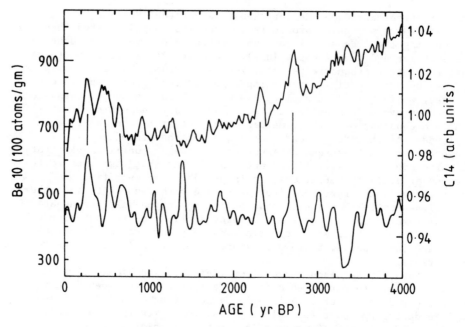

Fig. 3. (Upper curve) Variation in the ^{14}C concentration in the atmosphere as given by measurements in tree rings (Stuiver and Pearson, 1986; Pearson and Stuiver, 1986).
(Lower curve) Smoothed [10]Be concentration at Dome C, from Figure 2.

the calculated ages, these might correspond to the peaks at ∿ 950 and 1300 B.P. in the ^{14}C curve. The greater prominence in the [10]Be record could be due to their relatively short duration, and thus greater attenuation in the ^{14}C response. It is interesting to note that the Camp Century [10]Be record of Beer et al. (this volume) also show very significant peaks at these times.

There is no apparent ^{14}C analogue for the deep minimum in the Dome C curve at ∿ 3300 B.P. However, we are missing several samples for this period, so the smoothed curve may be misleading.

In general, then, we feel that the similarities between the Dome C [10]Be and tree ring ^{14}C records of Figure 3 are sufficiently promising that they support interpretation of short term variations in the former as being due to production rate changes, most probably due to variations in solar modulation. Unfortunately the absence of a firm independent time scale for the Dome C ice core leaves a certain latitude in the comparison of the two records which is not totally satisfying. Indeed, one might even turn the problem around and, assuming that the two curves are correlated, use the comparison to date the Dome C ice core. For example, we feel that the good correlation between the ^{14}C and [10]Be peaks at ∿ 2300 and 2700 B.P., discussed above, is strong support for the average precipitation rate used to establish the time scale of Figure 2, even though there may have been significant variations during that time.

We also find it very encouraging that there is perhaps an even greater similarity between the ^{10}Be records from Dome C and Camp Century (which also lacks a firm independent time scale) than between either of them and the ^{14}C record. If confirmed this would greatly strengthen the argument of using combined ^{10}Be ice core records to deduce production rate changes.

One of the potentially attractive aspects of using ^{10}Be in ice cores as a proxy indicator of solar activity, is that the stable isotope records (ΔH^2 or $\Delta^{18}O$) in the same ice represent important proxy indicators of paleo climate. Thus one can hope to address, without introducing uncertainties associated with separate time scales for the solar activity and climate records, the question as to what extent changes in solar activity may have influenced climate (Raisbeck and Yiou, 1980b). A preliminary comparison of the ^{10}Be data in Figure 1 with ΔH^2 data (J. Jouzel, private communication) for the same ice core, showed no significant correlation. If confirmed by a more detailed statistical analysis, such a result would be evidence against hypotheses in which solar activity has had an important influence on global climate over the past 4000 years.

LONG TERM VARIATIONS

Let us now consider longer term variations. As can be seen in Figure 3, the tree ring ^{14}C record shows an important long term trend, that is not apparent in the Dome C ^{10}Be record. This ^{14}C variation has often been attributed to a change in the intensity of the geomagnetic dipole, as deduced from independent paleomagnetic intensity measurements (Bucha, 1970; Damon et al., 1978). In addition to their inherent difficulty, as discussed by other authors in this volume, the problem with paleomagnetic intensity measurements is that, at best, they are usually only available for a few locations for any given time period. Since local geomagnetic records may be significantly influenced by non-dipole fields, it is very difficult to reconstruct an accurate secular record of the dipole field, which is the dominating influence on cosmogenic production rates (both because of its "global" nature, and because of its weaker fall off with distance from the earth's surface).

Although attempts have been made to estimate the paleo dipole intensity by averaging paleomagnetic records from a few different locations (McElhinny and Senanayake, 1982), there is considerable uncertainty as to their reliability. In fact, on the basis of the ^{10}Be record in the Camp Century ice core, Beer et al. (1984) argued that "there had not been a change in the geomagnetic dipole large enough to significantly affect atmospheric radioisotope production rates during the last 4000 years". It is evident from Figure 3 that the ^{10}Be record from Dome C also shows no significant long term trend over this time period. As discussed earlier, (Raisbeck and Yiou, 1985) however, it is not evident to what extent variations in the geomagnetic dipole field, which in this case are most influential on low latitude production, will be reflected in ^{10}Be concentration variations in polar precipitation. Thus, as discussed above for solar modulation, until a quantitative relationship can be established between ^{10}Be in polar ice and geomagnetic dipole variation, we believe it is premature to make any categorical conclusions regarding the paleomagnetic field on the basis of the ice core data.

Unfortunately calibration of the ^{10}Be response to geomagnetic intensity variations may be more difficult than in the case of solar activity variations. The reason is that the variation in the dipole intensity over the time period for which direct measurements of this parameter exist is rather small. It is for this reason that we have spent so much effort in trying to obtain a reliable ^{10}Be record from low latitude lake sediments, where the effects of geomagnetic variations on ^{10}Be production are expected to be more important.

PRE-HOLOCENE ^{10}Be VARIATIONS

One of the potential advantages of ^{10}Be compared to ^{14}C for studying cosmogenic production rate variations is that it permits investigation over a much longer time scale. Thus, although they fall outside the time domain chosen for this meeting, we would like to briefly mention some recently published ^{10}Be measurements (Figure 4) in older Antarctic ice (Raisbeck et al., 1987). While precipitation changes have been hypothesized as being the dominant cause of the systematically larger ^{10}Be concentration observed in glacial ice, compared to inter-glacial ice, there are at least two periods (at $\sim$ 35,000 and $\sim$ 60,000 B.P.) during the last ice age when unusually large ^{10}Be concentrations were observed, not correlated with any obvious climatic change, or corresponding variation in other parameters (Δ^{18}O, dust, CO_2, trace elements). Although the source of these variations is not yet established, we are presently inclined to favour a production rate origin, most likely of reduced solar modulation origin. One note-worthy feature is that at least the most recent of these "events" appears to have lasted of the order of 1000 years. Thus, if additional work confirms a solar modulation origin for these events, it would indicate that periods of reduced solar activity could last significantly longer than those observed during the Holocene.

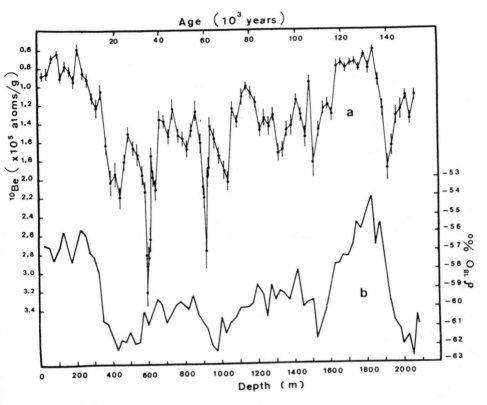

Fig. 4 ^{10}Be concentration (upper curve) and Δ^{18}O (lower curve)
in ice core at Vostok Antarctica (Raisbeck et al. 1987).
Note the large ^{10}Be concentrations at $\sim$ 35,000 and
60,000 B.P.

ACKNOWLEDGEMENTS

We would like to thank the organizers of this meeting both for the
invitation to participate and for their patience in awaiting this
manuscript.
     We thank J. Lestringuez, D. Deboffle and D. Bourlès for help
in preparing samples and making measurements with the Tandetron,
C. Lorius for making available samples from the Dome C ice core,
M. Pourchet for assistance in sampling, and J. Jouzel for making
available (and showing us how to use) the computer programme used
to generate Figures 1-3.
     Sample preparation was supported in part by the PNEDC.   Tandetron
operation is supported by the CNRS, CEA and IN2P3.

296

REFERENCES

Beer, J. et al., (1984), Nucl. Instr. and Meth., B5, 380.
Bucha, V. (1970) in Olsson, I.U. (ed.) Radiocarbon, variations and Absolute Chronology, Nobel Symposium, Vol. 12, Wiley, New York.
Damon, P.E., Lerman, J.C. and Long, A., (1978), Ann. Rev. Earth Planet Sci., 6, 457.
Eddy, J.A., (1976), Science, 192, 1189.
Gilet, F. and Rado, C., (1979), Antarctic J. U.S., 14, 101.
Lal, D., (1987), Geophys. Res. Lett., 14, 785.
Legrand, M. and Delmas, R.J., (1987), Nature, 327, 671.
Lorius, C., Merlivat, L., Jouzel, J. and Pourchet, M., (1979), Nature, 280, 644.
McElhinny, M.W. and Senanayake, J., (1982), Geomag. Geoelect., 34, 39.
Pearson, G.W. and Stuiver, M., (1986), Radiocarbon, 28, 839.
Petit, J.R., Jouzel, J., Pourchet, M. and Merlivat, L., (1982), J. Geophys. Res., 87, 4301.
Raisbeck, G.M., (1985), Proc. 19th Int. Cosmic Ray Conf., Vol. 9, 73.
Raisbeck, G.M. and Yiou, F., (1980a), in 'The Ancient Sun', Pepin, R.O., Eddy, J.A. and Merrill, R.B. (ed.), Pergamon New York.
Raisbeck, G.M. and Yiou, F., (1980b), in 'Sun and Climate', CNES.
Raisbeck, G.M. and Yiou, F., (1984), Nucl. Instr. and Meth., B5, 91.
Raisbeck, G.M. and Yiou, F., (1985), Ann. Glacial 7, 138.
Raisbeck, G.M., Yiou, F., Fruneau, M. and Loiseaux, J.M., (1978), Science, 202, 215.
Raisbeck, G.M., Yiou, F., Fruneau, M., Loiseaux, J.M., Lieuvin, M., Ravel, J.C. and Lorius, C., (1981), Nature, 292, 825.
Raisbeck, G.M., Yiou, F. and Livingston, D.A., (1986), in 'Global Change in Africa'.
Raisbeck, G.M., Yiou, F., Bourlès, D., Lorius, C., Jouzel, J. and Barkov, N.I., (1987), Nature, 326, 273.
Raisbeck, G.M., Yiou, F., Bourlès, D., Lestringuez, J. and Deboffle, D., (in press), Nucl. Instr. and Meth.
Stuiver, M. and Pearson, G.W., (1986), Radiocarbon, 28, 805.

TEMPORAL ^{10}Be VARIATIONS IN ICE: INFORMATION ON SOLAR ACTIVITY AND GEOMAGNETIC FIELD INTENSITY

[1,2]J. Beer, [2]U. Siegenthaler and [2,3]A. Blinov
[1]Institute for Medium Energy Physics, ETH-Hoenggerberg, CH-8093, Zurich, Switzerland.
[2]Physics Institute, University of Bern, CH-3012 Bern, Switzerland.
[3]Leningrad Polytechnic Institute, Polytechnicheskaya 29, Leningrad, 195251, USSR.

ABSTRACT. Temporal ^{10}Be variations in polar ice are caused by changes of the production rate and the following transport to the ice sheets. The production rate is determined by the cosmic ray flux and its modulation by helio- and geomagnetic shielding effects. The removal of ^{10}Be from the atmosphere is governed by the transport of aerosols to which ^{10}Be is attached. Atmospheric circulation and precipitation rates strongly influence this transport. The ^{10}Be record from Greenlandic ice cores exhibit variations on different time scales. In data from the last millenium, the 11 year solar cycle has been identified. For variations with characteristic times longer than a few centuries, there is the problem to distinguish between the different causes. A possible way to overcome this problem is to compare ^{10}Be with other tracers (^{14}C, ^{18}O) which behave differently. In fact a comparison between the ^{10}Be record from Camp Century, Greenland, and ^{14}C data from tree-rings reveals similar short-term variations which can be attributed to a variable sun. The agreement between the long-term variations of ^{10}Be with palaeomagnetic data, however, is less good and its interpretation is less conclusive.

1. INTRODUCTION

Today a steadily increasing number of parameters of the sun and the earth are being recorded directly and precisely. The sun is continuously monitored over the whole spectrum of observable wavelengths, and changes of its luminosity smaller than one permil can be detected by highly sensitive instruments mounted on satellites. The changes and distortions of the geomagnetic field are continuously registered all over the world by magnetometers. Going back in time the number of direct parameters recorded is decreasing: the systematic observations of sunspots was started in 1848 by Wolf, earlier data reaching back to about 1500 A.D. have been reconstructed. Monitoring and analysis of the geomagnetic field intensity was initiated by Gauss in 1832. Going further back in time one has to rely more and more on indirect

F. R. Stephenson and A. W. Wolfendale (eds.),
Secular Solar and Geomagnetic Variations in the Last 10,000 Years, 297–313.
© 1988 by Kluwer Academic Publishers.

and less reliable observations of sunspots and auroral events as described in historical reports or on palaeomagnetic information stored in natural archives. There is, however, an additional indirect source of information on the history of the solar activity and the geomagnetic field intensity: cosmic ray induced radioisotopes such as ^{14}C and ^{10}Be. In this paper we discuss how, in the case of ^{10}Be, the information is recorded and stored in archives and which are the main problems in interpreting this information. ^{10}Be data from polar ice cores are discussed in terms of solar activity and geomagnetic field intensity by comparing them with records of other proxy data.

2. INFORMATION FROM RADIOISOTOPES

Figure 1 schematically shows how radioisotopes are produced, distributed and stored and what kind of information they reflect. Cosmic ray particles (protons, alpha particles) impinging on the atmosphere produce by different nuclear interactions a large variety of radioisotopes. Since the atmosphere is composed mainly of nitrogen and oxygen the majority of the produced nuclei are lighter than oxygen (mass 16). Most of them decay due to their short half-lives before they even reach the surface of the earth. There are only two long-lived isotopes which are produced from N and O: ^{14}C ($T_{1/2}$ = 5730 y) and ^{10}Be ($T_{1/2}$ = 1.5 mio y). ^{14}C is generated by thermal neutrons knocking off a proton from nitrogen: $^{14}N(n,p)^{14}C$. ^{10}Be is produced by fast particles (p,n) knocking off many nucleons or small fragments (d,^4He) from the target nucleus (spallation reaction). The production rate of radioisotopes is directly related to the cosmic ray flux. There are three possible sources of production rate variations:

1. Primary cosmic rays. The primary cosmic ray flux may change due to e.g. supernova explosions. Only little is known about the effects of such events and they seem to be rather rare (Bhat et al., 1987; Raisbeck et al., 1987). Based on investigations of meteorites it has been found that the mean cosmic ray flux was constant within 20 - 30% over the last 1 - 10 million years (Geiss et al., 1962; Nishiizumi et al., 1980).

2. Solar modulation. Although this process is complex the general picture is quite simple: solar wind is carrying frozen-in magnetic fields which deflect especially the low energy cosmic ray particles and reduce therefore the radioisotope poduction rate. The solar modulation takes place in a large part of the solar system. Its existence can clearly be seen in the anticorrelation between the counting rate of neutron-monitors at sea level and the sunspot record.

3. Geomagnetic modulation. The geomagnetic field which is basically a dipole field prevents cosmic ray particles with too low rigidity (momentum) from penetrating into the atmosphere and producing radioisotopes (cutt-off rigidity). The shielding effect of the dipole field is strongly latitude dependent, having its maximum at the

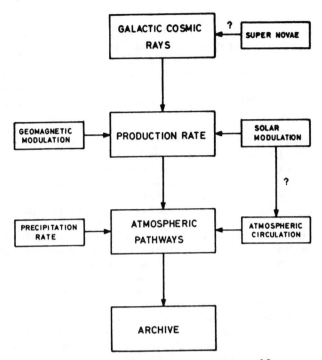

Fig. 1. Schematic diagram of production of [10]Be, its pathways
and different modulation effects involved.

equator (Lal and Peters, 1967).

Calculations show that production variations of 40 - 60% are
expected due to solar modulation (Blinov, 1988). In the case of geomagnetic field variations, the global change in the production rate
is < 50% for variation by a factor 2 of the field strength, and it
would increase by a factor 2-3 for an almost complete disappearance
of the field as expected during a magnetic reversal. Due to the
different production mechanism the modulation effects on ^{14}C and [10]Be
are slightly different. For 'normal' modulations the effects on [14]C
are larger by not more than 10% (Blinov, 1988).

After their production the radioisotopes are subject to mixing
and transport processes which strongly depend on their chemical
properties. [14]C is oxidized to [14]CO_2 and exchanges between atmosphere,
ocean and biosphere (Sigenthaler and Beer, 1988). Due to its long
atmospheric residence time [14]CO_2 can be considered as well mixed in
the atmosphere. The relatively long residence time of carbon in the
exchanging reservoirs leads to an attenuation of the production
variation, which is the stronger the shorter the time constant of
the modulation event is. [10]Be becomes attached to aerosols and follows
their pathway through the atmosphere before it is removed mainly by
wet precipitation. Since about 70% (Lal and Peters, 1967) of the
production of [10]Be takes place in the stratosphere, its mean residence
time in the atmosphere is about 1-2 y. The passage from stratosphere

through the tropopause into the troposphere and the final rainout
process are affected by the prevailing atmospheric conditions. For
individual precipitation events, large variations of the radioisotope
concentrations can be the result, but they are averaged out to some
extent over times of years or more. Because the exchange between
stratosphere and troposphere takes place mainly in the mid-latitudes,
the [10]Be fallout is latitude-dependent. Although the [10]Be fallout
pattern is not yet fully established (Finkel et al., 1977) it is
expected to be similar to the one observed for ^{137}Cs and ^{90}Sr having
its maximum between 40 - 60° (Sarmiento and Gwinn, 1986).

The last step, the deposition of aerosol bound [10]Be on the earth
surface (transfer function) is also not yet understood in detail.
Except for arid regions the fallout seems to occur mainly by precipitation (> 80%) and, within certain limits, to be proportional to the
precipitation (Junge, 1975). For regions with very low precipitation
rates the percentage of dry deposition is expected to be higher, leading
to higher concentrations. On the other hand, large precipitation
rates may dilute the radioisotopes and result in lower concentrations.

The above discussion shows that, in the case of [10]Be, the role
of atmospheric transport processes is not yet well understood, but
their effects can be large, at least as far as short-term variations
are concerned. For longer periods (>1 y) under climatically stable
conditions the transport and deposition effects should be more and
more averaged out. The situation is further complicated by the possible
coupling between the solar activity and the atmospheric transport
regime (Figure 1). If existent it may amplify or attenuate the direct
modulation effects on the production rate.

The information provided by the atmospheric ^{14}C and the deposited
[10]Be is only useful if it is stored stratigraphically undisturbed
in an archive. As far as the Holocene period is concerned, tree-rings
represent an ideal archive for [14]C. Each ring stores the atmospheric
[14]C/^{12}C ratio of the year of its formation. By analysing
dendrochronologically dated tree-rings it has been possible to reconstruct the atmospheric ^{14}C concentration over the last 9000 years
(Suess, 1970; Stuiver and Kra, 1986; Stuiver, 1988). In the case
of [10]Be the situation is less favourable. For the last 10,000 y,
ice sheets and lake sediments are the only archives providing the
necessary time resolution. The interpretation of lake sediments is
complicated by depositional effects, by the influx of additional [10]Be
(dust, soil from the drainage area) and by chemical and biological
processes going on within the sediment. Ice sheets sample the atmospheric fallout directly, but they are restricted to polar regions
or to high altitudes, and dating is not always easy.

3. INTERPRETATION OF RADIOISOTOPE DATA

The main problem in the interpretation of radioisotope data from an
archive is to separate the different causes determining the observed
concentration changes, especially to distinguish between production
variations and changes due to transport and depositions processes.

If there is no additional information available, a separation of the
different effects may be difficult or impossible. There are three
possibilities to overcome this problem. First, the record can be
compared with one of several other records which are influenced by
the same causes, but in a different way. Below, we will compare ^{14}C
and ^{10}Be that are produced similarly but are then distributed differently. The role of atmospheric processes for ^{10}Be could be studied
by constituents which are transported similarly, but which are not
produced by cosmic rays such as bomb-produced ^{137}Cs. Another approach
is to compare the 'modern' part of a record with direct observations
of the parameters to be reconstructed, e.g. the sunspot number, in
order to 'calibrate' the interpretation method. A calibration of
^{10}Be is still missing because most records do not include modern times
or are not detailed enough. In the last step an attempt should be
made to quantify the underlying processes by developing models and
comparing their results with the measured data. Thus, using semiempirical data on energy spectra and cross section values, it is
possible to calculate production variations (Blinov, 1988). However,
the transport and deposition processes are not yet included. General
circulation models (GCM), if supplied with an appropriate deposition
mechanism, should be adequate to describe these processes.

In the following we will discuss two ^{10}Be records from Greenland.
The Milcent record (Beer et al., 1983a) from central Greenland (70°18'N,
44°35'W) covers the period 1180 - 1800 A.D. with a time resolution
of 3 - 7 y (Figure 2a). The Camp Century core was drilled in northern
Greenland (77°10'N, 82°08'W) and covers about 100,000 y; the last
10,000 y of ^{10}Be data are shown in Figure 3a (Beer et al., 1987).
The time resolution varies from a few years around 1800 A.D. to about
100 y for the oldest part.

4. SOLAR ACTIVITY

The term 'solar activity' includes different phenomena related to
magnetohydrodynamic processes within the convection zone of the sun,
which appear at the surface in different forms such as sunspots and
flares. From the long sunspot record it is known that the number
of spots is waxing and waning with a period of about 11 years. An
important but still open question is, if this period keeps its phase
over long periods or not (Dicke, 1978; Newkirk, 1984). If yes, this
could mean that the 11-year or Schwabe cycle is driven by a clock
deep within the sun, if no, the cycle may be the result of periodic
processes in the convection zone with only a short memory. Another
interesting phenomenon was recognized when the sunspot record was
reconstructed back to 1600 A.D. During the period 1645 - 1715 A.D.
(Maunder minimum), almost no sunspots could be observed (Eddy, 1976).
^{14}C measurements on tree-rings revealed a clear increase in the
atmospheric ^{14}C concentration during that time, which can be
attributed to a higher production rate (Stuiver and Quay, 1980). Other
^{14}C peaks further back in time indicate that similar periods of quiet

302

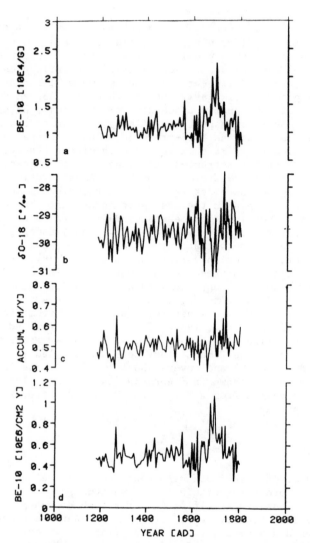

Fig. 2.   Parameters for the Milcent ice core:
(a) ^{10}Be concentration in units of 10^4 atoms/g (Beer
et al., 1983a).
(b) δ^{18}O values in permil (Hammer et al., 1978)
averaged to fit the time resolution of (a).
(c)   Accumulation rate (m/y) derived from the seasonal δ^{18}O
variations and corrected for the thinning of the annual
layer thickness.
(d) ^{10}Be flux (10^6 atoms/cm^{2}y) calculated from the ^{10}Be
concentration (a) and the accumulation rate (c).

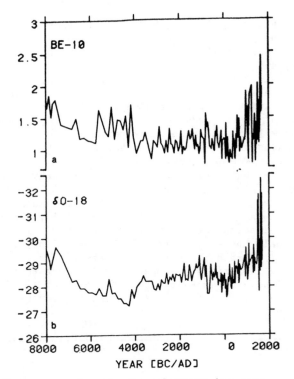

Fig. 3. Parameters for the Camp Century ice core
(a) ^{10}Be concentrations in units of 10^4 atoms/g (Beer et al., 1987a).
(b) δ^{18}O values in permil (Beer et al., 1987a).

sun occurred before (Stuiver, 1988).

As described, radioisotopes are supposed to reflect primarily variations of the solar wind. It is known that solar wind particles are responsible for auroral events in polar regions. In a first step we therefore compare qualitatively the Milcent ^{10}Be record with auroral data.

5. COMPARISON OF ^{10}Be AND AURORAE

Auroral phenomena are optical manifestations of ionization processes caused by solar wind particles penetrating into the upper atmosphere. Solar wind is connected with solar magnetic fields in the interplanetary space which shield the earth from galactic cosmic rays. Therefore, it can be expected that the frequency of aurorae is anticorrelated to the radioisotope production rate. Since it is impossible to derive the corresponding solar wind intensity from historically reported auroral events, only qualitative comparisons with ^{10}Be records are feasible. In Figure 4 the smoothed auroral record of Schove (1983) is compared with the Milcent [10]Be data. The comparison with another more detailed data set for the period 1590 - 1720 A.D. is shown in Figure 5 (Krivsky and Ruzickova, 1986). The general agreement between the two parameters is good. Most of the maxima and minima of the auroral record are also reflected in the [10]Be data. However, the amplitudes of the variations are not always well reproduced. The cross-correlation of the records reveals a phase shift of about 2 y, which corresponds well to the mean atmospheric residence time of [10]Be.

Similar comparisons between ^{14}C and solar activity have been reported earlier (Stuiver and Quay, 1980).

Several attempts have been made to detect the Schwabe cycle in the Milcent record. Although the time resolution (3 - 7 y) is not well suited for this purpose, periodicities between 9 and 12 y have been found (Cini Castagnoli et al., 1984; Beer et al., 1985; Attolini et al., 1988). Indications of the Schwabe cycle have also been observed in biannual [10]Be samples from Dye 3, Greenland, for the 20th century (Beer et al., 1983b). In the ^{14}C tree-ring record, the search for the solar cycle is seriously hampered by the fact that the ^{14}C production variations are damped by about a factor of 100 for a period of 11 y and therefore very difficult to detect (Siegenthaler et al., 1980: Povinec, 1983).

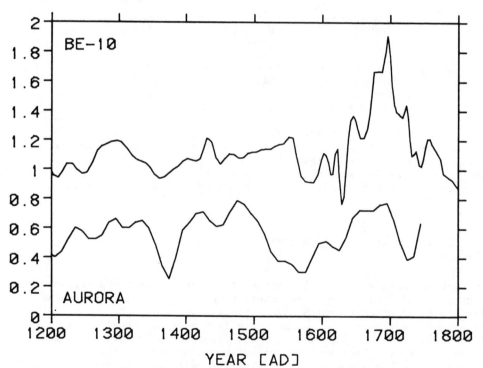

Fig. 4. Comparison of inversely plotted auroral data (Schove, 1983) with the smoothed [10]Be data of Figure 2a.

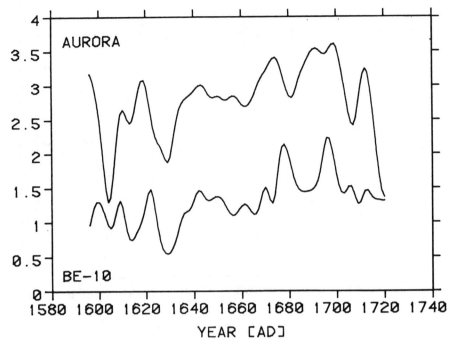

Fig. 5. Comparison of inversely plotted detailed auroral data (Krivsky and Ruzickova, 1986) with the smoothed ^{10}Be data of Figure 2a.

6. COMPARISON OF ^{10}Be and ^{14}C

Since ^{10}Be and ^{14}C are produced by similar processes, but behave geochemically rather differently, a comparison of the two radioisotopes should allow to distinguish between production rate variations and the effects of transport and system changes. However, in order to compare ^{14}C and ^{10}Be, a carbon cycle model has to be used to account for the exchange of CO_2 between atmosphere, ocean and biosphere. Assuming that the ^{10}Be concentration in ice directly reflects the atmospheric production rate, the expected ^{14}C variations can be calculated using the ^{10}Be record as an input function for the carbon cycle model. A detailed discussion of this procedure and its implications are given by Siegenthaler and Beer (1988) and Beer et al. (1987a). Figure 6 shows the calculated Δ^{14}C (solid line) when using the Milcent ^{10}Be record of Figure 2a for the period 1180–1800 A.D. and the box diffusion model of Oeschger (1975; Siegenthaler and Beer, 1988) together with the measured Δ^{14}C (dotted line) (Stuiver and Quay, 1980). This time period is characterized by three peaks

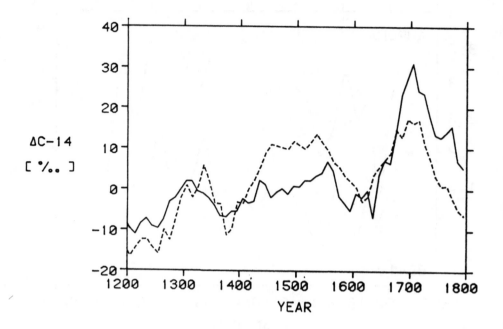

Fig. 6. Comparison of calculated $\Delta^{14}C$ (solid line) using a box-diffusion model (Siegenthaler and Beer, 1988) and the ^{10}Be data (Figure 2a) with measured data from tree-rings (dotted line) (Stuiver and Quay, 1980).

corresponding to the Wolf, Spoerer and Maunder minimum. The comparison shows, beside a generally good agreement, that the Wolf minimum (1300 A.D.) is reproduced quite well by the ^{10}Be record, whereas the Spoerer minimum (1500 A.D.) is somewhat too small and the Maunder minimum (1700 A.D.) too large. Possible reasons for these differences are climatic variations which have been neglected for the calculations. Since, there is no direct information about the climate of this period, we take the $\delta^{18}O$ record (Figure 2b) as a partial substitute for the temperature. As can be seen, the $\delta^{18}O$ values fluctuate statistically around the mean value without clear trends. Only during the Maunder minimum period, the values are slightly lower indicating lower temperatures. A relation between the $\delta^{18}O$ variations and the ^{10}Be concentration has been established (Beer et al., 1987a) by assuming that during cold periods less water is cycled and therefore the precipitation rate is reduced, which leads to higher ^{10}Be concentrations in snow. However, the $\delta^{18}O$ variations at Milcent are quite small, so that no significant climatic influence on the ^{10}Be concentration is to be expected. A more direct and reliable information on the precipitation rate can be derived from the excellent time scale constructed for the Milcent core (Hammer et al., 1978). Based on the seasonal variations of $\delta^{18}O$ it was possible to precisely determine the annual layer thicknesses. By correcting for the thinning of the layers by means of an ice flow model (Hammer et al., 1978) we calculated the accumulation rate as a function of time (Figure 2c). Although there is a slight indication of a lower accumulation rate during the Maunder minimum (consistent with the lowered $\delta^{18}O$), the accumulation rate can be considered as rather constant with a mean value of 0.51 ± 0.05 m. As a matter of fact, if the accumulation rate is used to calculate the ^{10}Be flux (Figure 2d) the resulting curve is almost indistinguishable from the concentration curve (Figure 2a). This suggests that the differences between ^{14}C in tree-rings and ^{10}Be at Milcent are not primarily caused by changes of the precipitation rate (Monaghan, 1987) but rather by transport effects or CO_2 system variations.

The same procedure for comparing ^{10}Be and ^{14}C was used in the case of the Camp Century record (Figure 3) (Beer et al., 1987a; Siegenthaler and Beer, 1988). Figure 7 shows the results which were obtained by assuming that the production before 7000 B.C. was higher by 20% than the mean value between 7000 B.C. and 1800 A.D. A detailed comparison (Beer et al., 1987a; Siegenthaler and Beer, 1988) reveals that the prominent $\Delta^{14}C$ peaks (Suess wiggles) are well reproduced. Even the long-term trend is under these initial conditions consistent with the measured $\Delta^{14}C$ record. This good agreement clearly indicates that the short-term variations are caused by changes of the production rate. It is very unlikely that the climate affects the two different systems just in the way necessary to produce the observed similarity. In addition a correction of the ^{10}Be record derived from the $\delta^{18}O$ values (Figure 3b) (Beer et al., 1987a) is of only minor importance and does not change the general picture. Finally, new records from Antarctica (Beer et al., 1987b; Raisbeck, 1987b) reveal similar structures. From all these findings and arguments it can be concluded that the ^{10}Be records indeed reflect to a large extent the solar activity.

7. COMPARISON OF ^{10}Be AND THE GEOMAGNETIC FIELD

There are three potential sources of information about the history of the geomagnetic field: the palaeomagnetic data from archeological finds and sediments, the $\Delta^{14}C$ record from tree-rings and the ^{10}Be profiles from ice cores. Palaeomagnetic data are based on the remanent magnetization acquired at the time of the fabrication of pottery or of sediment formation. They therefore reflect primarily the local magnetic field. For the sediment studies, the main problem in deriving the field intensity are secondary magnetizations of the sediment and changes in the sediment composition (Tarling, 1988). Figure 8a shows a compilation of a large collection of palaeomagnetic data (McElhinny and Senanayake, 1982). Whilst the palaeomagnetic data indicate local intensities, ^{14}C and ^{10}Be mainly reflect the dipole

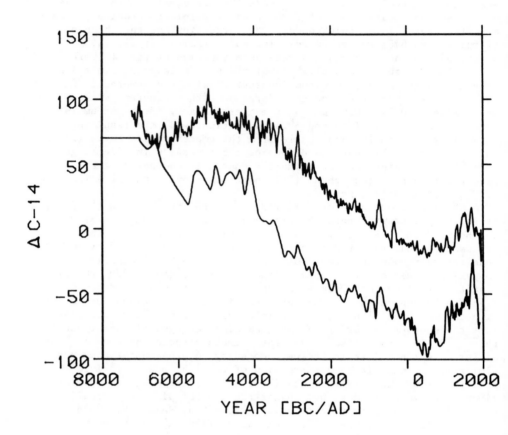

Fig. 7. Comparison of measured $\Delta^{14}C$ data (Stuiver and Kra, 1987) with calculated $\Delta^{14}C$ using the ^{10}Be concentration of Figure 3a as input for a box-diffusion model and assuming that the production rate before 7000 B.C. was higher by 20% than during Holocene (Siegenthaler and Beer, 1988).

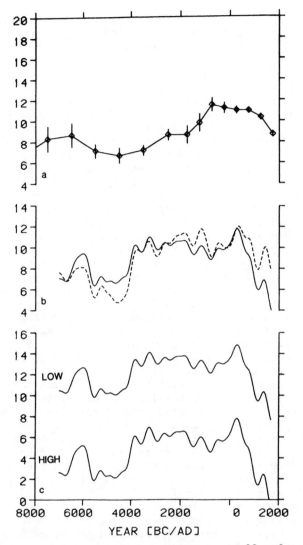

Fig. 8. Global dipole moment in units of 10^{22} Am^2.
(a) Compiled data from palaeomagnetic measurements (McElhinny
     and Senanayake, 1982).
(b) Calculated dipole moment using the smoothed ^{10}Be record
     (Figure 6a) and assuming a constant average level of solar
     activity (Blinov, 1988). The solid line shows the results
     for the smoothed original data, the dotted line after
     applying a δ^{18}O-climate correction (Beer et al., 1987a).
(c) The same as (b) but for low and high levels of solar activity.
     The calculations are based on the smoothed original ^{10}Be
     data (Figure 6a).

field. The intensity of quadrupole and higher order field components is relatively small and decreases quickly with increasing distance from the earth. It can therefore be neglected in a first order approximation of the modulation.

The ^{14}C long-term trend has often been attributed to a sinusoidally changing dipole field (Bucha, 1970; Damon and Linick, 1986). However, the long-term ^{14}C trend depends on the production rate over the preceding period of about 10,000 years. As shown in Figure 7 the ^{10}Be record, which looks rather flat (Figure 3a), is consistent with the Δ^{14}C curve if a production rate higher by 20% than the Holocene mean value is assumed before 7000 B.C. Since the tree-ring record does not yet reach back so far, the production rate during Wisconsin is not known well enough to draw final conclusions about the Holocene Δ^{14}C record. The ^{10}Be profiles on the other hand reflect production variations directly, but since, in contrast to solar modulation, the geomagnetic modulation is largest at the equator, polar ice cores might not reflect the average global production variations. The question how representative polar ^{10}Be profiles are is related to the problem of (1) how much of the deposited ^{10}Be originates from local production, and (2) how much the atmospheric mixing patterns are changing in time. In spite of this uncertainty, it is interesting to try to reconstruct the geomagnetic dipole moment by using the ^{10}Be data. This procedure is based on the following assumptions: the long-term ^{10}Be trend at Camp Century reflects the global production rate and is only determined by the geomagnetic field intensity. This means that solar modulation averaged over several hundred years is constant and there are no climatic effects. Based on these strong assumptions the history of the geomagnetic dipole moment can be re-constructed. Figure 8b shows the results for the original data (Figure 3a) and for the δ^{18}O-corrected data (dotted line). The comparison with the palaeomagnetic data (Figure 8a) reveals the following features: in both cases the profiles are characterized by an increase over the last 2000 y looking backwards in time, followed by a minimum between 6000 and 4000 B.C. Otherwise the agreement is rather poor, as stated earlier (Beer et al., 1984). To demonstrate the effect of the solar modulation the same calculations were repeated for the original smoothed ^{10}Be data for low (Φ = 100) and high (Φ = 900) levels of solar activity. As can be seen (Figure 8c) the changes of the solar activity levels strongly influence the calculated geomagnetic dipole moment. Thus no firm conclusions can be drawn on the role of the geomagnetic field for ^{10}Be (and ^{14}C) variations in the past. It is clear that geomagnetic variations must have influenced radioisotope production, but the palaeomagnetic data are affected with uncertainty. On the other hand, the observed long-term ^{10}Be variations, leaving aside the possible influence of climate, might also be due to solar modulation. For the long-term trends, the situation is therefore more complicated than for the short-term isotope fluctuations which can be traced back with some certainty to varying solar activity.

8. CONCLUSIONS

^{10}Be records from polar regions reveal short-term variations which correlate well with solar phenomena (aurorae, sunspots). The observed fluctuations of ^{10}Be and ^{14}C exhibit surprisingly good consistency. Since both isotopes are produced in a similar way, but have a different geochemical behaviour, this agreement shows that the isotope variations must have been caused mostly by changing production rate, which for short (∿ century) time scales were a consequence of a variable sun. Thus, ^{10}Be proves to be a useful tool to study the short-term history of solar activity over the last 10,000 years. In earlier time, the ice core ^{10}Be concentrations were strongly affected by considerably lower precipitation during the ice age. In the case of the geomagnetic field the situation is less favourable. All sources of information (palaeomagnetism, ^{14}C, ^{10}Be) agree in a decreasing trend of the dipole field over the last 2000 y and a minimum around 5000 B.C. Otherwise the agreement between palaeomagnetic data and the ^{10}Be Camp Century record is rather poor. Ice core ^{10}Be records are automatically biased towards high latitudes, so that uncertainty remains as to their representativeness of global conditions. For a more quantitative interpretation of the ^{10}Be data a better understanding of the transport and deposition processes is necessary. However, from the comparison of ^{10}Be and ^{14}C it seems that climatic effects during Holocene are of little importance, and smaller than might perhaps be expected from theoretical considerations.

ACKNOWLEDGEMENTS

We are thankful to all the people involved in the drilling of the ice cores, the sample preparation and the measurements, especially to H. Oeschger, W. Woelfli, M. Suter, G. Bonani, D. Lal and R. Finkel for interesting discussions, to S. Johnsen for providing us the Milcent ^{18}O values and to K. Haenni for designing Figure 1. This work was financially supported by the Swiss National Science Foundation.

REFERENCES

Attolini, M.R., Galli, M., and Nanni, T., 1988, this volume.
Beer, J., Siegenthaler, U., Oeschger, H., Andree, M., Bonani, G., Hofmann, H., Nessi, M., Suter, M., Woelfli, W., Finkel, R. and Langway, C, 1983a, 'Proc. 18th Int. Cosmic Ray Conf., Bangalore', 9, 317.
Beer, J., Andree, M., Oeschger, H., Stauffer, B., Balzer, R., Bonani, G., Stoller, C., Suter, M., Woelfli, W. and Finkel, R., 1983b, Radiocarbon, 25, 269.
Beer, J., Oeschger, H., Siegenthaler, U., Bonani, G., Hofmann, H., Morenzoni, E., Nessi, M., Suter, M., Woelfli, W., Finkel, R.C. and Langway, C. 1984, Nucl. Instrum. Meth., B5, 380.

Beer, J., Oeschger, H., Finkel, R.C., Cini Castagnoli, G., Bonino, G., Attolini, M.R. and Galli, M., 1985, M. Nucl. Instrum. Meth. B10/11, 415.

Beer, J., Siegenthaler, U., Bonani, G., Finkel, R.C., Oeschger, H., Suter, M. and Woelfli, W., 1987a, submitted to Nature.

Beer, J., Bonani, G., Hofmann, H.J., Suter, M., Synal, A., Woelfli, W., Oeschger, H., Siegenthaler, U. and Finkel, R.C., 1987b, 'Proc. fourth Int. Symp. on Accelerator Mass Spectrometry', in press.

Bhat, C.L., Mayer, C.J., Rogers, M.J. and Wolfendale, A.W., 1987, J. Phys. G: Nucl. Phys. 13, 257.

Blinov, A., 1988, this volume.

Bucha, V., 1970, in 'Radiocarbon Variations and Absolute Chronology', (Almquist & Wiksell, Stockholm), 501.

Cini Castagnoli, G., Bonino, G., Attolini, M.R., Galli, M. and Beer, J., 1984, Nuovo Cimento C7, 235.

Damon, P.E. and Linick, T.W., 1986, Radiocarbon 28, 266.

Dicke, R.H., 1978, Nature 276, 676.

Eddy, J.A., 1976, Science 192, 1189.

Finkel, R., Krishnaswami, S. and Clark, D.L., 1977, Earth Planet. Sci. Lett. 35, 199.

Geiss, J., Oeschger, H. and Schwarz, U., 1962, Space Sci. Rev. 1, 197.

Hammer, C.U., Clausen, H.B., Dansgaard, W., Gundestrup, N., Johnsen, S.J. and Reeh, N.J., 1978, J. Glaciol. 20, 3.

Junge, C.E., 1975, 'Proc. Grenoble Symp. on Isotopes and Impurities in Snow and Ice', IAHS-AUSH Publication 118.

Krivsky, L. and Ruzickova-Topolova, B., 1986, 'Proc. 12th Reg. Consult. on Solar Physics', in press.

Lal, D. and Peters, B., 1967, in 'Handbuch der Physik' 46, 551.

McElhinny, M.W. and Senanayake, W.E., 1982, J. Geomag. Geoelectr. 34, 39.

Monaghan, M.C., 1987, Earth Planet. Sci. Lett., 84, 197.

Newkirk, G., 1984, Nucl. Instrum. Meth. B5, 404.

Nishiizumi, K., Regnier, S. and Marti, K., 1980, Earth Planet. Sci. Lett. 50, 156.

Oeschger, H., Siegenthaler, U., Gugelmann, A. and Schotterer, U. 1975, Tellus 27, 168.

Povinec, P., 1983, Radiocarbon 25, 259.

Raisbeck, G.M., Yiou, F., Fruneau, M., Loiseaux, J.M., Lieuvin, M., Ravel, J.C. and Lorius, C., 1981, Nature 292, 825.

Raisbeck, G.M., Yiou, F., Bourles, D., Lorius, C., Jouzel, J. and Barkov, I., 1987, Nature 326, 273.

Raisbeck, G.M. and Yiou, F., 1988, this volume.

Sarmiento, J.L. and Gwinn, E., 1986, J. Geophys. Res. 91, 7631.

Schove, D.J., 1983, 'Sunspot cycles' (Hutchinson Ross Stroudsburg, PA).

Siegenthaler, U., Heimann, M. and Oeschger, H., 1980, Radiocarbon 22, 177.

Siegenthaler, U. and Beer, J. (1988) this volume.

Stuiver, M. (1988) this volume.

Stuiver, M. and Quay, P.D. 1980, Science, 207, 11.
Stuiver, M. and Kra, R.S., 1986, Radiocarbon, vol. 28, No. 2b.
Suess, H., 1970, in 'Radiocarbon Variations and Absolute Chronology',
   (Almquist & Wiksell, Stockholm), 595.
Tarling, D.H., 1988, this volume.

MODEL COMPARISON OF ^{14}C AND ^{10}BE ISOTOPE RECORDS

U. Siegenthaler and J. Beer,
Physics Institute, University of Bern, Switzerland.
and
Institute for Medium Energy Physics, Zurich, Switzerland.

ABSTRACT. The atmospheric concentrations of ^{14}C and of ^{10}Be are
influenced by variations in isotope production rate due to solar
or geomagnetic influence, but also by internal redistribution processes.
The various mechanisms are discussed for ^{14}C. While records of ^{10}Be
in Greenland ice and of ^{14}C in tree-rings for the past nearly 10,000
years cannot directly be compared because of different geochemical
behaviour, a comparison can be made using a carbon cycle model.
Short-term variations in both isotopes (with time scale $\cong$ 100 yr)
are highly correlated, which indicates that they are caused by
production rate variations, probably due to variations in the sun's
magnetic activity. The long-term trend for the two records is only
partly consistent with data on past variations of the geomagnetic
field. For ^{14}C, the possibility must be considered that the mean
production rate before the start of the tree-ring record was higher
than in the past 9000 years and that the observed long-term ^{14}C
decrease is thus due to radioactive decay of the "excess" ^{14}C produced
in the past.

1. INTRODUCTION

The rate of production of radioisotopes by cosmic ray interactions
in the atmosphere depends on the incoming flux of cosmic radiation.
This flux is modulated by magnetic fields carried by the solar wind
and by the geomagnetic field that both partly shield the earth from
the galactic cosmic radiation. Thus, radioisotope records contain
information about past variations of solar activity and geomagnetic
field. In addition, however, the concentration of a radioisotope
is influenced by processes on earth, like transport and deposition
of ^{10}Be and other aerosol-bound isotopes or changes in the global
carbon cycle for ^{14}C, as discussed in more detail in the accompanying
paper (Beer et al., 1988). The interpretation of the record of
one radioisotope alone is therefore not unequivocal; but the combined
consideration of two isotopes with different geochemical behaviour,
like ^{10}Be and ^{14}C, helps to distinguish between production rate

315

F. R. Stephenson and A. W. Wolfendale (eds.),
Secular Solar and Geomagnetic Variations in the Last 10,000 Years, 315–328.
© 1988 by Kluwer Academic Publishers.

variations and internal causes of concentration changes. This will be done in the following investigations based on a [10]Be record in ice from Greenland and on tree-ring ^{14}C data, both records covering about the last 9000 years (Figure 1). The [10]Be data were obtained from the ice cores from Camp Century and from Milcent station and measured by accelerator mass spectrometry (Beer et al., 1987, 1988). The ^{14}C record (Stuiver et al., 1986) combines high-precision data from several laboratories.

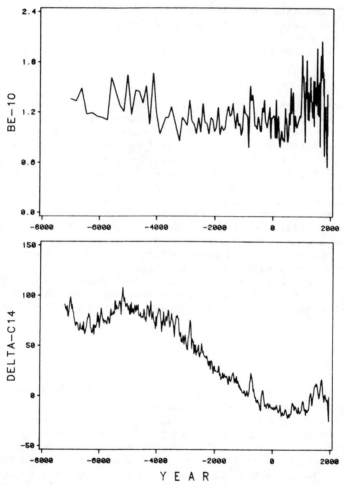

Fig. 1. [10]Be concentration in the Camp Century ice core, unit: 10^4 atoms per g of ice (from 1200 to 1800 A.D. also results from an ice core from Milcent, Greenland, are included), and tree-ring ^{14}C record (Stuiver, et al., 1986).

In order to compare the two isotopes, their different behaviour must be considered. [10]Be is attached to aerosols soon after its production and is deposited within a few years at most at the earth's surface, so that its concentration in precipitation reflects production changes directly, without attenuation or delay. On the other hand, a [10]Be record is not necessarily representative for global average variations. The atmospheric ^{14}C concentration, on the other hand, is rather homogeneous because of rapid mixing of the atmosphere, but it is related in a complex manner to production rate. Only 2 percent of the global ^{14}C amount resides in the atmosphere, the rest is distributed in the ocean and in the land biosphere. The exchange between the various carbon reservoirs strongly affects the relation between production rate and atmospheric concentration of ^{14}C, which must be taken into account by means of an appropriate model of the carbon cycle.

In the following, the influence of production rate variations and of internal system changes on atmospheric ^{14}C is first discussed schematically. Then [10]Be and ^{14}C are discussed in combination. It will be shown that the short-time ($\sim 10^2$ yr) variations are presumably due to solar influence. The cause of the long-time ($\sim 10^3$ yr) variations is more difficult to pin down; according to the available data, variations of the geomagnetic field may partly, but not fully, be responsible. For ^{14}C, the ^{14}C inventory on earth at the start of the tree-ring record, ca. 8000 yr B.P. - in other words the average production rate over the preceding 2 or 3 half-lives of [14]C - also influences the long-term trend.

2. MODEL RESPONSES OF ^{14}C TO CHANGES IN PRODUCTION OR IN THE CARBON CYCLE

For this study, a carbon cycle model is used (see Figure 2) consisting of an atmosphere subdivided into troposphere (80 per cent of mass) and a stratosphere (20 per cent of mass), an ocean subdivided into a mixed layer of 75 m depth and a deep sea (3654 m deep) mixed by one-dimensional eddy diffusion with constant diffusity (box-diffusion model) (Oeschger et al., 1975), and a four-box biosphere (see Siegenthaler and Oeschger, 1987).

The dynamic parameters describing ocean-atmosphere exchange were determined from the distribution of natural [14]C (eddy diffusivity $K = 4005$ m^2/yr; atmospheric residence time of CO_2 with respect to exchange with the ocean $\tau_{am} = 7.95$ yr). The assumed atmospheric CO_2 concentration is 280 ppm. The steady-state ^{14}C/C ratios (in the following just called ^{14}C concentrations) corresponding to the CO_2 fluxes indicated in Figure 2 are: troposphere 100%, stratosphere 102.1%, biosphere (average) 99.0%, mixed layer 95.4% and deep sea (average) 84.4%. The ^{14}C production rate is assumed such as to balance the global desintegration rate by radioactive decay; following Lal and Peters (1967) 56 per cent occurs in the stratosphere, 44 per cent in the troposphere.

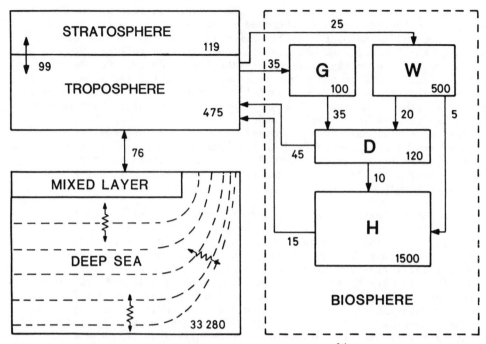

Fig. 2. Carbon cycle model used for calculating ^{14}C concentrations.
Reservoir sizes are in Gt C (gigatons of carbon,
$1Gt = 10^9$ t $= 10^{15}g$), fluxes in Gt C/yr.

What are the factors that determine the atmospheric ^{14}C concentration? Obviously, it is proportional to the global production
rate Q. Furthermore, it depends on the size of the different carbon
reservoirs and on the exchange fluxes between and within them.
Siegenthaler et al. (1980) discussed this in detail. They found for
the sensitivity of the atmospheric ^{14}C concentration R_a to (small)
changes in the system parameters:

$$\frac{dR_a}{R_a} = \frac{dQ}{Q} - 0.11 \frac{dK}{K} + 0.047 \frac{d\tau_{am}}{\tau_{am}} - 0.050 \frac{dN_a}{N_a} - 0.009 \frac{dN_b}{N_b}$$

$$+ 0.22 \frac{dh_{ds}}{h_{ds}} \quad , \tag{1}$$

where K = eddy diffusivity, τ_{am} = atmospheric residence time of CO_2,
N_a, N_b = total amounts of carbon in atmosphere and biosphere, h_{ds} =
depth of deep sea.

This result was obtained by Siegenthaler et al. (1980) based
on a model that differed in details (parameter values, structure of
biosphere) from that used here, but the effect on the numerical results
in equation (1) should be marginal.

Equation (1) permits the importance of various changes to be judged. Clearly, the isotope production rate Q is of dominant influence. The potential role of the other, internal, parameters may be discussed by comparing ice-age and Holocene conditions. The atmospheric CO_2 concentration, and therefore N_a, was about 30% lower (200 ppm instead of 280 ppm) during the ice age (e.g. Neftel et al., 1982). The corresponding atmospheric ^{14}C change according to equation (1) was + 1.5%; physically, it was caused by the larger air-sea transfer resistance due to the smaller exchange flux (which is proportional to the atmospheric CO_2 concentration). The atmospheric CO_2 residence time may have been reduced due to stronger winds leading to a relative increase of the air-sea gas exchange; if we, somewhat arbitrarily, assume a decrease of τ_{am} by 20 per cent, R_a would have been lower by 1%.

The influence of a modified biosphere mass is, according to equation (1), negligible. The ice age sea level was about 100 m lower than now, h_{ds} thus about 3 per cent smaller; this did not affect ^{14}C in a measurable way. The potentially most important internal change is that in the vertical exchange rate within the ocean, represented in the model by the eddy diffusivity K. Since most of the ^{14}C decay on earth, which in steady state has to balance the production rate, takes place in the ocean, a reduced oceanic turnover rate affects not so much the $^{14}C/C$ ratio in the deep sea, but rather raises it in the upper ocean and in the atmosphere. ^{14}C produced in the atmosphere is then less easily transported to the deep ocean and "piles up" in the atmosphere. Studies on deep sea sediments have yielded clear evidence that deep water formation in the North Atlantic, which today accounts for roughly half of the deep ocean ventilation, was significantly reduced during the ice age. It is difficult to judge the corresponding change in the eddy diffusivity K, since K is only a model parameter summarizing the different mechanisms of vertical exchange. A reasonable estimate, probably at the upper limit, may be a decrease of K by a factor of two. This would have lead to a ^{14}C age of the bottom water in the model, relative to the mixed layer, of 1880 years, instead of 1020 years for the Holocene value, K = 4005 m^2/yr. Indeed, first ^{14}C measurements on surface-dwelling and bottom-dwelling foraminifera sediments from the South China Sea indicate that the surface-bottom age difference there may have been several hundred years larger during late glacial time than during the Holocene (M. Andrée, personal communication). If K decreased to half its standard value, atmospheric ^{14}C increased by 9.1 per cent (calculated from the exact model equations, not the linearized equation (1)).

We conclude that during the ice age, atmospheric ^{14}C may, for an unchanged production rate, have been something like 5 to 10% higher than during the Holocene, essentially because of reduced deep water ventilation. However, the isotope records discussed here are only from the Holocene during which environmental conditions remained much more stable than during the glacial-postglacial transition. Therefore, variations of the steady-state ^{14}C concentration due to carbon cycle changes during the past 9000 years were probably small. This is supported by the finding that the ^{14}C concentration difference between surface-dwelling and bottom-dwelling foraminifera picked from deep-sea sediments from the South China Sea was more or less constant in that time period (Andrée et al., 1986).

So far, we have only discussed the steady state ^{14}C distribution on earth. It is instructive to consider also the transient variations. There is evidence that the rate of deep water formation changed at the glacial-postglacial transition. After a sudden change in oceanic mixing (eddy diffusivity), the ^{14}C concentration has adjusted to its new value within about 1000 years (Figure 3, solid line). Therefore, the tree-ring record (Figure 1) that extends back to about 9000 calendar years B.P. cannot be influenced by an event that took place at or before the glacial-postglacial boundary, 11,000 calendar years B.P. (For the dating of this boundary, given by the end of the Younger Dryas cold phase, see Beer et al., 1987.) Specifically, the long-term ^{14}C decrease from 7000 to 0 B.C. cannot be an after-effect of the glacial-postglacial transition. In addition, as discussed above, ^{14}C dates on foraminifera give direct evidence that the oceanic surface-to-bottom ^{14}C difference did not decrease during the Holocene.

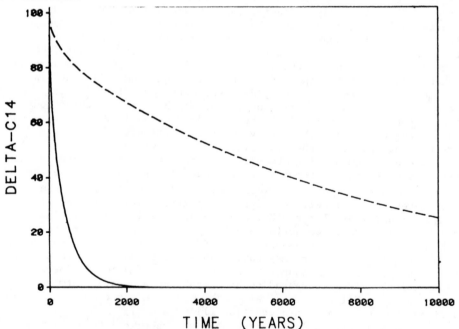

Fig. 3. Response of atmospheric ^{14}C to a sudden change of the eddy diffusivity from 2000 to 4000 m^2/y at time 0 (solid line) and to a sudden lowering of the ^{14}C production rate (dashed line). Vertical axis: ^{14}C concentration excess (in permil) above the final steady state value.

Because of the carbon exchange between different reservoirs, atmospheric ^{14}C reflects production variations only with attenuated amplitude. This is illustrated by Figure 3: 1000 years after a sudden step in the production rate, the atmospheric concentration has changed by only 17 per cent, compared to the change in the final steady state. The reason is that the global ^{14}C inventory adjusts to the new production rate with the ^{14}C half life of 5730 years; the dashed curve in Figure 3 therefore is approximately an exponential with this half life. All production rate variations on time scales shorter than several ^{14}C half lives are therefore considerably attenuated.

Another way to discuss this is by considering periodic, sinusoidal production rate variations. Figure 4 shows the attenuation, defined as the amplitude ratio of the relative atmospheric ^{14}C variations, $\Delta R_a/R_a$, and the relative production rate variations, $\Delta Q/Q$, and the phase lag between production and concentration in degrees. For a period of 11 years (sunspot cycle), the attenuation factor is 0.01; for this reason this cycle is difficult to detect in [14]C.

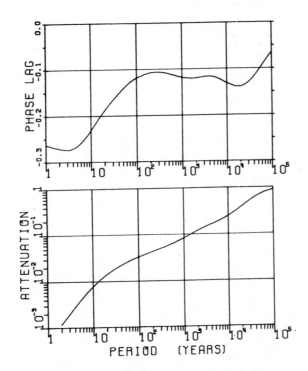

Fig. 4. Phase lag (as a fraction of one period) and amplitude attenuation factor of the atmospheric ^{14}C concentration in response to sinusoidal production rate variations, calculated for the carbon cycle model of Figure 2.

3. SHORT-TERM VARIATIONS OF ^{10}BE AND ^{14}C: THE SOLAR SIGNAL

As the foregoing discussion shows, the records of the two isotopes cannot be compared directly. Rather, we calculate from ^{10}Be a synthetic ^{14}C record which can then be compared with the tree-ring data. For doing this, we assume that the ice core ^{10}Be data truly reflect the variations of the global production rate, i.e. that they are not influenced by changing atmospheric transport or varying rate of snowfall (which would, for a constant ^{10}Be flux, cause varying concentration in the snow). Further, we assume that the production rates of ^{14}C and of ^{10}Be are always proportional to each other. Then the ^{10}Be record is used as input to the carbon cycle model, representing the ^{14}C production rate variations. Figure 5 shows the comparison between the thus calculated atmospheric ^{14}C concentration and the tree-ring record; as usual both are expressed as Δ^{14}C (in ‰), i.e. as the deviation from the ^{14}C/C ratio of a standard. Since the proportionality factor between ^{10}Be concentration in Greenland ice and the global ^{14}C production rate is not known, the absolute values of the model-calculated concentration are not fixed, and the curves are normalized such as to have the same average value between 0 and 1000 A.D. as the tree-ring record. Two calculated curves are shown in Figure 5; they differ by a factor of 1.2 in the assumed ^{14}C production rate before 7000 B.C. which is not known. This initial condition strongly affects the long-term trend, as is understandable from the discussion in the last section and as considered further below.

In order to test the sensitivity of the results, a model run was carried out with an eddy diffusivity of 7110 m^2/yr, the value obtained by a calibration with bomb-produced ^{14}C (Siegenthaler and Oeschger, 1987), instead of 4005 m^2/yr as obtained from the distribution of natural ^{14}C. This change has a very minor influence; the amplitude of the calculated short-term ^{14}C excursions is changed by a few per cent only.

Comparison of the observed and calculated curves in Figure 5 reveals strong similarities regarding the short-term fluctuations. In order to permit a better comparison, the long-term trend was removed from the tree-ring record and the calculated ^{14}C curve (upper model curve in Figure 5). The detrending was performed with a binomial ($\cong$ Gaussian) high-pass filter with a standard deviation of 140 years; then the data were slightly smoothed with a 5-point binomial low-pass filter (separation of the interpolated data points: 20 years). The result is shown in Figure 6 for the period 3000 B.C. - 1800 A.D. There are numerous fluctuations of ca. 200 yr duration, and the similarity between the two curves is striking.

A cross-correlation calculation yields a maximum correlation coefficient of 0.58 for a 0 to 10 yr lag between the two time series (Beer et al., 1987). In order to test the significance of the correlation coefficient, the number of degrees of freedom must be known. Considering the uneven spacing of the ^{10}Be data and the filtering, we estimate that there are about 150 degrees of freedom between 3000 B.C. and 1800 A.D., for which already a correlation

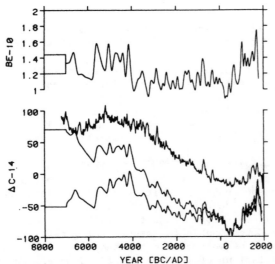

Fig. 5. Top: ^{10}Be concentration of Figure 1 (unit: 10^4 atoms/g)
used as model input function representing ^{14}C production
rate. Center: Tree-ring Δ^{14}C data (Stuiver et al., 1986).
Bottom: model-simulated Δ^{14}C history based on the ^{10}Be
data, for two different initial conditions (cf. the two
straight lines before 7000 B.C.). The model curves were
shifted by 70‰ to permit easier comparison with the
measured Δ^{14}C data.

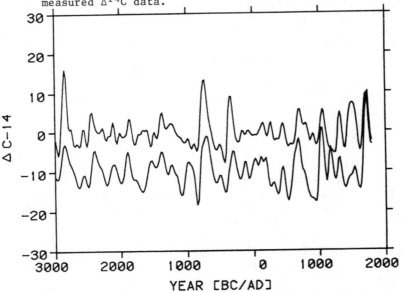

Fig. 6. Detrended measured (upper curve) and model-calculated
Δ^{14}C histories of Figure 5. The close similarity indicates
that the fluctuations of ^{10}Be and ^{14}C are both due to
the same cause, i.e. varying production rate.

coefficient of 0.264 is different from zero with 99.9 per cent probability. With r = 0.58, the two time series are thus correlated with
extremely high significance.

The good agreement between observed and calculated ^{14}C variations
strongly points to a common cause of the fluctuations in both isotopes,
namely variations of the isotope production rate due to modulation
of the cosmic ray flux. Indeed, it is very hard to conceive of another
mechanism that would lead to simultaneous concentration variations
of both isotopes that, in addition, are consistent in amplitude as
shown by Figure 6.

For ^{14}C, the "wiggles" of 100-200 yr duration have been discussed
for a long time (e.g. Suess, 1970; Damon et al., 1986), and a good
correlation with sunspot numbers indicates that, at least for the
past few centuries, the ^{14}C changes occurred in response to varying
solar modulation of the isotope production rate (Stuiver and Quay,
1980). It then seems reasonable to assume that generally, the short-
term isotope fluctuations were caused by changing solar activity,
especially since variations of the geomagnetic dipole with characteristic times of < 1000 yr are small (McElhinny and Senanayake, 1982).

One may wonder whether the isotope production fluctuations were
deviations from an average in both directions, or whether they
correspond mainly to periods of increased or decreased cosmic ray
flux. The Maunder sunspot minimum, ca. 1645-1715 A.D. (Eddy, 1976),
was a period of very low solar activity with correspondingly high
isotope production rates, as seen in Figure 6 and especially in the
figures of the accompanying paper (Beer et al., 1988). Inspection
of Figure 6 suggests that the isotope fluctuations are due mainly
to positive excursions of varying amplitude; such excursions may have
been caused by periods of a quiet sun of the Maunder minimum type.
If this hypothesis is correct, then the distribution of the detrended
^{10}Be and ^{14}C concentration values should be asymmetric, with a tail
at positive values. Figure 7 shows frequency histograms for the detrended ^{10}Be and ^{14}C data for the period 3000 B.C. to 1800 A.D. They
are indeed asymmetrical for both isotopes. A test for non-normal
distribution of the data can be made by considering the skewness,
defined by $\langle (x-\langle x \rangle)^3 \rangle / \sigma^3$, where the angular brackets denote mean
values and σ is the standard deviation (Sachs, 1978, p.253). (The
values for the skewness depend in our case somewhat on the way of
detrending). The skewness for the ^{10}Be data is 0.756, that for the
tree-ring ^{14}C data is 1.268; both are significantly different from
zero (and therefore the distributions different from normal) with
a probability of $\geq$ 99.9 per cent.

Thus, the data give evidence that from time to time the sun goes
through a period lasting 100-200 years of considerably reduced
activity with low sunspot numbers and low frequency of aurorae (see
Beer et al., 1988). During such a period, the solar activity is
reduced and consequently the galactic cosmic radiation is shielded
less, leading to higher isotope production rates in the atmosphere.

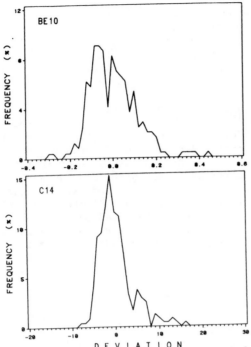

Fig. 7. Frequency histograms for the detrended ^{14}C record of
Figure 6 (lower curve; unit: permil) and the analo-
gously detrended ^{10}Be record (upper curve; unit: 10^4 atoms/g).
Both distributions are asymmetrical, suggesting that
periods of differently strong concentration increases
are superimposed on a slowly varying baseline.

4. LONG-TERM ISOTOPE TRENDS: GEOMAGNETIC SIGNAL?

Assuming that the ice core ^{10}Be concentrations are always proportional
to the isotope production rate and taking them as model input, we
have obtained the two simulated ^{14}C curves shown in Figure 5. They
differ in the initial condition for the production rate before 7000
B.C., corresponding to a ^{10}Be concentration of 1.44×10^4 atoms/g (upper
curve) and 1.20×10^4 atoms/g (= average ^{10}Be value 7000 B.C. to 1800
A.D.: lower curve). The pre-7000 B.C. production rate cannot be taken
from the ice core ^{10}Be data, because these were, at that time,
obviously influenced by lower snowfall rates during glacial and early
postglacial time (Beer et al., 1987).

The model-calculated curves in Figure 5 show in their long-term
trend similar features to the tree-ring ^{14}C record, namely a maximum
around 5000 B.C. and an increase after 500-1000 A.D. (The calculated
curves are shifted by $-70\%_0$ in order to avoid confusion with the
tree-ring record). The model curve with the higher initial condition
fits quite well the observed trend from 6000 B.C. on. Thus, a possible

326

explanation for the slow ^{14}C decrease from 5000 to 0 B.C. is that
the weighted mean production rate during the time before 7000 B.C.
had been about 20% higher than in the period since, while afterwards
the production was proportional to the ^{10}Be record.

The global isotope production rate Q depends on the strength
M of the geomagnetic dipole field as $Q/Q_o = (M/M_o)^{-0.5}$ (Elsasser et
al., 1956, Blinov, 1988). Based on roughly antiparallel trends of
the tree-ring ^{14}C data and the time history of the geomagnetic dipole
field, the long-term ^{14}C trend has often been attributed to geomagnetic
variations (Bucha, 1970; Damon and Linick, 1986). The correlation
between the archaeomagnetic data and the Greenland ^{10}Be results is
not very good (Beer et al., 1988). Here we compare the geomagnetic
intensity record, as compiled by McElhinny and Senanayake (1982),
with ^{14}C by calculating the atmospheric concentration corresponding
to the geomagnetic variations, assuming that the solar activity remained
constant on average. As for the ^{10}Be-based calculation, the ^{14}C trend
depends on the initial condition; Figure 8 shows the result for two
different initial conditions. The upper calculated curve is partly similar

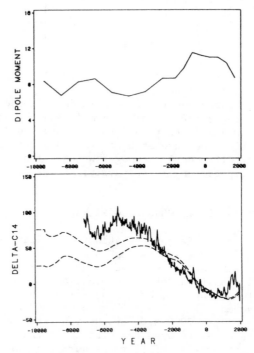

Fig. 8. Top: Geomagnetic dipole variations based on archaeomagnetic data (McElhinny and Senanayake, 1982); (unit: 10^{22}
A m^2). Bottom, solid line: tree-ring ^{14}C record (Stuiver
et al., 1986); dashed line: model-calculated ^{14}C history
for the prescribed geomagnetic variations and two
different initial conditions (cf. the two straight lines
before 9500 B.C.).

to the tree-ring record. The maximum ca. 5000 yr B.P. and the minimum 0-1000 A.D. are both simulated, but they significantly lag behind the tree-ring data. Thus, the geomagnetic dipole variations may well be responsible for some features of the long-term ^{14}C trend. However, the decrease 5000-0 B.C. cannot be explained by the geomagnetic data which go back to 9500 B.C. alone; instead, as discussed above, a relatively high initial ^{14}C inventory on earth at the start of the model calculation is necessary to simulate the concentration decrease between 5000 and 0 B.C. Interestingly, archaeomagnetic data seem indeed to point to low dipole intensity 20,000-30,000 yr B.P. (McElhinny and Senanayake, 1982).

Generally, it must be emphasized that the long-term ^{14}C variations do not depend on the instantaneous production rate only, but on the history of the isotope production during two to three ^{14}C half-lives preceding the start of the tree-ring record. It is therefore highly desirable to obtain high-quality information on isotope concentrations, geomagnetic field intensity and solar activity from the time before 10,000 B.P.

ACKNOWLEDGEMENTS

Thanks are due to A. Blinov, H. Oeschger, M. Stuiver and W. Wölfli for valuable discussions. This work was supported by the Swiss National Science Foundation.

REFERENCES

Andree, M., Oeschger, H., Broecker, W., Beavan, N., Klas, M., Mix, A., Bonani, G., Hofmann, H.J., Suter, M., Woelfli, W., and Peng, T.-H., 1986, Climate Dynamics 1, 53.
Beer, J., Siegenthaler, U., Bonani, G., Suter, M., Woelfli, W., Oeschger, H., and Finkel, R.C., 1988, Nature, in press.
Beer, J., Siegenthaler, U., and Blinov, A., 1988, This volume.
Blinov, A., 1988, This volume.
Bucha, V., 1970, In: 'Radiocarbon variations and Absolute Chronology', 501.
Damon, P.E., and Linick, T.W., 1986, Radiocarbon 28, 266.
Eddy, J.A., 1976, Science 192, 1189.
Elsasser, W., Ney, E.P., and Winckler, J.R., 1956, Nature, 178, 1226.
Lal, D., and Peters, B., 1967, In: 'Handbuch der Physik', vol. 46/2, 551. Springer-Verlag.
McElhinny, M.W., and Senanayake, W.E., 1982, J. Geomag. Geoelectr., 34, 39.
Neftel, A., Oeschger, H., Schwander, J., Stauffer, B., and Zumbrunn, R., 1982, Nature 295, 220.
Oeschger, H., Siegenthaler, U., Schotterer, U., and Gugelmann, A., 1975, Tellus 27, 168.
Sachs, L., 1978, Angewandte Statistik, Springer-Verlag.
Siegenthaler, U., and Oeschger, H., 1987, Tellus 39B, 140.

Siegenthaler, U., Heimann, M., and Oeschger, H., 1980, <u>Radiocarbon</u>, 22, 177.

Stuiver, M., and Quay, P.D., 1980, <u>Science</u>, 207, 11.

Stuiver, M., Kromer, B., Becker, B., and Ferguson, C.W., 1986, <u>Radiocarbon</u>, 28, 969.

Suess, H.E., 1970, In: 'Radiocarbon variations and Absolute Chronology', 595. Almquist & Wiksell, Stockholm.

THE DEPENDENCE OF COSMOGENIC ISOTOPE PRODUCTION RATE ON SOLAR ACTIVITY
AND GEOMAGNETIC FIELD VARIATIONS

A. Blinov
Physics Institute, University of Bern
Permanent address: Department of Cosmic Research,
Leningrad Polytechnic, Politechnicheskaya 29,
Leningrad 195251, USSR.

ABSTRACT. In order to interpret the measurements of cosmogenic isotope
concentrations in various terrestrial archives it is important to
understand their production by nuclear reactions induced by the
galactic cosmic rays in the atmosphere. In this paper the global
production rates of ^{10}Be and ^{36}Cl are calculated using a hadronic
cascade model. Primary cosmic ray flux variations due to heliomagnetic
and geomagnetic modulation and the resulting changes of the ^{10}Be and
^{36}Cl production rates are discussed and compared with ^{14}C variations.
Several simple approximations describing the relationship between
production rate, a solar activity parameter and geomagnetic dipole
moment are given. The accuracy of the developed model is estimated
by analysing the main uncertainties, and possible future improvements
are proposed.

1. INTRODUCTION

The recent development of accelerator mass-spectrometry (AMS) has
made it possible to measure with high precision long-lived cosmogenic
radioisotopes in terrestrial archives, such as polar ice cores and
deep sea sediments, covering a time period of several million years.
The most promising results were obtained for ^{10}Be ($T_{\frac{1}{2}} = 1.5 \times 10^6$ years)
and ^{36}Cl ($T_{\frac{1}{2}} = 3.08 \times 10^5$ years) (Beer et al., 1983; Raisbeck & Yiou,
1984). However, the interpretation of the experimental data in terms
of geophysics or cosmic ray physics requires a mathematical description
of the nuclear interaction processes leading to the atmospheric isotope
production. Especially interesting phenomena are the solar and
geomagnetic modulation of the primary cosmic ray flux. Many studies
have been performed for estimating the global ^{14}C production rate, in
addition a large number of papers have been devoted to atmospheric
neutron flux calculations (Lingenfelter & Ramaty, 1969; Korff et al.,
1979). The relatively good agreement between the theoretically
predicted and the measured ^{14}C inventory on Earth proved the reliabiliy
of the models. Two different approaches to calculate the cosmic-ray
induced production rates in spallation reactions have been used (Lal

F. R. Stephenson and A. W. Wolfendale (eds.),
Secular Solar and Geomagnetic Variations in the Last 10,000 Years, 329–340.
© 1988 by Kluwer Academic Publishers.

& Peters, 1967; O'Brien, 1979). In this paper we are going to discuss a different approach, which is based on an analytical description of the hadronic cascade. We shall compare our results for ^{14}C production with those obtained by Castagnoli & Lal (1980) using their approximation for the solar modulation of the cosmic ray flux.

2. ATMOSPHERIC PRODUCTION OF ^{10}Be AND ^{36}Cl

The radionuclides ^{10}Be and ^{36}Cl are produced in the atmosphere by the following cosmic-ray induced nuclear reactions (1-4):

$$^{40}Ar \ (N, \ ^{5}X) \ ^{36}Cl \tag{1}$$

$$^{36}Ar \ (n,p) \ ^{36}Cl \tag{2}$$

$$^{14}N \ (N, \ ^{5}X) \ ^{10}Be \tag{3}$$

$$^{16}O \ (N, \ ^{7}X) \ ^{10}Be \tag{4}$$

where n and p are neutron and proton, N is a nucleon, ^{5}X and ^{7}X are any possible combinations of particles with total mass equal to 5 and 7 atomic units respectively. The reactions (1) and (3) play the dominant role, so in order to simplify the calculations we shall consider them in the following as the only production channels. It has been shown by Mlotek (1981) and Levchenko & Blinov (1984) that in the case of medium energies (100 MeV $< E_N <$ 30 GeV) of the interacting nucleons the kinetic Boltzmann equations used for describing the propagation of the nucleon component in the atmosphere are formally similar to the Schrödinger equations and for their solution one can apply Feynman's diagram method. The solution in the case of a thick target gives the yield of cascade interaction products as the sum of the yields from each step of the cascade development. Presuming a small contribution to the total production rate from the secondary pion-induced reactions, we can write the equation for the isotope production rate "Y(E)" by a primary cosmic ray proton with kinetic energy "E" in the form

$$Y(E) = Y_1(E) + \sum_{i=2}^{I} \sum_{j}^{p,n} Y_{ij}(E) \tag{5}$$

where

$$Y_1(E) = \frac{K_r \ Sig_r(E)}{K_{tot} Sig_{tot}(E)} \tag{6}$$

$$y_{i,j} = \frac{K_r}{K_{tot}} \int_{E_t}^{E} Y_{(i-1),j}(E') n_j(E',E) dE' \tag{7}$$

K_r/K_{tot} is the relative target gas concentration in air, $Sig_r(E)$ the excitation function for the production reaction, $Sig_{tot}(E)$ the energy dependence of the total inelastic cross section for a proton in air, $n_j(E;E)$ the total multiplicity of the secondary particles with energy "E'", produced by a primary aprticle of the sort "j" with the energy "E", E_t the threshold energy for the considered reaction, I the maximum number of nuclear interactions in the atmosphere for the primary cosmic ray particle.

Unfortunately some of the nuclear parameters involved in the above expressions are only partly unknown. Since the experimental data on neutron spallation cross sections are sparse, most authors (see, for example Perl (1974)) do not distinguish between protons and neutrons in the expressions. This assumption is correct for high energies ($E > 10$ GeV), but is less precise just above threshold. In our model we also assume these cross sections to be equal.

Unknown excitation functions $Sig_r(E)$ can be obtained using either the semiempirical charge-distribution mass-distribution (CDMD) formula of Rudstam (1966) or, which we prefer, the more detailed Silberberg-Tsao approximation (1973), although without experimental data it is difficult to decide which one of them comes closer to reality. In Figure 1 and Figure 2 the calculated excitation functions for reactions (1) and (3) are shown. One can see an important difference both in the absolute value of the cross-section approximations and especially in the form of the excitation functions for the above-threshold energy.

To account for the energy losses of protons due to ionization and excitation the proton spectrum $N_0(E)$ has to be modified:

$$N_{mod}(E) = \int_{E_t}^{E} N_0(E') \exp[(T(E')-T(E))/L(E')] dE' \tag{8}$$

where $T(E)$ is the characteristic length for ionization losses, $L(E)$ is the characteristic length for nuclear inelastic processes for protons in air and $N_0(E)$ represents the primary spectrum. $L(E)$ is given in tabulated form by Medvedev et al. (1981) and in analytical form by Hagen (1976).

For the energy distribution of the secondary particles the following factorized expression was used

$$n(E',E) = \frac{dN(E',E)}{dE'} = C(E')B(E) \tag{9}$$

with the normalization conditions:

$$B(E) \int_{0}^{E} C(E') dE' = n(E) \tag{10}$$

$$B(E) \int_{0}^{E} E'C(E')dE' = \varepsilon(E) \qquad (11)$$

where n(E) is the total multiplicity of secondary nucleons and $\varepsilon(E)$ is the total energy of secondary nucleons.

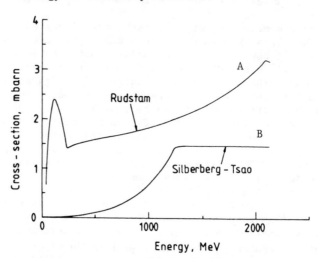

Fig. 1. Approximation of the $^{14}N(N,5X)^{10}Be$ excitation function
A – Rudstam formula
B – Silberberg-Tsao formula

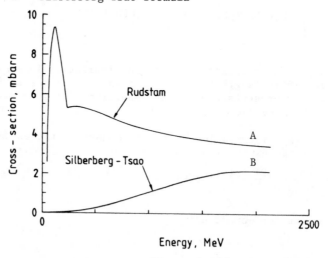

Fig. 2. Approximation of the $^{40}Ar(N,5X)^{36}Cl$ excitation function
A – Rudstam formula
B – Silberberg-Tsao formula

3. SOLAR ACTIVITY MODULATION

The influence of the level of solar activity on the primary cosmic ray energy spectrum and flux in the vicinity of the Earth is most pronounced for energies below 30 GeV/nucleon. In general a higher level of solar activity leads to a higher density of the solar wind magnetic field, which deflects cosmic ray particles. However, this simplified picture does not fit all the experimental data, for example there is no direct correlation between the solar wind velocity and the number of sunspots manifesting the level of solar activity. Although the modulation process is not yet understood in detail, the experimental data clearly show that there is an anticorrelation between the level of solar activity and the intensity of galactic cosmic rays.

The mathematical approximations of the solar modulation process are usually based on a number of parameters such as the diffusion tensor, the dimensions and form of the modulation envelope, the regular and irregular solar wind velocity etc. (Toptygin, 1985). In our calculations we use the one parameter approximation proposed by Castagnoli and Lal (1980). It is in a good agreement with the result of cosmic ray measurements during the last two solar cycles, moreover it gives us the opportunity to compare our results with the ^{14}C calculations of Castagnoli & Lal (1980). According to Castagnoli & Lal (1980) the dependence of the primary proton spectrum on the solar modulation parameter ϕ (MeV) can be described as

$$N_{op}(E,\phi) = A \frac{E(E + 2E_o)(E + \phi + m)^{-\gamma}}{(E + \phi)(E + \phi + 2E_o)} \quad cm^{-2}sec^{-1}sr^{-1}MeV^{-1} \quad (12)$$

where E is the kinetic energy in MeV, E_o is the proton rest energy in MeV, $A = 9.9 * 10^4$, $m = 780*exp(-2.5*10^{-4}E)$, $\gamma = 2.26$. The parameter ϕ varies from 100 MeV, which corresponds to the absence of modulation, to 900 MeV, which simulates the highest level of modulation ever observed. The energy dependence of the interplanetary cosmic proton flux on the level of solar activity is shown in Figure 3.

The isotope production rate in the absence of the geomagnetic field can be written using (5), (8) and (12) as

$$W(\phi) = \int_{E_t}^{E_{max}} Y(E')N_{opmod}(E',\phi)dE' \quad (13)$$

The alpha-particle primary flux was taken into account by re-normalization of the constant A in (12), considering it as a flux of unbound nucleons with the same energy spectrum.

The energy dependence of the function $Y(E')*N_{opmod}(E',\phi)$ for ^{10}Be and ^{36}Cl is plotted in Figures 4 and 5. As can clearly be seen, the most important energy interval for spallation reactions ranges from several GeV to several tens of GeV.

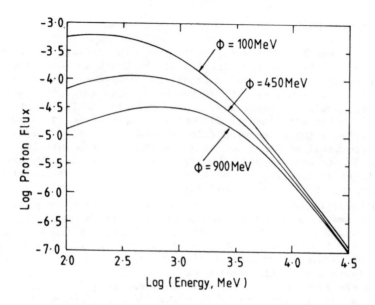

Fig. 3. The energy dependence of the interplanetary cosmic proton flux as a function of the solar activity level (Castagnoli and Lal, 1980).

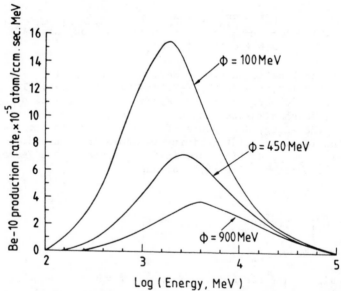

Fig. 4. The energy dependence of the ^{10}Be production rate in the absence of geomagnetic field for three solar activity levels.

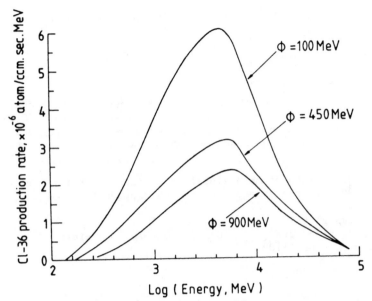

Fig. 5. The energy dependence of the ^{36}Cl production rate in the absence of geomagnetic field for three solar activity levels.

4. GEOMAGNETIC MODULATION OF THE PRODUCTION RATE

When calculating the global mean production rate of an isotope it is necessary to consider the latitude dependence of the geomagnetic field intensity. Neglecting the nondipole component of the geomagnetic moment and its longitude variation, we use the well known expression for the cut-off rigidity

$$R(\alpha) = R_o \cos^4(\alpha) \tag{14}$$

where α is the geomagnetic latitude, R_0 corresponds to the equatorial cut-off and $r_0 = 14.9$ GV. The value of R_0 is directly proportional to the geomagnetic dipole moment.

We can introduce the parameter "p"

$$p = M(T)/M(0) \tag{15}$$

where $M(T)$ is the geomagnetic moment at the time T and $M(0)$ is the modern average value, $M(0) = 8.0*10^{25}$ Gs* cm^3. The expression (14) can be rewritten for the cut-off energy dependence using (15) and a relationship between energy and magnetic rigidity as

$$E(\alpha,p) = -E_o + [E_o^2 + p^2 R_o^2 \cos^8(\alpha)]^{1/2} \tag{16}$$

The global mean production rate <Q> is then determined from

$$<Q>(\phi,p) = \int_0^{\pi/2} \sin(\alpha)\,d\alpha \int_{E(\alpha,p)}^{E_{max}} Y(E')\,N_{opmod}(E',\phi)\,dE' \qquad (17)$$

The final expression reflects the relationship between the global parameters. The results are summarized in Tables 1 and 2. For ^{14}C calculations we used the curve for the $Y(E)$ published by Castagnoli & Lal (1980) and, as expected, derived the same values for the production rate variations.

Table 1. Comparison of mean global production rates for ^{10}Be and ^{36}Cl in the atmosphere and their production ratio.

Mean global production rate, atom $cm^{-2}\,sec^{-1}10^3$		$<Q>^{36}Cl/<Q>^{10}Be$	Reference
^{10}Be	^{36}Cl		
25	0.87	0.0345	O'Brien (1979).
19	1.1	0.0567	Lal & Peters (1967)
14	2.2	0.157	Oeschger et al. (1969)
26	1.9	0.074	this work

Table 2. The maximum isotope production variations as a function of helio- and geomagnetic modulation (in this table and later on the sign δ corresponds to the ratio $Q(\phi,p)/Q(450,1)$ for each isotope).

$M(T)/M(0)$	ϕ,MeV	$\delta^{10}Be$	$\delta^{36}Cl$	$\delta^{14}C$
0	450	2.0	1.6	2.6
2	450	0.69	0.74	0.65
1	100	1.3	1.2	1.42
1	900	0.78	0.83	0.72

The results of Table 2 are also included in Figures 6 - 9. Simple approximate formulae can be constructed for the ^{10}Be and ^{36}Cl variation dependences as was done for ^{14}C (Lingenfelter and Ramaty, 1969; Castagnoli and Lal, 1980).

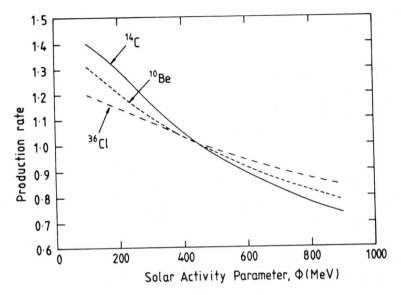

Fig. 6. The dependence of the relative cosmogenic isotope (for ^{14}C, ^{10}Be and ^{36}Cl) production rate Q/Q_o on the level of solar activity.
$Q(\phi = 450 \text{ MeV}) = Q_o$

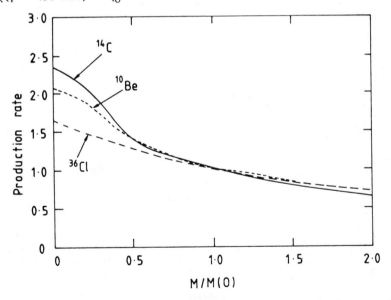

Fig. 7. The dependence of the relative cosmogenic isotope (for ^{14}C, ^{10}Be and ^{36}Cl) production rate Q/Q_o on the geomagnetic moment variations.
$Q (p = 1) = Q_o$

338

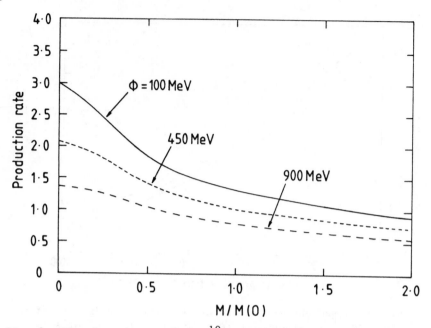

Fig. 8. The dependence of the ^{10}Be production rate on the geomagnetic dipole moment value for three solar activity levels.

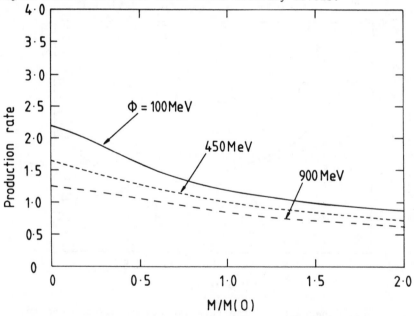

Fig. 9. The dependence of ^{36}Cl production rate on the geomagnetic dipole moment value for three solar activity levels.

$$\delta\ ^{14}C(\phi = 450\ MeV) = p^{-0.5} \qquad 0.5M_o < M < 1.5M_o \qquad (18)$$

$$\delta\ ^{10}Be(\phi = 450\ MeV) = p^{-0.48} \qquad 0.5M_o < M < 1.5M_o \qquad (19)$$

$$\delta\ ^{36}Cl(\phi = 450\ MeV) = p^{-0.36} \qquad 0.5M_o < M < 1.5M_o \qquad (20)$$

The production variations of ^{10}Be and ^{36}Cl estimated for small changes of the geomagnetic field can be well determined by these formulae

$$\delta\ ^{14}C(p = 1.0) = 3.2 - 1.2E\text{-}3\ \phi \qquad (21)$$

$$\delta\ ^{10}Be(p = 1.0) = 4.2\ \phi^{-0.24} \qquad (22)$$

$$\delta\ ^{36}Cl(p = 1.0) = 2.6\ \phi^{-0.16} \qquad (23)$$

The accuracy of the results is directly proportional to the accuracy of the excitation function values and proportional to the accuracy of the multiplicity values with a factor 1.2. These are the main reserves for the model improvement. We think that precise measurements of the excitation functions both for proton and neutron induced reactions are of great importance and should be carried out as soon as possible.

5. CONCLUSION

The model for atmospheric isotope production discussed here combines the physical backgrounds with practical simplicity and can therefore be applied to the interpretation of long-time scale records of cos-mogenic isotopes. It can also be useful for the separate investigation of global stratospheric and local tropospheric isotope production, phenomena which can play an important role for equatorial and high-latitude areas.

REFERENCES

Beer, J., Andree, M., Oeschger, H̄. et al., 1983, Radiocarbon, 25
    (2), 269.
Castagnoli, G., Lal, D., 1980, Radiocarbon, 22, 2, 133.
Hagen, F.A., 1976, Ph.D. Thesis, University of Maryland.
Korff, S.A., Mendell, R.B., Merker, M., Light, E.S., Verschell, H.J.,
    1979, Final report, New York University.
Lal, D., Peters, B., 1967, 'Encyclopedia of Physics', 46/2, 551.
Lingenfelter, R.E., Ramaty, R., 1969, In: 'Radiocarbon variations
    and absolute chronology', 12 Nobel Symposium, New York, 514.
Levchenko, V.A., Blinov, A.V., 1984, In: 'Abundances of the isotopes
    in the environment and astrophysical phenomena', Leningrad, 61,
    (in Russian).

Medvedev, Y.A., Stepanov, B.M., Trukhanov, G.Y., 1981, 'Nuclear-physics constants of neutron interactions. Reference book', Moscow (in Russian).

Mlotek, V.I., 1981, Izvestia Acad. of Sc. USSR, Ser. Phys. 45, 1215, (in Russian).

O'Brien, K., 1979, J. Geoph. Res., 84, A2, 423.

Oeschger, H., Houtermans, J., Loosli, H., Wahlen, M., 1969, In: 'Raidocarbon variations and absolute chronology', 12th Nobel Sumposium, New York, 471.

Perl, M.L., 1974, 'High energy hadron physics', J. Wiley & Sons, New York.

Raisbeck, G.M., Yiou, F., 1984, Phys. Res., B5, 91.

Rudstam, G., 1966, Z. Naturforsch A., 21, 1027.

Silberberg, R., Tsao, S., 1973, Astrophys. J. Suppl. Ser., 220 (1), 25, 315.

Toptygin, I.N., 1985, 'Cosmic rays in interplanetary magnetic fields', D. Reidel, Dordrecht.

SOLAR IMPRINT IN SEA SEDIMENTS: THE THERMOLUMINESCENCE PROFILE AS
A NEW PROXY RECORD.

G. Cini Castagnoli and G. Bonino
Istituto di Fisica Generale dell'Università and
Istituto Cosmogeofisica del CNR Corso Fiume,
4-10133 Torino, Italy.

ABSTRACT. A new method for tracing the solar behaviour in the past
is proposed. We show that the relative thermoluminescence signals
from the different layers of a recent mud carbonate sediment core,
measured as a function of time in the last 8 centuries (after dating
the core by radiometric methods and checking the dates by exceptional
volcanic events) carries the imprint of the secular variations of
solar activity and of the 11 year cycle. The TL signals are in phase
with the solar activity as monitored by Sunspots and Aurorae.
Moreover they are in antiphase with the ^{14}C production rate variations
and the ^{10}Be time profile in the Greenland Milcent ice core. This
supports the possibility of using the TL profiles as solar proxy
data.

1. INTRODUCTION

The natural thermoluminescence (TL) of polyminerals, so sensitive
to light and ionizing agents, has been used as a new physical method
of investigating the possibility of tracing the imprint of solar
effects in recent uniform marine sediments (sedimentation rates
0.05 - 0.1 cm/year). The procedure for the natural TL measurements
is based on the treatment of adjacent layers of the core. The TL
analysis is performed if in the core (uniform mud, deposited in undisturbed sites, negligibly bioturbated) a constant sedimentation
rate is measured.
    We present here the results obtained in an Ionian Sea core for
the last 8 centuries. The sedimentation rate is determined by radiometric measurements of ^{210}Pb and refined by tephroanalysis. The
potentiality of the method is evident from the results obtained:
in particular by the presence of the 11 year cycle in the time profile
of the TL data. Moreover other features and periodicities appear
in the TL profile which can be ascribed to Solar-Terrestrial effects.
    The TL variations due to the imprint of the Maunder, Spörer
and Wolf solar events are shown in comparison with those obtained in
the concentrations of the cosmogenic isotopes ^{10}Be in the Milcent

F. R. Stephenson and A. W. Wolfendale (eds.),
Secular Solar and Geomagnetic Variations in the Last 10,000 Years, 341–347.
© *1988 by Kluwer Academic Publishers.*

ice core (Beer et al. 1983) and in atmospheric [14]C (Stuiver and Quay 1980). The different nature of the Maunder, Wölf and Spörer variations is shown and discussed by a further comparison of these series with the global temperature index $\delta D\%_0$ measured in the Antarctica Dome C core during the same period (Benoist et al. 1982).

We infer that the TL properties of the sediment are predominantly induced by agents in phase with solar activity as monitored by sunspots and aurorae. This supports the point of view that the TL level is controlled by the equilibrium radiation characteristic of the environment at the time of the deposition of the material forming the sediment.

2. TL SERIES IN AN IONIAN SEA SEDIMENT

The TL intensity profile in the time range 1100-1900 A.D. was obtained from 207 adjacent layers, equally spaced and sampled at 2.5 mm width of the recent Ionian Sea core GT 14 (lat. 39°45'55", long. 17°53'30") drilled in 1979 A.D. by means of a gravity corer at a water depth of 160 m on the continental platform.

The dating of the core was established by the following procedures (Cini Castagnoli et al. 1987): (a) the sedimentation rate was determined by the measurement, as a function of depth, of the ^{210}Pb excess, due to the atmospheric fall-out of this radioisotope, (taking into account the in-situ decay of [226]Ra), (see e.g. Krishnaswami et al. 1971); (b) microscopic observations were made on powdered samples in order to investigate the presence of volcanic ashes in different layers; (c) concentration measurement of ^{137}Cs, in the top layers (the fall-out of this radioisotope derives from the nuclear atmospheric explosions having a maximum in 1963-64 A.D.) for checking the lack of perturbations at the top.

Following the procedure (a), a sedimentation rate s = 0.06 cm/year has been found, with an accuracy of about 5%. This corresponds to Δt = 4 ± 0.2 years per layer.

The procedure (b) allowed a refinement of Δt to the value 3.87 ± 0.04 years. The least-square fit of the excess ^{210}Pb activity with respect to the exponential decay function has a correlation coefficient r = 0.98. This high value of r ensures that the sedimentation rate in the considered layers is about constant and the monotonic decrease of the activity also for the first few centimeters ensures that bioturbation is negligible. In addition, the measurements of ^{137}Cs activity show a maximum in the layer 0-1 cm, as expected.

The TL signals were obtained by polymineralic samples following a standard procedure already described (Cini Castagnoli and Bonino, 1987). For each layer the average TL intensity was obtained from four measurements (standard deviation < 5%) on samples of sieved powder (< 44 µm) of 15 mg. The glow curves were recorded in a neutral atmosphere of ultra pure nitrogen (flow 5 litres/min) at a heating rate of 5° C/s. The photomultiplier used is a Hamamatsu R269 coupled to a Corning 5-60 filter. The TL levels were measured at 340°C.

   The TL profile as a function of depth was transformed as a
function of time on the basis of the above described dating procedure.
In Fig. 1, the TL series for the time interval 1100-1900 A.D. is
shown. Variations of the TL specific intensity of about 10 per
cent take place between different layers; these variations are not
randomly distributed.

3. TL PERIODICITIES

The TL series has been analysed in order to test the presence of
periodicities. Two methods have been adopted: power spectral density;
superposition of epochs.
   The standard discrete Fourier transform applied to the detrended
series of Figure 1 is shown in Figure 2. The spectrum reveals

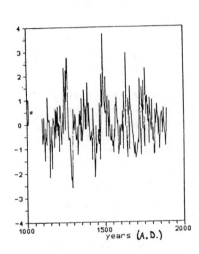

Fig. 1. Thermoluminescence
series of the Ionian Sea
core GT14.

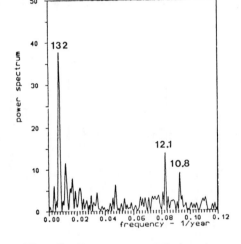

Fig. 2. Power spectral density
of the detrended TL series.

periodicities of 132, 12.08, 10.8 years similar to those exhibited
by solar activity as monitored by sunspots and aurorae. In fact
the 132 yr period is very similar to the period recently found in
the record of historical aurorae (Attolini et al. 1987). The period
corresponding to the beating of the two higher frequencies (corresponding
to 12.08 yr and 10.8 yr) is 11.4 yr which appears in the sunspot
record between 1824 A.D. and 1903 A.D. (Attolini et al. 1985). The
same periodicity is present in the ^{10}Be record in the Milcent ice
core in the time interval 1180-1450 A.D. (Beer et al. 1985), while
during the time interval 1500-1700 A.D. the 10.8 yr period appears
(Cini Castagnoli et al. 1984).
   Moreover two small peaks, at 84 years (corresponding to the
Gleissberg cycle), and at 22 years (corresponding to the magnetic

solar cycle), appear in Figure 2.

In order to investigate the amplitude and phase of the 132 yr period, the method of superposition of epochs has been adopted. The detrended time series is partitioned into consecutive subseries of length T = 132 yr which are then superposed on a single interval which is divided into 10 subdivisions. The mean and the standard deviation of the experimental data falling in each subdivision are computed. The result is shown in Figure 3. The curve drawn through the points is the least-square fit with a sinusoidal wave of period 132 yr. The peak to peak amplitude of the effect is $\sim$ 6 TL arbitrary units: this corresponds to an oscillation of about 10% around the TL average level of $\sim$ 60 (a.u.). The significance level of the wave with respect to a random distribution evaluated by means of the deviations of the 10 binned data from their average value corresponds to about 5 standard deviations.

The times of the minima of the 132 yr period can be expressed by (1166 + n132) A.D. where n is a positive integer.

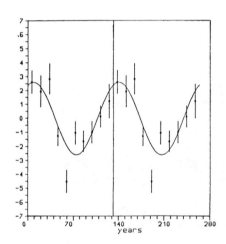

Fig. 3. Superposition of epochs of the TL detrended data for a periodicity of 132 years.

4. COMPARISON OF THE TL PROFILE WITH ^{14}C, ^{10}Be AND δD/H

The TL time series has been compared with the following proxy data series:
- ΔQ_m of ^{14}C (production rate variations from the mean level in the time interval 1000 – 1860 A.D.) (Fig. 4a) evaluated by Stuiver and Quay (1980) from the data measured in Douglas fir wood, adopting the four-reservoir box-diffusion model of Oeschger et al. (1975) with a carbon reservoir of 60 years biospheric residence time.
- ^{10}Be concentration in the polar ice Milcent core (70°18'N, 44°35'W) Greenland (Beer et al. 1983) (Fig. 4b).
- Deuterium (expressed in δD‰ vs standard mean ocean water, SMOW, with a D/H ratio equal to 155.76 10^{-6}) climatic record over the last 2.5 Kyears from Dome C (74°39'S, 124°10'E) Antarctic ice core 1979 (Benoist et al. 1982) (Fig. 4d).

Observations and comments on the main features of the four proxy data series in correspondence with the Maunder, Spörer and Wölf solar events are the following;
(1) During the Maunder Minimum of solar activity (1645-1715 A.D.) ^{10}Be and ^{14}C have maxima, while TL has a minimum, δD shows evidence for a global mean low temperature. Moreover the TL curve is generally

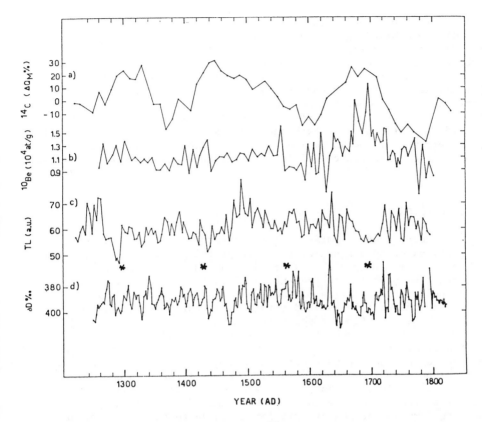

Fig. 4. (a) Percentage deviation of ^{14}C production rate relative
to the average level in the 1000-1860 A.D. interval
(from Fig. 8 of Stuiver and Quay, 1980);
(b) [10]Be in Milcent core;
(c) TL series;
(d) $\delta D\%_0$ vs standard mean ocean water, from Dome C Antarctic
ice core.

smooth.
(2) During the Wölf minimum (1270-1340 A.D.) the relationships
between the series of TL, ΔQ_M of ^{14}C and [10]Be are similar to those
described above for the Maunder Minimum.
(3) The behaviour of solar activity during 15th and 16th century
(to which the Spörer minimum belongs) is still a debated subject
(see e.g. Jiang and Xu 1986): a) to a TL minimum around 1440 A.D.
corresponds a [10]Be maximum in 1437 A.D. and a ΔQ_M of ^{14}C maximum
in 1440-1450 A.D.; b) the TL minimum around 1560 A.D. corresponds
to the [10]Be maximum in 1555 A.D. as a second episode of low solar
activity, which in the ^{14}C production rate is not evident; c) the
TL maximum around 1500 A.D. corresponds to a temperature increase

shown by the δD curve at the same time.

The understanding of the origin and role of the 132 yr wave in addition to the role of the 11.4 yr cycle in modulating the series is essential in order to explain the different behaviours pointed out during the Spörer period as well as to explain the different characteristics of the Maunder-like and Spörer-like events. We can point out however that the minima of the 132 yr period (marked as stars in Fig. 4C), persistent in the whole TL series, are always coincident with maxima in both the radiocarbon and ^{10}Be series, while the occurrence of the maximum of this wave around 1500 A.D. is the main point of discrepancy. This could find an explanation in climatic effects.

5. CONCLUSIONS

A new method of investigating solar-terrestrial relationships in the past, based on a study of the time profile of the relative thermoluminescence (TL) signals in a recent sea sediment core has been proposed. We have considered the possibility that the TL properties of the polimineral samples of the sediment (carbonate mud) is controlled by the equilibrium radiation level characteristic of the environment at the time of deposition.

The TL profile shows the effects of the Maunder, Spörer and Wölf solar minima: changes in the Sun and in the solar wind larger than any seen since space observations began had already been inferred from the behaviour of the concentrations of different cosmogenic iotopes stored in different Earth reservoirs. Information can be drawn from the TL measurements as well; indeed certain features of the long term variations are similar in shape to those of cosmogenic isotopes, but opposite in sign. The TL signal seems to record directly the solar output. Further investigation is in progress in order to be able to make a quantitative analysis.

Moreover the TL periodicites are similar to those of solar activity, as monitored by Sunspots and Aurorae.

In addition, climatic influences seems to appear in the TL series. The natural TL measured in adjacent layers of appropriate recent marine sediments can be adopted as a new proxy record of solar activity in the past.

ACKNOWLEDGEMENTS

We are grateful to Prof. C. Castagnoli for helpful discussions and to Mr. P. Cerale and Mr. A. Romero for technical assistance during the experiment.

REFERENCES

Attolini, M.R., Galli, M. and Cini Castagnoli, G., 1985, Solar Phys.,
    95, 391-395.
Attolini, M.R., Cecchini, S., Galli, M. and Nanni, T., 1987, Nature,
    in press.
Beer, J., Siegenthaler, U., Oeschger, H., Andrée, M., Bonani, G.,
    Suter, M., Wolfli, W., Finkel, R.C. and Langway, C.C., 1983,'Proc.
    18th Int. Cosmic Ray Conf'. (Bangalore, India), 9, 317-320.
Beer, J., Oeschger, H., Finkel, R.C., Cini Castagnoli, G., Bonino,
    G., Attolini, M.R. and Galli, M., 1985, Nucl. Instrum. Meth., B10/11
    415-418.
Benoist, J.P., Jauzel, J., Lorius, C., Merlivat, L. and Pourchet
    1982, Annals of Glaciology, 3, 17-22.
Cini Castagnoli, G., Bonino, G., Attolini, M.R., Galli, M. and
    Beer, J., 1984, Il Nuovo Cimento, 7C, 235-255.
Cini Castagnoli, G. and Bonino G., 1987, in "Solar-Terrestrial
    relationships and the Earth environment in the last millenia",
    G. Cini Castagnoli editor, (E. Fermi Int. Summer School of Physics
    85), North Holland, in press.
Cini Castagnoli, G., et al., 1987, in preparation.
Jiang, Y. and Xu, Z., 1986, Space Science, 118, 159-162.
Krishnaswamy, S., Lal, D., Martin, J.M. and Meybeck, M., 1971, Earth
    and Planetary Sci. Letters, 11, 407-414.
Oeschger, H., Siegenthaler, U., Schotterer, U. and Gugleman, 1975,
    Tellus, 27, 168-192.
Stuiver, M. and Quay, P.D., 1980, Science, 207, 11-19.

SECULAR VARIATIONS OF THE GEOMAGNETIC FIELD
- THE ARCHAEOMAGNETIC RECORD

D.H. Tarling
Department of Geological Sciences
Plymouth Polytechnic
Plymouth PL4 8AA
England

ABSTRACT. The ability of geological and archaeological materials to
retain a record of the geomagnetic field at specific times enables
a record of past changes of the geomagnetic field to be compiled
Palaeointensity determinations can only be effectively made with
fired clays, although sedimentary records can provide continuous
records of generalized relative intensity changes. With notable
exceptions, sediments are noisier recorders of directional changes
than fired materials. Using baked clays, the direction and intensity
of the past geomagnetic field should be determinable to about 1°
and 1-3%. Unfortunately there are methodological uncertainties that
mean that the real errors may be of the order of 5° in direction
and well over 10% in palaeointensity. The main factor in both methods
appears to be magnetic distortion resulting from magnetic interactions
within grossly inhomogeneous structures and is particularly serious
in the areas that cooled slowly. The current British secular variation
record suggests periodicity of some 200 years in the rates of change
in direction and 400 years in the departures from the axial geocentric
dipole field direction.

INTRODUCTION

Observatory records of the direction of the geomagnetic field
began in London and Paris towards the end of the 16th century, while
intensity records commenced in the mid 19th century. The spatial
variation of the present, and future, geomagnetic field can now be
defined using satellite-based magnetometers, as exemplified by MAGSAT
which surveyed between 50°N/S for 9 months (October 1979 to June
1980) and provided the first, almost global, instantaneous record
of the geomagnetic field at heights of 350 to 650 kms. However,
while the spatial resolution of the extant geomagnetic field is now
resolvable, direct measurements of secular variations of the geomagnetic
field are still severely limited in both space and time when compared
to their periodicities of some 10^{3-4} years.

F. R. Stephenson and A. W. Wolfendale (eds.),
Secular Solar and Geomagnetic Variations in the Last 10,000 Years, 349–365.
© 1988 by Kluwer Academic Publishers.

The ability of geological and archaeological materials to retain
a memory of the intensity and direction of the geomagnetic field
at specific times is now well documented (Aitken, 1974, 1983; Tarling,
1983), as are most of the techniques and methods employed (Collinson,
1983). Such factors will therefore only be briefly discussed, mainly
in the context of the remaining sources of error. Instrumentation
now allows the measurement of the direction of the remanent magnetization, in most materials to ± 1° (Collins, 1983; Tarling, 1983),
yet the directions from different samples from the same site can
differ by > 5° and can only in part be attributed to the difficulty
in orienting individual samples in the field (Thellier & Thellier,
1952; Weaver, 1962; Tarling et al., 1986). Similarly, the intensity
of magnetization can be measured to about ± 1% (Collinson, 1983;
Walton, 1987), but different samples from the same site yield apparently
reliable determinations of the past intensity of the geomagnetic
field that may well be inconsistent to > 50% (Walton, 1987). In
both cases, therefore, the main sources of inaccuracy are not fundamentally instrumental, but arise mostly from either invalid, or
poorly constrained, assumptions about the mode of acquisition of
remanence under natural conditions or from the inadequacy of the
techniques by which such geomagnetic parameters are derived from
the measurements. It is therefore necessary to consider these factors
prior to considering the available data for the last 10,000 years.
However, as some major data bases are discussed elsewhere in this
volume (Creer, Verosub and Dubois), such consideration will be mostly
restricted to the archaeomagnetic record for western Europe.

THE ARCHAEOMAGNETIC METHOD

All methods assume that most geological and archaeological materials
can acquire a remanent magnetization which is directly related to
the geomagnetic field at a specific time and that this remanence
can be isolated and used to determine the original geomagnetic parameters. Four fundamental processes are involved in the acquisition
of remanence – thermal, chemical, depositional, and viscous.

(i) Thermal

Magnetically isotropic ferromagnetic minerals, as they cool, become
magnetized in the direction of the ambient field and acquire a
remanent intensity of magnetization that is related to the strength
of the ambient field in linear proportion to their low-field magnetic
susceptibility. The direction of remanence in such materials will
be parallel to the ambient field, and can thus be measured directly,
while the determination of the previous geomagnetic field intensity
is only slightly more complicated. It is based on the argument that,
assuming no change in the original magnetic susceptibility, the
intensity of magnetization acquired by a particular sample in a known
(laboratory) field over a specific temperature range is exactly
proportional to the intensity of natural remanence which it originally

acquired over the same temperature range. The observed ratio between the natural and laboratory intensities is thus identical to the ratio between the laboratory and original field strengths (Thellier & Thellier, 1942).

(ii) Chemical

As crystallites of magnetic minerals grow within a magnetic field, they acquire a remanence. If the material containing them is magnetically isotropic, then the direction of remanence acquired is the same as the applied field, and the intensity of remanence is related to the original field strength in the same way as thermal remanence. Indeed, thermal and chemical remanences are generally considered to have extremely similar magnetic properties. In the context of the last 10,000 years, remanence of this origin has not been widely used to determine past geomagnetic field directions or strengths, and is usually encountered as a source of inaccuracy (discussed later).

(iii) Depositional

Already magnetized particles, as they are deposited and shortly after deposition, are capable of being aligned by the ambient geomagnetic field. In the absence of other forces, such alignment is parallel to the ambient field and is proportional to the strength of the field if it is < 0.2 mT (Collinson, 1974). Under ideal conditions, the direction of their remanence should therefore reflect the original geomagnetic field direction (Lund, 1985) and laboratory re-deposition experiments in known fields should allow the determination of the previous geomagnetic field strength.

(iv) Viscous remanence

The original magnetization in all substances containing magnetic minerals will very gradually decay, but as they continue to lie within the geomagnetic field they also acquire a time-dependent magnetization. Conveniently, the grains that most readily lose their original magnetization are those that most readily acquire a viscous remanence, so that the viscous component of the total natural remanence of an archaeological sample can be readily removed by partial demagnetization in weak alternating magnetic fields, generally < 10 mT, or by heating and cooling (in zero ambient magnetic field) from relatively low temperatures, c. 100–120°C. In general, therefore such loss and acquisition of remanence is not significant as long as the properties of the remanence at such low demagnetization levels are excluded from subsequent analysis - as is standard procedure.

SOURCES OF ERROR

A. General

If later chemical changes (weathering) occur, these commonly affect
the magnetic minerals directly (Barbetti et al., 1977) and may also
result in the formation of new magnetic minerals from previously
non-magnetic, iron-bearing minerals. The chemical remanences associated
with such changes will often have similar or even greater stability
to partial demagnetization than the original remanence (Grommé et
al., 1979). At the moment it is safer to exclude samples where such
changes are suspected, but this can be difficult to quantify. Sediments are, however, much more sensitive to chemical changes subsequent
to deposition than are fired materials.

In fired materials, such weathering effects can be reduced by
acid washing (Sakai & Hirooka, 1986), but it is difficult to be certain
that only the weathering effects are removed and that original magnetic
minerals are not also chemically changed. There is also a possibility
of long term chemical changes occurring as different minerals exsolve
and approach chemical equilibria in an ambient environment which
may differ significantly from that in which the remanence was
originally acquired. Even if such conditions could be duplicated
in the laboratory, there are still significant time-dependent chemical
changes involved. Such chemical changes may also be triggered by
the initial laboratory heating and comparison of various magnetic
parameters determined during subsequent heating runs will not
distinguish those changes initiated during the initial heating from
those occurring before sampling. (Thermally induced chemical changes
also prevent heat treatment of unconsolidated sediments).

In fired materials, laboratory induced changes in the magnetic
mineralogy may be detected by monitoring changes in their magnetic
parameters - such as susceptibility or coercivity spectra. However,
not all experimentalists monitor these properties and, even if monitored,
significant changes can take place that are not always detected (Coe,
1967; Coe & Grommé, 1974; Prévot et al., 1983; Walton, 1983).
Furthermore, if all of the chemical changes occur during the initial
laboratory heating, comparisons after cooling may not distinguish
such changes from the original properties.

Examination of the vector content of the remanence, as a function
of temperature, is a routine part of directional studies. This may
well be a more sensitive gauge of the changes during laboratory
treatment than measuring individual bulk properties. It also
establishes whether only a single vector is involved in the study.
Unfortunately, most determinations of palaeointensity do not involve
component analysis, with the notable exception of the Oxford
laboratory, and there are clear dangers in such procedures. Clearly,
combined palaeointensity and palaeodirectional studies are desirable
to indicate their mutual reliability.

B. Fired Materials

(i) Magnetic Anisotropy

At a domain scale, each magnetic grain is strongly anisotropic, but their orientations are assumed to be random within the sample, so that the bulk anisotropy is zero. The sample anisotropy is rarely measured directly and isotropy is generally assumed unless there are clear visual lineations of grains or colouration. However, the presence of a visual anisotropy is not necessarily diagnostic of the presence, or absence, of anisotropy within the small percentage (< 5%) of the magnetic fraction of the total sample that actually carries the original remanence. On the same basis, even if measured, the measured anisotropy may be similarly misleading as this will normally be dominated by the total properties. Nonetheless, measured anisotropies rarely exceed 5% and so this effect is likely to be less than 1° in direction and 1% in intensity. The only real test is to determine how closely a new remanence, acquired under identical conditions to those originally pertaining, reflects the direction and intensity of an applied field. Unfortunately, such conditions cannot usually be duplicated in laboratory experiments. (In the near future, the various other forms of remanence (Stephenson & Molyneux, 1987) may enable examination of the effective anisotropy of the remanence-carrying grains without thermally-induced chemical changes).

(ii) Inhomogeneity

Most materials, in archaeological and geological environments, are commonly inhomogeneous on a range of levels. (Thellier's original studies (1942, 1981) utilized large samples, partially in order to increase the moment being measured, but also to reduce the effects of inhomogeneity). Modern instrumentation is insensitive to inhomogeneity within an individual specimen so that inhomogeneity in grain size, composition and distribution should not contribute significantly to noise if they occur at scales less than that of a specimen. However, "standard" specimen volumes vary from 0.3 to > 100 cm^3 and so different scales of inhomogeneity are involved, depending on the magnetometer systems used. The highest degree of directional scatter in fired materials is associated with the occurrrence of the greatest gross inhomogeneity (Tarling et al., 1986), i.e. on scales much greater than that of an individual specimen. This means that the effect of large-scale inhomogeneity must be significant, but probably indirect, i.e. inhomogeneity alone does not cause the deviations, but appears to create an environment within which other influences can operate more effectively.

(iii) Cooling Rates

The simpler theoretical models for the acquisition of thermal remanence normally assumes instantaneous cooling in weak fields comparable to

that of the Earth. However, different intensities of magnetization
can be acquired in the same field, depending on the cooling rate
(Weaver, 1970; Walton, 1980; Dodson and McClelland-Brown, 1980).
This apparently corresponds to the acquisition of a "high-temperature
viscous" component which is normally additional to the thermal
remanence. The number of studies of these effects are still few, but
such acquisitions are considered to be exponentially dependent on time,
but also dependent on grain size and composition. The exponential
nature of such effects means that they are likely to be similar for
most archaeological materials that have cooled for longer than 24 hours.
Fox and Aitken (1980) consider that the correction for the difference
between laboratory cooling rates (using 3 mm sized specimens) and
those commonly associated with pottery kilns is usually in the region of
5% and corrections of this magnitude are routinely made by some
research workers (Wei et al., 1987). Such a correction is clearly
significant in palaeointensity determinations, but is smaller than
the reported intensity discrepancies of > 50% based on materials
characterized by almost identical correction factors.

(iv) Magnetic interactions/refraction

As materials cool, they become magnetized in a field that is the
summation of the ambient geomagnetic field and any other extraneous
field. In some circumstances, such an extraneous field could be
associated with a large, external, strongly magnetized object, but
the square law decrease in the strength of the magnetic field of
an object, as a function of its distance away, means that such effects
will usually be localized and have only rarely been suggested (Hoye,
1982). However, this could reflect the paucity of studies of
sufficient quantity for times when the true direction of the geomagnetic field is known.

On a smaller scale, the magnetic grains of blocking temperature
> 20°C below that of other grains will be cooling in the magnetic
fields of already magnetized grains within the structure itself (Neél,
1949; Stacey, 1963; Dunlop & West, 1979). These internal magnetic
fields comprise induced components, mostly associated with grains
of high susceptibility, and remanent components associated with the
already acquired remanences. These self-demagnetizing field effects
are commonly termed 'refraction', the theory for which is relatively
well understood (Dunlop & Zinn, 1980; Abrahamsen, 1986). The
observed dispersion in direction and poor repeatability in palaeointensity can be largely explained in terms of a refraction by a
thin sheet of different permeability. Dunlop & Zinn (1980) obtained
some ± 15% range in palaeo-field intensities for samples from the
floor of a 19th century kiln and angular deviations from the mean
of up to 7.3°. These were explained in terms of both tilting of
the bricks and magnetic refraction for samples with a vertical intensity
> 2 x 10^{-2} mA/m – although the observed dispersions in samples of
weaker vertical intensity could not be explained. Most archaeological
samples of fired materials from middle to high latitudes exceed this

intensity and this suggests that such magnetic interactions could
be a dominant factor.

(v) Comments

A variety of factors operate that can account for some of the observed
dispersions in both directions and palaeointensities, but most of
these do not explain all of the observed dispersions nor the observed
correlations. The 'refraction' effect is the most effective, but
does not account for the observed dispersions in weakly magnetized
materials. Furthermore, it predicts that the greatest effects should
be associated with the most strongly magnetized materials. This
does not appear to be the case as the greatest scatters appear to
be associated with sites showing (a) the greatest large-scale inhomogeneity and (b) the slowest cooling (Tarling et al., 1986). The
implication is that much of the observed dispersion in direction
and variation in palaeointensity is caused by the field distortions
associated with relatively large areas of the structures that have
become magnetized in the direction of the ambient field by fast
cooling. Other areas, while homogeneous within themselves and with
their own characteristic cooling rates in laboratory experiments,
are affected by cooling in the field of these other sections. In
other words, the problem arises from the different scales on which
inhomogeneity occurs and hence the resultant magnetic interactions.
Observations based solely on the individual samples can thus be misleading. However, it seems strange that these effects should not
be more dependent on the intensity of magnetization than appears
to be the case.

It also seems probable that thermally-induced chemical changes
also occur more readily within the slower-cooled regions, and so give
rise to changes in the magnetic mineralogy as a function of the time
taken to cool. This latter effect would not be repeatable in subsequent experiments as the changes would be towards their equilibrium
conditions, which would then be chemically stable to subsequent heating,
but would mean that the laboratory thermal behaviour does not exactly
mimic the original cooling behaviour - as required for palaeointensity
determinations. The situation could, in fact, be even more complicated
as the less stable magnetic minerals may well undergo deuteric alterations subsequent to the cooling of the archeological body. However,
this effect is only likely to be significant when partially
unexsolved, but metastable, magnetic minerals have been incorporated
into the structure. In most instances such deuteric changes will
complete during the cooling process.

Some recent investigations have been concerned with magnetizations
acquired as a result of the formation of new magnetic minerals during
thermal heating at temperatures in excess of 500°C (Xu et al., 1986).
The observed effects are variable, with some examples of remanences
being acquired parallel to the field of the already magnetized grains,
but, more commonly, randomly (Tarling, 1983). The effects almost
certainly depend on the distance away from the already magnetized
grains where the newly forming minerals grow. If in intimate atomic

contact, super-exchange forces probably allow a duplication of the original field direction thereby enhancing the original direction, while at greater distances, the magnetization will be acquired in various orientations, depending on the grain distribution pattern. This model is analogous to that invoked for magnetic 'refraction', but clearly needs modification to consider different scales within which the phenomenon could operate. One major problem is that most of the observed effects are only detected by thermal demagnetization (Dunlop & Zinn, 1980; Tarling et al., 1986) while most British archaeomagnetic results have been obtained using A.F. demagnetization procedures. Luckily, however, the sampling method adopted in these studies has largely been the disk method (Tarling, 1983) whereby the surface materials have been preferentially sampled and hence are most likely to have rapidly cooled and been less affected by such complex processes.

DEPOSITED MATERIALS

The main reason for the deviations frequently observed in sedimented materials arises from the assumption that only the aligning force of the geomagnetic field is present during deposition. Early re-deposition experiments clearly demonstrated the occurrence of a deviation in inclination due to gravitational effects on particles, and that this is enhanced if deposition takes place on a slope (King, 1955; Rees, 1961, 1965). However, it is thought that this effect is of limited importance in slowly deposited sediments and that this inclination 'error' often disappears as a result of post-deposition re-orientation of the magnetic particles (Irving & Major, 1964). Nonetheless, water currents and other depositional factors can cause major deviations, and several long sedimentary cores are usually required, with appropriate statistical filtering, to define the changing direction of the geomagnetic field in most lakes. In general, the reversals in the changes in declination and inclination can usually be recognized (Creer, 1985), but it is difficult to quantify the observed swings. However, if quiet depositional environments are carefully selected, as exemplified by Creer (this vol.) and Verosub et al. (1986 & this vol.), then multiple cores, when averaged, appear to give a reasonable record of directional changes in the geomagnetic field (Thompson, 1982; Creer, 1985); such cores have the major advantage that they provide a record that can extend well over 10,000 years for these specific locations. However, even care-fully selected records can be affected to indeterminate amounts by biological activity, chemical changes, water movements through the sediments (particularly if the water table traverses through the sequence - Nöel 1980), and the like.

Nonetheless, in particularly favourable conditions, the funda-mental inaccuracy is in the dating of sediments as ^{14}C is expensive and frequently unreliable in sediments which commonly contain old carbon. Between such horizons, intermediate dates are obtained by extrapolation, usually assuming constant sedimentation rates. The significance of such dating inaccuracies obviously depends on the precision with which other parameters show variations over similar time-scales. Where calibration can be made with archaeomagnetic observations, the lake sediment dating may well have uncertainties of only a few decades, but where such calibrations are not available, particularly for sediments older than 2,000 years, the dating uncertainties become larger than the known errors of > 100 years associated with the ^{14}C dating under these circumstances.

Despite the theoretical and laboratory evidence that the palaeointensity of the field can be determined from comparison of the observed intensity of natural remanence and that acquired during laboratory deposition - at least for fields up to 0.2 mT (Collinson, 1974), the difficulty of reproducing the natural redepositional conditions, particularly the rate of deposition and degree of flocculation of the detrital grains, has inhibited most palaeointensity studies. Relative changes in past geomagnetic intensities can be estimated by comparison of the normalized natural intensities of remanence, as a function of depth. (The normalization allowing for changes in the composition and concentration of magnetic particles). Where marker horizons can be calibrated against archaeomagnetic observations, then such determinations may be a good indication of the real trends in the geomagnetic field intensity, although it is doubtful if numerical values of the past geomagnetic field can yet be given.

Although there have been few studies (Nöel et al., 1979), cave sediments provide an opportunity for geomagnetic determinations using sediments in which there is little or no bioturbation, the physical depositional conditions are often clearly constrained, and chemical changes are largely inhibited by the even, generally low temperature and low organic content. Unfortunately dating provides a greater problem because of the absence of organic carbon and the common occurrence of old carbon. Nonetheless, such sediments do offer an ideal opportunity for establishing a better understanding of the magnetization processes, and hence derivation of past geomagnetic parameters. In terms of dating, varved sediments, such as those available in Scandinavia and Canada, provide well dated records, but while these were studied in the early days of palaeomagnetism, subsequent studies have been limited in scale.

SUMMARY OF ERRORS

While much of the above comments, on both fired and deposited materials, have been largely concerned with directional studies, this mainly reflects that there has been much greater study of the 'errors' in this parameter than has been undertaken in palaeointensities. The main reason for this is that most palaeointensity studies only measure scalar quantities, and therefore provide no information on the nature and composition of the vector of remanence. However, it must be emphasized that most causes of directional deviation are equally, and often of greater, importance to palaeointensity determinations

and this must raise serious questions about the absolute reliability
of even highly repeatable palaeointensity determinations. At this
stage, the 'unknown' causes for directional deviations in fired
archaeological materials can exceed 1-2° and those in palaeointensity
determinations are probably of the order of ± 5%. Such errors, that
are mostly only indicated by lack of consistency between different
studies, must be regarded as methodological errors and additional
to the quoted repeatability accuracy. It is more difficult to assess
the real uncertainties in the use of sediments as this depends
critically on largely unknown, past depositional conditions, and
on the validity of computer or subjective rejection criteria. An
error estimate of about 10° would be generous in some circumstances,
and excessive in others, while it is not considered that quantitative
palaeointensity evaluations can yet be made.

It must be emphasized that both approaches - fired materials
and sediments - must be considered mutually complementary. Both
are discontinuous in space, being restricted to particular localities,
but the lake sediments provide moderately continuous records at a
few localities. Archaeological sites usually cover a wider spatial
range, but individually provide brief, very discontinuous records
for any one location so that some form of geomagnetic model is required
to allow direct comparison of individual observations. It is also
relevant that palaeointensity determinations on fired materials can
be undertaken on virtually any fired materials. There is therefore
a vastly greater choice of material for study, often including well
dated pottery, while directional studies of archaeological materials
requires them still to be in situ since they were last fired. Directional studies are therefore more limited in the materials available
and are particularly sensitive to site disturbances.

GEOMAGNETIC EVALUATION

At this stage, the inaccuracies of archaeomagnetic observations are
such that small amplitude changes in either intensity or direction
of the geomagnetic field cannot be considered established for any
period before 1600 A.D. Only sequences of consistent directional
or intensity changes can be used for comparison with other solar
and terrestrial phenomena during the last 10,000 years. Such considerations preclude, for example, the likelihood of detecting geomagnetic field changes associated with sun-spot cycles during this
century. However, it is reasonable to expect that improving the
understanding of the processes of magnetization is likely to enable
much more precise field intensity and directional determinations
in the near future. The major consideration will then be the uncertainty of the actual archaeological dating which is commonly of
the order of ± 100 years. The improvement in archaeomagnetic dating
will soon lead to the improvement of such errors, but this will occur
on the assumption that secular variation changes are gradual in both
space and time, i.e. the directional and intensity determinations
from disparate sites must be corrected to a central location (see

below) and then placed in a more appropriate chronological sequence by means of various best-fitting curve procedures. Such dating improvements will therefore result in some smoothing of the archaeomagnetic observations which may well be real, but still obscuring short term, small amplitude fluctuations, i.e. periodic functions of less than some 10-20 years are likely to be suppressed as such improvements are implemented.

The spatial correction of archaeomagnetic data involves some model of the geomagnetic field. Previous corrections have usually assumed that the inclination values can be corrected using the standard dipole formula (tan inclination = 2 tan latitude), but no correction was made for the spatial variation of declination. During the last ten years, the British directional data have been corrected to a central location on the assumption that the regional geomagnetic field can be largely simulated by an inclined geocentric dipole, the pole of which is defined by the observed site direction. This model has been tested for the 1985.0 IGRF relative to a central site within Britain, Meriden, at 52.43°N 1.62°W. After applying this correction, the average solid angular deviation for sites 550 kms away is 0.88 ± 0.32° and the maximum observed deviation is only 1.35°. (For sites 220 kms away, the average deviation is only 0.36 ± 0.13°.) This suggests that the maximum error involved in making this correction for sites 550 kms apart is still comparable with the measurement errors. However, it also means that if measurement errors continue to improve, there will still be errors of this magnitude within the secular variation spectrum - unless a more effective correction can be determined.

Within such constraints, the geomagnetic field direction during the last 2,000 years or so is gradually becoming better defined (Clark et al. submitted) and there are now indications of its probable direction back to c. 750 B.C. (Figure 1) - although further revision is still in progress. These data can now be used to estimate the rate of directional change of the geomagnetic field over this period and other properties of geomagnetic interest, such as the average departure from direction of the axial geocentric dipole field. In view of the uncertainties outlined above in terms of specific determination of the directions and their age, the data have been selected and evaluated to provide estimates of the field direction at 25 year intervals back to 50 B.C. To partially reflect the uncertainties, these estimates were only calculated or estimated to the nearest 0.5° in most cases, although more precise values were used when available.

(a) Angular Change

Analysis of the rate of angular change of the field direction (Figures 2 and 3) shows a modal rate of 0.05°/year, but with distinct variations, ranging from 0.01 to over 0.2°/year. These variations are not irregular, but show evidence of regular periodicity, of the order of about 200 years, superimposed on longer wavelengths. There may well be shorter period variations present, but the data are not

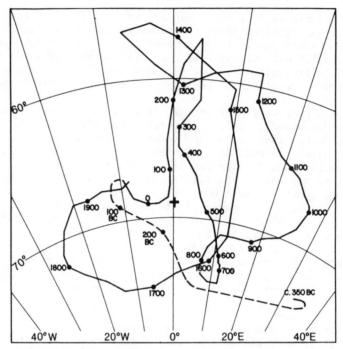

Fig. 1. The British Archaeomagnetic Curve. Derived from well-
dated archaeological materials, generally ± 25 years,
for which more than 5 samples have been used, the magnetization of which has been partially demagnetized (mostly
by alternating magnetic fields).

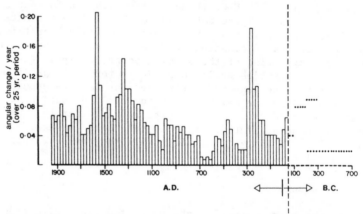

Fig. 2. The rate of geomagnetic directional change in Britain.
Individual directions were determined or interpolated
at 25 year intervals, using the curve given in Figure
1, and the solid angular change determined.

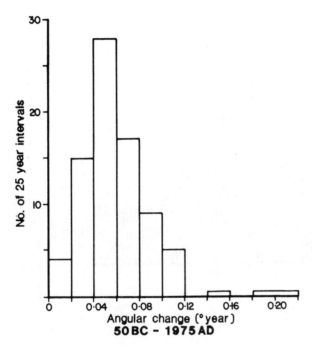

Fig. 3. The frequency of annual directional change rates. See
        Figure 2.

considered sufficiently reliable for such analyses at this time.

(b) Departure from the Axial geocentric dipole.

The average field direction (50 B.C. to 1975 A.D.) differs significantly from that of the axial geocentric dipole field. The
observed mean direction is 4.1°, + 67.5°, with α_{95} = 1.3°. This is
2.1° away from the axial geocentric dipole field direction, 0.0°,
+ 68.96°, and statistically different from it at a > 95% probability level. However, the overall mean direction for this 2025
year period is closer to the axial geocentric dipole field than that
based on observatory records since 1600 A.D. (347.7°, + 71.6,
α_{95} = 2.1°) which is some 4.9° away from the axial geocentric dipole
field. The frequency pattern (Figure 4) suggests that the field
only rarely approaches the axial geocentric dipole direction, and
there are indications of two possible preferred distributions, centred
mainly on a departure of 5°, with another distribution possibly
centred around 9°. The angular departure from the axial geocentric
dipole direction, as a function of time (Figure 5) shows a clear
sinusoidal variation with a periodicity of about 400 years, and
indications of other periodicities, e.g. c. 1000 years. At this
preliminary stage, there does not appear to be any obvious relationship
between the departure from the axial geocentric dipole field and

362

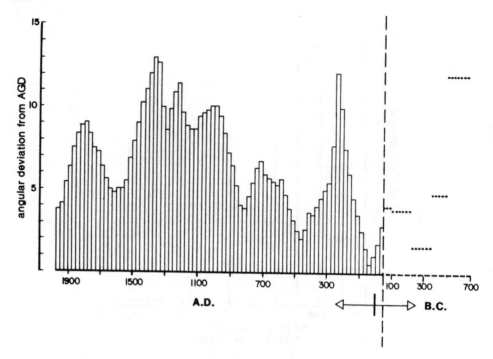

Fig. 4.   The angular distance from the axial geocentric dipole
          field as a function of time.   Based on the same data
          as Figure 2, but determining the solid angular
          distance from the axial geocentric dipole field
          direction.

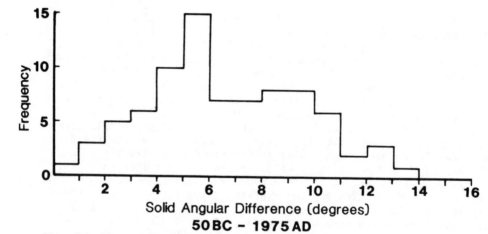

Fig. 5.   The angular separation from the axial geocentric dipole
          field.   See Figure 4.

the rate of secular change in direction, but this requires further examination.

At this stage, it is not considered practical to compare such analyses with directional observations from other areas of the world. The French data (Thellier, 1981), after applying the same inclined dipole correction, is broadly similar to that of Britain, but many key French data are poorly constrained in age. The most extensive European record, including both directional and intensity determinations, is that established for S.E. Europe as a result of very detailed and extensive research by Kovacheva (1983), mainly in Bulgaria. Unfortunately, these data are not sufficiently numerous for the last 2,000 years for direct comparison and the data have been presented as century averages that obscure some of the relevant shorter term variations of geomagnetic interest. Nonetheless, it should be emphasized that this work can be regarded as a definitive example of combined palaeointensity and directional studies.

Clearly, comparison is still required with the increasingly well defined Japanese data, and that from southwestern North America (Dubois, this vol.). However, the greater need is to synthesize the directional findings with those for intensity. Since Bucha's (1970) suggestions, it has generally been considered that there are long term, global changes in the dipolar field that are responsible for, amongst other things, the variation in ^{14}C production in the ionosphere. However, such early data were poorly reliable, individually, and of very localized derivation. There is increasing evidence that the global picture is far from established, despite the subsequently vast increase in palaeointensity determinations (Aitken et al., 1984, 1985; Evans, 1986; Kovacheva & Kanarchev, 1986; Sakai & Hirooka, 1986). Certainly the most recent Chinese data (Wei et al., 1987) are reported as not being globally consistent with patterns determined elsewhere. However, until palaeointensity determinations are more consistent, using different methods, it seems premature to attempt to analyse intensity variations that are less than 20% in absolute difference, unless consistent trends can be established. However, this presupposes that brief, rapid geomagnetic changes do not occur - when such changes are already known from observatory records. Hence there is a huge potential in such studies, but the uncertainties, at this stage, preclude all but the most consistent geomagnetic trends during the last 10,000 years. However, the possibility that the 200-year cycle in palaeointensity in Greece (Liritzis, 1986), may correlate with the British 200-year cycle in the rate of angular change may not only be geomagnetically significant, but would suggest that the errors in both techniques may not be as serious as indicated from detailed studies of certain specific sites.

ACKNOWLEDGEMENTS

I am particularly grateful for consultations with various colleagues, in particular Dr. A.J. Clark, and Profs. R.A. Hide and D. Walton.

REFERENCES

Abrahamasen, N., 1986, In 'Twenty five years of Geology in Aarhus',
ed. J.T. Möller, Geoskrifter 24, 11.
Aitken, M.J., 1974, 'Physics and Archaeology' 2nd Edn., Clarendon
Press, Oxford,p.291.
Aitken, M.J., 1983, In 'Geomagnetism of Baked Clays and Recent
Sediments', eds. K.M. Creer, P. Tucholka and C.E. Barton, Elsevier,
Amsterdam, 78.
Aitken, M.J. and Hawley, H.N., 1971, Archaeometry 13, 83.
Aitken, M.J., Allsop, A.L., Bussell, G.D., and Winter, M.B., 1984,
Nature 310, 306.
Aitken, M.J., Allsop, A.L., Bussell, G.B., and Winter, M., 1985,
Nature 314, 753.
Barbetti, M.F., McElhinny, M.W., Edwards, D.J., and Schmidt, P.W., 1977,
Phys. Earth. Planet. Ints., 13, 346.
Bucha, V., 1970, Phil. Trans. Roy. Soc., Lond., A269, 47.
Clark, A.J., Tarling, D.D. and Nöel, M., J. Archaeo. Sci. (submitted).
Coe, R.S., 1967, Geomagn. Geoelect., 19, 157.
Goe, R.S., and Grommé, C.S., 1973, J. Geomagn. Geoelect., 25, 415.
Collinson, D.W., 1974, Geophys. J. R. Astr. Soc., 38, 253.
Collinson, D.W., 1983, 'Methods in Rock Magnetism and Palaeo-
magnetism', Chapman & Hall, London, p.503.
Creer, K.M., 1985, Geophys. Surveys, 7, 125.
Dodson, M.A., and McClelland-Brown, E.A., 1980, J. Geophys. Res., 85,
2625.
Dunlop, D.J., and West, G.F., 1969, Revs. Geophys. Space Phys., 7, 709.
Dunlop, D.J., and Zinn, M.B., 1980, Canad. J. Earth Sci., 17, 1275.
Evans, M.E., 1986, J. Geomagn. Geoelect., 38, 1259.
Fox, J.M.W., and Aitken, M.J., 1980, Nature, 283, 462.
Grommé, S., Mankinen, E.A., Marshall, M., and Coe, R.S., 1979, J.
Geophys. Res., 84, 3553.
Hoye, G.S., 1982, Archaeometry, 24, 80.
Irving, E., and Major, A., 1964, Sedimentology, 3, 135.
King, R.F., 1955, Mon. Not. R. astr. Soc., 7, 115.
Kovacheva, M., 1983, In 'Geomagnetism of Baked Clays and Recent
Sediments', eds. K.M. Creer, P. Tucholka, & C.E. Barton,
Slsevier, Amsterdam, 110.
Kovacheva, M. and Kanarchev, M., 1986, J. Geomag. Geoelect., 38, 1297.
Liritzis, Y., 1986, Earth Moon Planets, 34, 235.
Lund, S.P., 1985, Geophys. Res. Letters, 12, 251.
Neél, L., 1949, Ann. Geophys., 5, 99.
Nöel, M., 1980, Geophys. J. R. astr. Soc., 62, 15.
Nöel, M., Homonko, P., and Bull, P.A., 1979, Trans. Brit. Cave Res.
Assoc., 6, 85.
Prévot, M., Mankinen, E.A., Grommé, C.S., and Lecaille, A.,
J. Geophys. Res., 88, 2316.
Rees, A.I., 1961, Geophys. J. R. astr. Soc., 5, 235.
Rees, A.I., 1965, Sedimentology, 4, 257.
Sakai, H., and Hirooka, K., 1986, J. Geomagn. Geoelect., 38, 1323.

Stacey, F.D., 1963, Adv. Phys., 12, 46.
Stephenson, A., and Molyneux, L., 1987, Geophys. J. R. astr. Soc.,
  90, 467.
Tarling, D.H., 1983, 'Palaeomagnetism', Chapman & Hall, Lond., 379.
Tarling, D.H., Hammo N.B., and Downey, W.S., 1986, Geophysics 51,
  634.
Thellier, E., 1981, Phys. Earth Planet. Sci., 24, 89.
Thellier, E., and Thellier, O., 1942, C.R. Acad. Sci., Paris, 214, 382.
Thellier, E., and Thellier, O., 1952, C.R. Acad. Sci., Paris, 234, 1464.
Thompson, R.m, 1982, Phil. Trans. R. Soc., Lond., A303, 103.
Verosub, K.L., Mehringer, P.J., and Waterstraat, P., 1986,
  J. Geophys. Res., 91, 3609.
Walton, D., 1980, Nature, 286, 245.
Walton, D., 1983, In 'Geomagnetism of Baked Clays and Recent Sediments',
  eds. M.K. Creer, P. Tucholka, & C.E. Barton, Elsevier, Amsterdam, 90.
Walton, D., 1987, Revs. Geophys. (in press).
Weaver, G.H., 1962, Archaeometry 4, 23.
Weaver, G.H., 1970, Archaeometry, 12, 87.
Wei, Q-Y, Zhang, W-N, Li D-J, Aitken, M.J., Bussell, G.D., and Winter,
  M., 1987, Nature 328, 330.
Xu, T-C., Tarling, D.H., Eustance, N.B., and Hijab, B.R., 1986,
  Geophys. J. R. astr. Soc., 87, 305.

THE GEOMAGNETIC FIELD OVER THE LAST 200 YEARS

F.J. Lowes
School of Physics
University of Newcastle upon Tyne
Newcastle upon Tyne, NE1 7RU
England.

ABSTRACT. Detailed direct observations of the geomagnetic field were at their peak in 1980, but go back about 200 years with decreasing accuracy and resolution. This paper describes the main features of the field, and particularly the complexities of its secular time variation, so that the older, indirect, and often smoothed, data from archaeomagnetism and palaeomagnetism can be put into context.

The temporal and spatial power spectra of the field are used to give a first-order separation between those parts of the field which come from electric currents in the fluid conducting core (the main field), from the magnetisation of crustal rocks, and from the iono-sphere and magnetosphere. The main field is then discussed in more detail.

1. INTRODUCTION

We only have reasonably good direct observations of the geomagnetic field for about 200 years. This is a very small fraction of the more than 10^9 years for which the field has existed, so we cannot be certain that the present field is typical. However, this meeting is about the last 10,000 years, so the extrapolation is not quite so extreme. In fact what evidence we have suggests that, at least over this 10,000 year scale, the present field is not untypical.

No matter how careful the techniques used, the geomagnetic "data" presented in other papers at this meeting are based on deductions from inevitably "noisy" very sparse and patchy information. For example, archaeomagnetism gives essentially point data, the points being irregularly scattered in both space and time; lake sediment data, though continuous in time, have been subject to uncertain smoothing. If we are going to interpret these data sensibly we need to have some feel for the underlying process of which these data are a very imperfect sample.

The purpose of this paper is therefore to summarise our more detailed knowledge of the more recent field, and particularly of its time variation, so that the older "measurements" can be put into context.

F. R. Stephenson and A. W. Wolfendale (eds.),
Secular Solar and Geomagnetic Variations in the Last 10,000 Years, 367–379.
© 1988 by Kluwer Academic Publishers.

2. INTERNAL AND EXTERNAL SOURCES

If we observed the field at one point for a (very) long time we would
see variation on a very wide range of time scales. Figure 1 shows
schematically the sort of power spectrum we would see; it has contributions having typical time scales ranging from millions of years
(reversals) down to seconds (pulsations) and beyond. However there
appears to be a kink in the spectrum at a period of a few years, and
we now know that this separates fields from two sources. The longer
period variations come from the Earth's core (and to a very minor extent
its crust), its sources are "internal", while the shorter period
variations come from the ionosphere and the magnetosphere, "external"
sources.

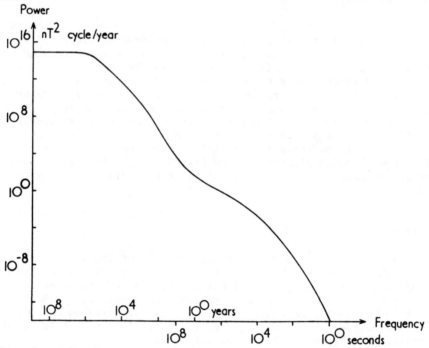

Fig. 1.   Schematic temporal power spectrum of the geomagnetic field.
          (For convenience the Frequency axis has actually been
          labelled in terms of period.) In addition to this con-
          tinuous spectrum there are spectrum "lines" at 1 day, 1 year,
          ∿ 11 years, and harmonics of these.

For workers interested in the internal field the high-frequency
external field is just noise (though of course it is precisely that
part of the field, coming from the solar-terrestrial interaction, which
is the subject of other papers at this meeting). In principle, given
sufficient vector field observations over the Earth's surface, it is
possible to use spherical harmonic analysis to separate the internal
and external parts of the field, but in practice the data are not good

enough. So we eliminate (most of) the external field by using a low-pass filter, conventionally by using (calendar) annual means; we then have what we call the secular variation, where "secular" is used in the sense of "slow". There is a minor complication in that the ring current, produced by the longitudinal drift of the charged particles in the radiation belts and which varies with time, is not easily separated from the internal dipole; there is also a small but significant 11-year modulation produced by the sunspot cycle. (For "instantaneous" observations, such as from a short-lived satellite, we try to use data from "quiet" conditions, and accept the remaining external field as increased noise.)

3. THE FIELD OF INTERNAL ORIGIN

Having rejected the high-frequency "noise", let us now look at how the field varies over the Earth at any one time. To a first approximation the field resembles that of a roughly central dipole, being horizontal near the equator, and vertical (and twice as strong) near the poles. But as can be seen from Figure 3, (which shows only the vertical part of the vector field, and on a projection which inevitably distorts the almost spherical Earth's surface) this is only a crude approximation even if we think of the dipole as being inclined at about 11° to the geographic axis. It is convenient to subtract (vectorially) the inclined dipole approximation from the total field to give the "non-dipole" field – whether this subtraction has any physical significance is discussed below. To allow simple comparisons of field magnitudes, from now on I will always quote r.m.s. vector magnitudes, the mean being taken over the sphere. So the dipole field, which varies from about 30,000 to 60,000 nT, has a r.m.s. value of 44,000 nT, while the non-dipole field adds about 10,000 nT, to give the total field of 45,000 nT.

Even without the smoothing imposed in producing Figure 3, because of its small scale it would not be possible to show all the small wiggles of the real contours. However, just as we can characterise the variation in time by means of a (temporal) power spectrum, which shows the contribution of different periods (or frequencies) to the total variation, then so also can we produce a spatial "power" spectrum showing the contribution of different length scales (spatial frequencies) to the total spatial variation. In fact the circumference of the Earth gives a fundamental "period", so we have the (spherical) spatial analogue of a Fourier series.

Figure 2 shows a recently obtained spatial spectrum. Again the spectrum (essentially) decreases monotonically, so smoothing such as that used in Figure 3 does not produce significant distortion. Also again there are two parts, corresponding to different source regions. Up to about n = 13 (wavelengths down to 1300 km) we are looking at the "main" geomagnetic field, that coming from electric currents in the fluid conducting core 2900 km below us; these currents produce the vast majority of the surface field.

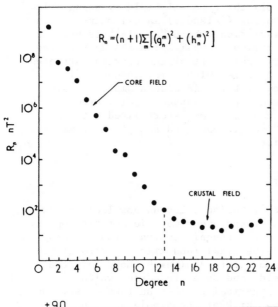

$$R_n = (n+1)\sum_m \left[(q_n^m)^2 + (h_n^m)^2 \right]$$

CORE FIELD

CRUSTAL FIELD

Fig. 2. Spatial power spectrum of the main geomagnetic field. The ordinate is (the logarithm of) the m.s. field produced by all harmonics of degree n. (After Langel, 1987).

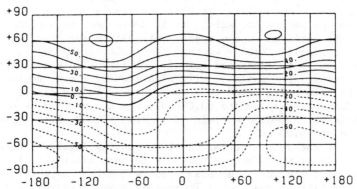

Fig. 3.   Vertical component of the main geomagnetic field (that coming from electric currents in the core of the Earth) at the Earth's surface R_e = 6371 km in 1980. (Fields of wavelength less than 45° omitted.)
Contour interval 10 μT; full contours for field inwards, dashed contours for fields outwards.
Cylindrical equidistant (Plate Caree) projection.

From about n = 15 upwards we are looking at the field produced by the magnetisation of crustal rocks (and archaeomagnetic sites, and lake sediments etc.!) of the outer 20 km or so of the Earth which is above the Curie temperature isotherms. (The temperature increase with depth is about 30°C/km, and above about 600°C the rocks are non-magnetic.) This crustal magnetisation contributes typically 200 nT to the surface field.

Consider first this crustal magnetisation. Some of this is permanent and its directions reflect some past field; for example, the sea-floor "striping" produced by sea floor dykes and lavas produced near the mid-ocean spreading centres, and the (often small) stable components of magnetisation of rocks and sediments used in palaeo-magnetism. But the vast majority of the magnetisation is induced, with its direction being parallel to the present main field; however, because of the small scale variability of the susceptibility, the surface field resembles that of random sources. To a main-field geo-magnetician the field of this crustal magnetisation is again "noise"; we would like to remove it by low-pass (spatial) filtering, but this is difficult to do for surface data (and archaeomagnetic and palaeo-magnetic field data inevitably include the local crustal field). Fortunately satellite observations are at an altitude such that the height itself gives very considerable filtering.

Returning to the main field, clearly there must be some, as yet unknown, physical mechanism which imposes such a smooth spectrum. However, note that the n = 1 dipole point is well above the line, suggesting that there might indeed be different mechanisms for the production of the dipole and non-dipole fields, and a justification for our, otherwise arbitrary, separation. But this still leaves the problem of which dipole we subtract. If we average the whole field over 10^7 years (and ignore the reversals) we obtain an (almost) axial central dipole. But if we average for only 10^3 years we obtain an (almost) central dipole inclined at about 10° to the axis. If we are looking at cosmic ray trajectories we might wish to approximate the whole field by an inclined dipole displaced about 500 km from the centre. We need to be clear about what we are doing in producing a "non-dipole" field.

Another problem concerns the choice of surface, i.e. radius, on which we look at the field. The figures I have been quoting refer to the surface of the Earth. However, as far as the main magnetic field is concerned this surface is not "visible" as there is no significant contrast in permeability and no significant magnetisation and electric currents. The really important surface is the core-mantle boundary (CMB), which is the outer boundary of the electric currents producing the main field. Figure 4 (note the change in contour interval from Figure 3) shows an approximation to the radial field at this surface; as the higher harmonics (shorter wave-lengths) of the field (which are less well known) increase more rapidly as we approach the source, at this surface the dipole field no longer dominates (and the relative uncertainty of the field is a lot larger than at the surface). Conversely we might be interested in cosmic rays being deflected as they approach the Earth; in this case the field at, say, R = 2 R_{earth} would be more relevant. Figure 5 shows that at this distance the field is much more nearly (inclined) dipolar. However, our observations are in fact all at or near the Earth's surface, so I will revert to describing what we see there.

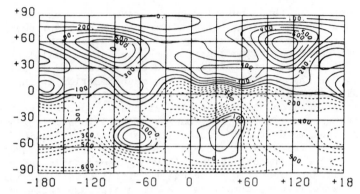

Fig. 4.   Approximation to the vertical component of the main geomagnetic field at the core-mantle boundary, R = 0.545 R$_e$.
Contour interval 100 μT, otherwise as Figure 3.

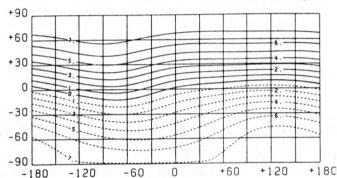

Fig. 5.   Vertical component of the main geomagnetic field at R =
2 R$_e$.
Contour interval 1 μT, otherwise as Figure 3.

Table 1 summarises the various contributions to the surface field.
For 1980, because of the global satellite survey by MAGSAT, we knew
the main field (that from the core) to about ± 20 nT.   However, as
this Workshop emphasises, the geomagnetic field varies very rapidly
by geophysical standards (if not by solar standards) at about
80 nT/yr.   Unfortunately we know this 80 nT/yr only to about ± 20 nT/yr,
so our knowledge of the Earth's magnetic field is rapidly getting
poorer again; we need a satellite survey about every 10 years to
maintain 0.1% accuracy – ESA and NASA please note!

4.   THE NON-DIPOLE FIELD

If we filter out the field due to crustal magnetisation, and subtract
the inclined central dipole we are left with the surface "non-dipole"
field shown in Figure 6.   This non-dipole (ND) field has features which
are of continental length scale; the other field components show

Table 1

Magnetic fields at the Earth's surface

(a) Main geomagnetic field (from the core) (r.m.s. over surface)

	Total field	=	Dipole	+	Non-dipole
	45,000 nT		44,000 nT		10,000 nT
	80 nT/year		(5-40) nT/year		70 nT/year

(b) Crustal field   ∿   200 nT

(c) External field   ∿   30 nT on quiet days

                          >   1000 nT on disturbed days

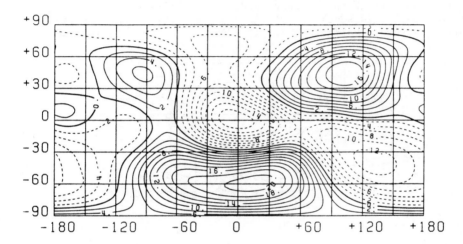

Fig. 6.   Vertical component of the non-dipole field at the Earth's
          surface.
          Contour interval 2 µT, otherwise as Figure 3.

similar scale features but with different patterns. As this field
is produced in the core we would not expect any direct correlation
with the surface continents; in fact at present the ND field is
particularly small over the Pacific. Of course, individual features
of the ND field can change and move quite rapidly (see below), but
there is some (disputed) evidence that this lack of ND field over the
Pacific has existed for perhaps a million years. If this feature has
been long-lived it might suggest some link between the pattern of the
(fluid) convection currents in the core which produce the field, and
the (solid state) convection in the mantle which (perhaps) controls
the positions of the continental plates; such a link could perhaps
be thermal and/or topographic anomalies on the CMB.

5. THE SECULAR VARIATION FIELD

The first time derivative of the field of internal origin is called the secular variation (SV) field; Figure 7 shows the 1980 field, but remember that we do not know it at all well.

The dipole part of the main field varies only slowly; the 44,000 nT is decreasing at about 35 nT/yr, about 10% century, at present, but this decrease has varied in the range 5 - 40 nT/yr this century. The corresponding change in dipole direction is a few tenths of a degree.

The 10,000 nT non-dipole field has a larger time variation (relatively much larger) of about 70 nT/yr, or 70%/century.

As most of the secular variation comes from changes in the non-dipole field it too has features of continental extent. Again there is no direct correlation with the actual continents, except for (at present) being small over the Pacific. Nor is there any _direct_ correlation with the ND field itself (but there is indirect correlation through the westward drift - see below).

Although this secular variation is only poorly known it is clear that it is _very_ complicated. I will discuss two particular aspects of it.

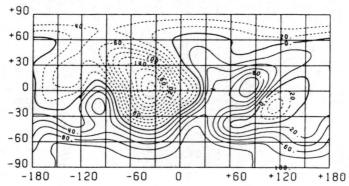

Fig. 7.   Vertical component of the secular variation field at the Earth's surface.   Contour interval 10 nT/year, otherwise as Figure 3.

6. THE WESTWARD DRIFT

If we look at charts of the ND field for succcssive epochs we see that some individual features are created and destroyed, and some features move significantly, on a time scale of 10 - 100 years. It was realised quite early that individual features _on average_ tended to drift westwards, though with much scatter. We would like to quantify this average, or global, "westward drift", but it is not possible to do this uniquely without introducing arbitrary assumptions. The first assumption is usually to consider only the non-dipole part of the field, because the dipole part seems to have very much less drift.

One approach, introduced by Bullard et al. (1950), is to assume that the whole ND field drifts westward, but at the same time changes its pattern, so that at each point

$$\frac{\partial \underline{B}}{\partial t} = \frac{D\underline{B}}{Dt} + \omega \frac{\partial \underline{B}}{\partial \lambda}$$

where $\partial \underline{B}/\partial t$ is the total time variation seen by the stationary observer, $D\underline{B}/Dt$ is the time variation which would be seen by an observer moving with the field at angular velocity ω with respect to the stationary observer, and $\partial \underline{B}/\partial \lambda$ is the longitudinal gradient of the (ND) field. We can observe $\partial \underline{B}/\partial t$ and $\partial \underline{B}/\partial \lambda$, but do not know $D\underline{B}/Dt$ or ω. In practice ω is usually estimated by least squares' fitting $\omega \partial \underline{B}/\omega \lambda$ to $\partial \underline{B}/\partial t$ over the surface; this involves the quite arbitrary assumption that nature is such as to minimise $D\underline{B}/Dt$, a most unlikely situation! Also, the value of ω obtained depends on what property of the field is used, and at what radius! However, at present the field appears to drift westward on average at about $0.2°$/year, making the two contributions to the observed secular variation about equal in magnitude (but remember that they add vectorially).

An alternative approach is that introduced by Yukutake and Tachinaka (1968) who argued that the ND field has two parts, one stationary and one drifting, with both parts having fixed patterns. So we have

$$\underline{B} = \underline{B}_{drift} + \underline{B}_{stat}$$

$$\frac{\partial \underline{B}}{\partial t} = \omega \frac{\partial \underline{B}_{drift}}{\partial \lambda}$$

whence

$$\frac{\partial^2 \underline{B}}{\partial t^2} = \omega \frac{\partial}{\partial \lambda} (\frac{\partial \underline{B}}{\partial t})$$

On this model the secular variation rotates bodily, and its westward drift gives ω directly; unfortunately $\partial^2 B/\partial t^2$ is very poorly known. Inevitably Yukutake's ω is larger than Bullard's value; we find about $0.4°$/year, with the stationary and drifting parts of the ND field having comparable magnitude.

Both approaches give about an equally good (or bad) fit to the surface observations at any one time, and we do not know which is the best way of modelling the situation over a significant period of time - certainly both models would have to be made more complicated to give a reasonable fit. It is not inconceivable that the apparent drift is simply the result of chance correlation between the SV and ND fields.

Our two hundred years of direct global observation is probably comparable to the relevant time constants, so would not be long enough to give a reasonable test, and although there is older data it is much more scattered. Similarly, although workers using the inclination/declination profiles of lake sediments from different continents claim to see time-displaced correlation which would correspond to the westward drift (e.g. Creer and Tucholka, 1982), unfortunately their data inevitably represents very few points on the Earth's surface.

Although 0.2°/year corresponds to a "period" of 1800 years, it is unlikely that individual features would persist so long, so I would not expect to see any significant periodicity as such.

If there is a real westward drift then the place to look for it would be at the core/mantle boundary, where, assuming that for short time scales the magnetic field is "frozen" to the core fluid, we would be looking at the surface motion of this fluid. Unfortunately, we do not know the SV field very well at the surface, so we know it even less well at CMB, particularly as the SV field has a flatter spatial spectrum than the ND field. However, Gubbins and Bloxham (Gubbins and Bloxham, 1985; Bloxham, 1986) have recently produced a series of charts of the total field (dipole + non-dipole) at the CMB for epochs from 1715 to 1980. They are talking in detail about drifting and stationary sources of magnetic flux, but there is still discussion on the accuracy of their charts.

7. TIME VARIATION OF THE SECULAR VARIATION

Our knowledge of the secular variation comes mainly from the records of about 120 magnetic observatories (unfortunately, very unevenly distributed over the Earth's surface). Figure 8 shows the annual means from a not untypical observatory. There is some instrumental noise, and also, particularly for the X component, significant solar cycle signal (mainly through the ring current). To obtain the secular variation we have to find the first derivative of these "curves"; ideally we should try to remove the solar signal first, but in practice this is rarely done. Clearly, even at one epoch it is difficult to decide what set of gradients to draw; direct differencing of the annual means can give a quite misleading impression of the general trend.

But it is also obvious that one straight line for each curve is a quite inadequate representation of the data. Historically some workers used a series of straight lines as a purely empirical help to charting. More recently the more usual approach has been to use a Taylor series solution

$$\underline{B}(t) = \underline{B}(t = 0) + \left[\frac{\partial \underline{B}}{\partial t} \right]_o (t-t_o) + \tfrac{1}{2} \left[\frac{\partial^2 \underline{B}}{\partial t^2} \right]_o (t-t_o)^2 + \ldots$$

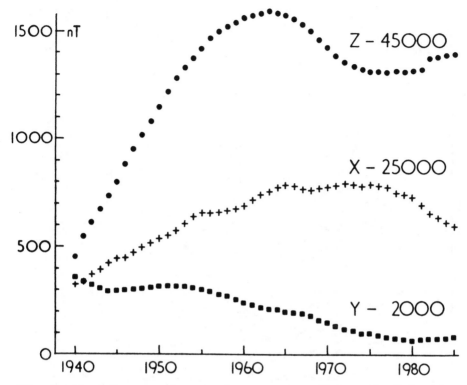

Fig. 8.   Annual means of cartesian components X (north), Y (east),
Z (down) at Tashkent magnetic observatory.

but of course $\partial^2 \underline{B}/\partial t^2$, the secular acceleration, is given in effect by
second differences, so is even less well known than the secular
variation.   Also, while perhaps a reasonable way of representing
past variation, a Taylor series to second or third order is a
notoriously unstable way of extrapolating to the future, as the
compilers of world declination charts for navigators know to their
cost!

In fact, the situation is even worse. Many workers believe that
there is now convincing evidence that the field is not sufficiently
smoothly varying to allow a power series representation, except for
discrete intervals.   In particular they argue (e.g. Malin et al.,
1983) that at about 1970 the data from many observatories indicated
a "jerk", a step change in $\partial^2 \underline{B}/\partial t^2$ (Figure 9).   There is controversy
as to how sharp this event was (e.g. Backus et al. 1987), and as to
whether it was truly global, but it certainly did exist over a large
area.

So remember, when you look, for example, at the magnetic record
preserved in a lake sediment, you are looking at an inevitably
smoothed (and possibly distorted) version of a very complicated time
variation.

378

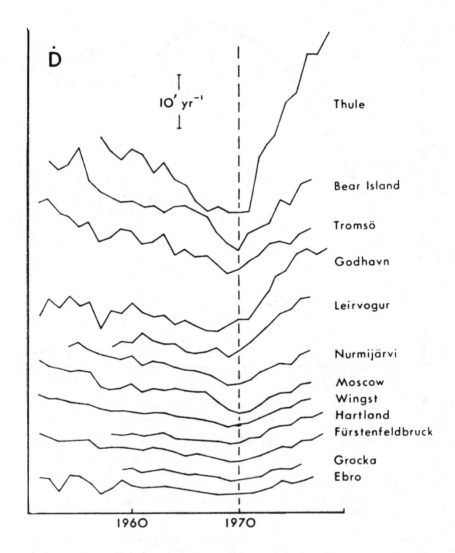

Fig. 9.   Secular variation in declination D showing the "jerk"
in 1970 at European observatories.   (From Malin et al.,
1983.)

REFERENCES

Bloxham, J., 1986, J. Geophys. Res., 91, 13954.

Backus, G.E., Estes, R.H., Chin, D. and Langel, R.A., 1987, J. Geophys. Res., 92, 3615.

Bullard, E.C., Freedman, C., Gellman, H. and Nixon, J., 1950, Phil. Trans. R. Soc. Lond. A., 243, 67.

Creer, K.M. and Tucholka, P., 1982, Phil. Trans. R. Soc. Lond. A., 306, 87.

Gubbins, G. and Bloxham, J., 1985, Geophys. J. R. astr. Soc., 80, 695.

Langel, R.A., 1987, 'Geomagnetism', J. Jacobs (Editor), Academic Press, Chapter 4.

Malin, S.R.C., Hodder, B.M. and Barraclough, D.R., 1983, 'Publ. Ebro Obs., Jubilee Vol'., 239.

GEOMAGNETIC FIELD AND RADIOCARBON ACTIVITY THROUGH HOLOCENE TIME

K.M. Creer
Department of Geophysics
University of Edinburgh
Edinburgh, U.K.

ABSTRACT. The properties of baked clays and lake sediments as recorders of geomagnetic variations through pre-historic times are compared. Archaeomagnetism has provided a substantial body of intensity data but few directional data while lake sediment palaeomagnetism has provided continuous, smoothed records of direction but no reliable intensities.
    Lake sediment records of direction, constructed for several regions in both northern and southern hemispheres, extend through the entire Holocene. They are characterized by concentration of energy in periods of a few thousands of years which are shown to be due to variations in the nondipole field.
    Reliable archaeointensity data are derived mainly from a rather restricted geographic region comprising Greece and surrounding countries. The 8000 year long record shows a weak long wavelength trend contaminated by shorter periods. By analogy with the lake sediment results, the shorter periods are attributed to nondipole field variations. The overall trend of the record has been attributed to variations in dipole moment and has been associated with a similar long period of about 12000 years carried by the 8500 year long bristlecone pine record of temporal variability of concentration of atmospheric radiocarbon. Spectral analysis of this record has also revealed weak sporadic variations with period about 2000 years. It is proposed that these are caused by a drifting quadrupole field which causes a weak spatial assymmetry in cosmic ray influx and maintains the resulting spatial variation in radiocarbon concentration in the atmosphere against mixing as it slowly drifts westward.

1. THE HISTORIC FIELD

Geomagnetic measurements have been made at continuously recording observatories since the early 19th century. Satellite data have provided a more uniform coverage of the globe in the last two decades. Readings of dated geomagnetic declinations recovered from ships' logbooks, supplemented by occasional inclination readings taken at ports of call by mariners cover the last few centuries.

F. R. Stephenson and A. W. Wolfendale (eds.),
Secular Solar and Geomagnetic Variations in the Last 10,000 Years, 381–397.
© 1988 by Kluwer Academic Publishers.

Nowadays spherical harmonic anlyses of observatory data are normally carried out up to order about n = 10. These analyses can be extended up to about n = 20 by incorporating satellite data, though this is interesting mainly for the study of short wavelength, near surface magnetic anomalies, rather than for studies of the field produced directly by the geodynamo. Gauss was able to extend his original analysis of the 1834AD field only as far as n = 4, being limited by a sparse data set and also by having to make calculations by hand. Analyses of pre-observatory, <u>directional</u> data have been taken as far as n = 4 (Barraclough, 1974), though without intensities Gauss co- efficients can only be calculated relative to the axial dipole g_1^o.

The results of such analyses show that about 90% of the geomagnetic field as observed at the Earth's surface can be described by a <u>geocentric dipole</u> tilted at about 11.5° to the rotation axis. The <u>best-fitting dipole</u> is described by the coefficients of the three terms of order n = 1 in the spherical harmonic expansion. These are usually expressed as the Gauss coefficients g_1^o (axial component), g_1^1 and h_1^1 (equatorial components along the 0° and 90°E meridians respectively). Here the subscripts and superscripts refer respectively to order and degree. The field remaining when the <u>dipole field</u> has been subtracted from the <u>observed field</u> is called the <u>nondipole field</u> (NDF). Contours of the NDF define several <u>foci</u> with linear dimensions of roughly 2000km across. While it is convenient for many practical purposes to separate the observed field into dipole and nondipole parts, it must be stressed that these two parts of the field are <u>not</u> physically distinct from each other.

TABLE 1

GAUSS-COEFFICIENTS DEFINING DIPOLES, QUADRUPOLES AND OCTUPOLES FOR THE 1980 IGRF

	n	m	g_n^m	h_n^m
dipoles	1	0	-30001	
	1	1	-1950	+5634
quadrupoles	2	0	-2038	
	2	1	+3035	-2136
	2	2	+1652	-179
octupoles	3	0	+1293	
	3	1	-2156	-38
	3	2	+1244	+261
	3	3	+851	-235

The <u>quadrupole</u> terms of the spherical harmonic expansion are represented by the five Gauss coefficients of order n = 2: g°_2, g^1_2, h^1_2, g^2_2 and h^2_2. Together with the geocentric dipole, these define an <u>eccentric dipole</u> which through the last century has been displaced from the geocentre by about 300km.

The magnitudes of Gauss coefficients of orders n = 1 to 3 are listed in Table 1. The magnitudes of higher order coefficients decrease, relative to the quadrupole, by a factor of about 10 at n = 5 and by a factor of almost 100 by n = 7.

1.1 Secular Variations

Variations in strength and direction of the geomagnetic field occur over a very wide range of time-scales. The more rapid variations, which have both regular (e.g. diurnal) and irregular (storm) character- istics, are caused by electric currents in the ionosphere. Variations of mean annual values of the geomagnetic elements (declination, inclination, intensity etc.) are called <u>secular variations</u> (SV). These are characterized by time-scales of tens and hundreds of years and their principal characteristics can be represented by:-
(i) westward drift of the NDF at an overall average (though not spatially uniform) rate of about 0.2° per year;
(ii) waxing and waning of the strength of individual NDF foci;
(iii) a slow decrease in the strength of the main dipole from about 8.5 10^{22} Am^2 in about 1834 to 7.0 10^{22} Am^2 in 1980, (the mean rate is about 5% per century);
(iv) westward drift of the equatorial component of the main dipole at a rate of about 0.08° of longitude per year.

Yukutake and Tachinaka (1968, 1969) separated the NDF into drifting and standing parts. The drifting part consists mainly of low harmonics (n < 3). When plotted, a strong quadrupolar symmetry is visually apparent (see Creer, 1981), with one positive and one negative focus in each of the northern and southern hemispheres: this north/south symmetry is such that the sign changes across the equator along any meridional band. The drift rate is about 0.3° per year, faster than the averaged value of about 0.2° per year for the complete NDF. The standing part of the NDF exhibits a more complex pattern.

2. THE PRE-HISTORIC FIELD

2.1 Methods and Techniques

We have to rely on the inadvertent consequences of routine activities of ancient civilizations or on natural geological processes, rather than purpose made instruments, as sources of geomagnetic records prior to about the 14th century AD. <u>Archaeomagnetism</u> relies on the thermo- remanent magnetization acquired by pottery, kiln walls and other relicts left behind by the more advanced of the ancient civilizations. <u>Lake sediment palaeomagnetism</u> depends on the alignment of the magnetic grains contained in sediments in the ambient geomagnetic field direction, during or shortly after deposition in still water. The reader is referred to The Geomagnetism of Baked Clays and Recent Sediments (GBCRS-1983), for a broad-based multi-authored account of methods, techniques and results of archaeomagnetism and lake sediment palaeo-magnetism.

2.2 Targets

The objectives of archaeomagnetic and lake sediment palaeomagnetic research have to be restricted to investigations of the grosser scale properties of the ancient (archaeo- and palaeo-) geomagnetic field because the coverage of the global surface will always be sparse and irregular, because pottery and sediments are at best low fidelity recorders, and because there were neither mechanical clocks nor written diaries to provide an accurate time scale. Examples of the kind of question that can realistically be posed are as follows.

(ii) Is westward drift a persistent or a transient property of the geomagnetic field?
(ii) What is the typical lifetime of a nondipole field focus?
(ii) What is the relative importance of drifting and standing nondipole sources?
(iv) Is the geomagnetic spectrum stationary?
(v) How rapidly has the geomagnetic field changed in direction and intensity over thousands of years?
(vi) To what extent do published reports of archaeomagnetic intensity, and lake sediment directional, excursions really represent abnormal changes? (See, for example, Verosub and Banerjee, 1977; Creer, 1985).
(vi) Has the ratio between the intensities of the dipole and nondipole parts of the field varied appreciably over thousands of years?

3. LAKE SEDIMENT PALAEOMAGNETISM

3.1 Depositional Remanent Magnetization

Lake sediments acquire a natural remanent magnetization when the axes of magnetic carrier grains (usually iron oxides of detrital and/or biological origin) become aligned statistically in the direction of the ambient geomagnetic field on falling through the water to settle on the lake floor. This Depositional Remanent Magnetization (DRM) may register falsely shallow magnetic inclinations, especially if the grains are not spherical (King, 1955). The alignment of magnetic grains along the ancient field direction may also be degraded by water currents on the lake bottom, slumping etc.

3.2 Post Depositional remanent Magnetization

Another process, Post Depositional Remanent Magnetization (PDRM) can take place in newly deposited, oozy sediment while its water content is still high enough to allow magnetic grains to rotate in the interstices of the matrix. The alignment along the ambient geomagnetic field direction may be aided by bioturbation, opposed by surface tension due to bubbles of gas arising from chemical and biological action, and opposed by the matrix grains whenever they are pushed aside by the magnetic grains as they rotate.

Laboratory investigations have shown that the effectiveness of the PDRM process is influenced by the size of the magnetic grains relative to the matrix and by the water content (Tucker, 1981 and section 1.3 of GBCRS-1983). In particular, grain rotation is no longer possible when the water content is less than about 70%. In the natural lake environment, the water content of a sediment is progressively reduced by the increasing overburden of accumulating younger sediments. The indications from laboratory experiments are that the critical water content should be reached at sediment depths of around 15 - 20 cm, thus defining a magnetic lock-in zone. The length of time so represented depends on the rate of sediment deposition: e.g. for 1mm/yr, it would be 150-200 years.

Laboratory experiments are completed in times which are orders of magnitude shorter than the time taken for sediments to acquire their PDRM. Hence bacteria have no time to play any role in the magnetization process as they seem to in the lake environment. Nevertheless, the concept of a lock-in time will still hold, imposing a lower period, (i.e. high frequency) cut-off to the recorded signal which must therefore be accepted as a somewhat smoothed version of the actual geomagnetic signal. While attempts have been made to normalize measured values of intensity to allow for variability and concentration of magnetic minerals, grain size etc. as a function of depth, using e.g. anhysteretic remanent magnetization (Levi and Banerjee, 1976) or PDRM (Tucker, 1981), none of these can be relied upon to yield secure values of palaeointensity of the geomagnetic field. Hence the lake sediment palaeomagnetic data base is composed exclusively of directions.

3.3 Decay and oveprinting of recorded signal

Subsequent to deposition, the reducing action of bacteria may cause ferric iron to be reduced to ferrous iron (e.g. haematite may be partly reduced to magnetite). It may also cause the formation of sulphides (Canfield and Berner, 1987). Chemical changes such as these may cause the original PDRM to be progressively lost with increasing age and it may create a new component of magnetization called chemical remanent magnetization (CRM). This explains why lakes yielding good quality palaeomagnetic SV records are rare. Nevertheless, such lakes do exist, though a persistent effort is required to find them.

3.4 Type curves depicting secular variations

Mackereth (1971) published a convincing record of declination variations spanning Holocene time, recovered from the bottom sediments of Lake Windermere. This demonstration that unconsolidated sediments can carry a record of geomagnetic declination variations led to a growing interest in palaeomagnetic work on lake sediments.

Progress has been slower than initially anticipated because only a small proportion of lakes yield good quality records. Lakes should be small, with minimum fluctuations of water level. In such lakes, deposition rates, averaged over decades to centuries, are found to have been fairly uniform through millennia. Cores should be taken from the central part of the flat floor to avoid the direct effects of slumping of sediment from the slopes around the perimeter.

The procedure is to construct type curves depicting declination and inclination variations. Logs from individual cores from each lake must be transformed to a common depth scale for that lake. They should then be stacked or merged, but it may be necessary to take out any long wavelength trends introduced by twisting or bending of the core tube during the coring operation. After within-lake stacking, transformation to a time scale is carried out using the available age control information (radiocarbon dates, palynology, etc). Finally, whenever possible, logs from cores from several lakes within the region are stacked as described by Creer (1982), Creer and Tucholka (1982a) and Creer et al. (1983).

Such type curves have been shown to be valid for geographical regions 1000km or more across (Creer, 1985). The test of reliability is repeatability, first at the level of inter-core correlation within a given lake, and then at the inter-lake level within the geographic region. As yet, stacked type curves have been constructed for only three different regions, viz. UK (Turner and Thompson, 1979), east central N. America (Creer and Tucholka, 1982a), and Japan (Hyodo and Yasakawa, 1986). These are shown in Figure 1. High quality curves comprising paired measurements taken down cores from two lakes in Australia (Barton and McElhinny, 1982) and measurements merged from three cores from each of two lakes in Argentina (Creer et al., 1983) are shown in Figure 2.

3.5 Comparison between type curves for different regions

(Creer and Tucholka, 1982b and c) compared phase relationships and spectra of the inclination and declination type curves for UK and east central North America. Both inclination type curves exhibit a sequence of maxima and minima. The correlation coefficient is maximized by a phase shift of about 650 years which is compatible with westward drift at a rate of 0.13° per year. However the phase relationship between the pair of declination curves is quite different. The oscillations are approximately in phase through the last 5000 years while for the earlier half of the Holocene they are almost exactly out of phase.

A standing NDF source situated in the northern hemisphere of the Earth's core at a longitude between UK and North America (i.e. somewhere under the Atlantic) would produce oppositely directed east-west horizontal fields in these two regions in agreement with the declination results prior to 5000 years ago (Figure 1). This standing source must have been replaced by a similar source located at a longitude either to the east or to the west of both regions, so producing a

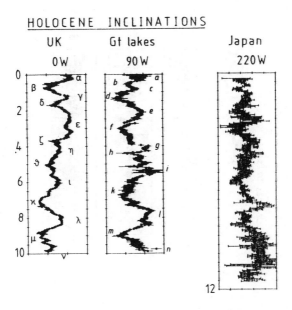

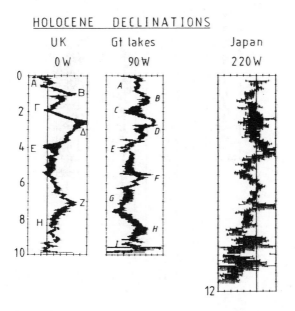

Fig. 1. Stacked type curves for inclination (above) and for
declination (below) for UK (Turner and Thompson, 1979),
east-central North America (Creer and Tucholka, 1982a)
and Japan (Hyodo and Yaskawa, 1986).

388

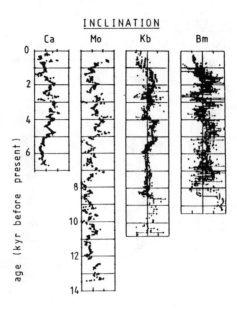

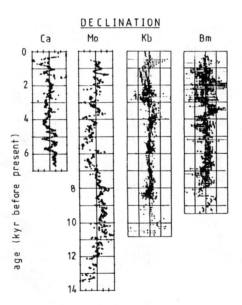

Fig. 2. Inclination and declination logs compiled from multiple
coring, reading from left, from Lakes Campanario and
Moreno in Argentina (Creer et al., 1983) and from Lakes
Keilambate and Bullenmari in Australia (Barton & McElhinny (1982).

horizontal field component (of different intensities) in the same
(i.e. either east or west) sense at both places.

Creer and Tucholka (1982c) proposed a model based on the structure
of the present field (section 1.2), consisting of drifting and standing
NDF sources to explain the apparently conflicting results provided
by the inclination and declination logs. Radial dipole or horizontal
electric current loop type NDF sources perturb the vertical component
of the main field (and hence the inclination) most strongly when they
are vertically beneath the observer and they perturb the horizontal
component of the main field when they are some tens of degrees of
arc away (Creer, 1983 and Epilogue GBCRS-1983). Since it is improbable
that an observation point chosen at random should be sitting directly
above a standing NDF source, while it is certain that drifting NDF
sources will pass beneath the observation point from time to time,
it follows that drifting sources will have a stronger influence on
the pattern of inclination variations than on the pattern of declination
variations. The opposite will be the case for standing sources. Thus
the combination of dominance of drifting sources on inclination and
standing sources on declination can be attributed to observer/geomagnetic
source geometry.

When we try to extend the inclination correlations over a greater
distance, as far as Japan, we find that the maxima and minima identified
along the UK and North American curves cannot be traced eastwards
across Asia or westwards across the Pacific with confidence. Taking
a drift rate of 0.13° per year as estimated above, the expected phase
lags would be 1100 years to UK from Japan and 1000 years to Japan
from east-central North America, while it would take about 2800 years
for a given NDF source to make one complete rotation relative to a
given observation point on the Earth's surface. Noting that the size
and intensity of nondipole foci of the historic field have grown and
decayed through centuries, we should expect the signature recorded
by a given focus as it passes beneath an observation site to change
appreciably as it drifts from site to site. We should also expect
the signature to be affected by standing sources because the phase
and amplitudes of the signal from these will be different every time
the drifting source providing the basic signature passes beneath a
given site. The results shown in Figure 1 suggest that while the
signature is still recognizable after drifting about a quarter of
one revolution relative to the surface (UK to North America), it has
usually changed so much as to be almost unrecognizable after about
half a revolution (Japan to UK and North America to Japan). The
extent to which the signature changes through one complete revolution
is also reflected by the changes in shape of successive maxima and
minima along the inclination curve for a given site.

Because of the dominance of standing sources on the declination
variations, we should not expect curves to correlate well over distances
larger than the range of influence of a single NDF source.

3.6 Spectra

Techniques of spectral analysis suitable for applying to palaeomagnetic
data have been discussed by Barton (1982, 1983). Spectra of the UK
and North American type curves were obtained using autocorrelation
and maximum entropy methods by Creer and Tucholka (1982c). The spectra
are not stationary and most of the energy resides in periods of a few
thousands of years. The time series were divided into halves at 5000
years because of the pronounced change in relationship between the
patterns of the UK and North American declination type curves (section
3.5). The spectra for declination and inclination are different (Table
2). The dominant period in the inclination record is 2200 - 2400
years which must be associated with the westward drifting sources,
giving a drift rate of about 0.15° per year (similar to that found
in section 3.5 above), though this period is replaced in the UK record
from 5000 years BP to the present by a shorter period of 1050 years.
The declination variations are characterized by longer periods of
3000 to 4000 years which differ by about 30% for the two regions,
and which decreased for both regions by about 20% in the younger half
of the record. Only the strongest spectral peaks which were stable
over a range of prediction error filter values from a half to a third
of the number of equally spaced data points in the time series are
given in Table 2.

TABLE 2

RESULTS OF SPECTRAL ANALYSIS OF UK AND NORTH AMERICAN TYPE CURVES

DECLINATION

Age Interval (yrs)	Region	Dominant Periods (yrs)		
0 - 5000BP	UK	3500		
0 - 5000BP	NA		2600	1050
5000-9000BP	UK	4100		
5000-9000BP	NA	3200		

INCLINATION

Age Interval (yrs)	Region	Dominant Periods (yrs)		
0 - 5000BP	UK	3500		1050
0 - 5000BP	NA		2200	
5000-9000BP	UK		2200	
5000-9000BP	NA		2400	

The most commonly occurring periodicities are sporadic in character and they are best explained by a combination of drifting and stationary nondipole sources and not by variations in dipole orientation or intensity.

4. ARCHAEOMAGNETISM

4.1 Method

Baked clays acquire a thermoremanent magnetization (TRM) on cooling down through the Curie temperatures of their magnetic constituents (usually haematite, T_c = 670°C). The strength of the TRM is proportional to the strength of the magnetic field in which the cooling took place. In the archaeological context, this is the ancient geomagnetic field since the anomaly in local geomagnetic intensity caused by a kiln is less than the error of measurement of archaeointensity. The linear relationship holds for fields up to a few hundred microtesla (μT), which comfortably exceeds the geomagnetic field strength measured anywhere on the Earth's surface at the present time.

Orientated samples are essential for the recovery of the direction of the ancient field. These can be collected from kiln walls. However, archaeomagnetists have tended to concentrate their efforts on determining intensity rather than direction for the practical reason that the most reliable results are obtained using samples from pieces of pottery. Directions cannot usually be recovered from pottery because their orientations when being baked cannot be reconstructed. Therefore the archaeomagnetic data base, in contrast to the lake sediment one, consists mainly of intensities.

Since archaeomagnetic data provide unattenuated spot values of the ancient field, they can be used to calibrate the amplitudes of the smoothed lake sediment records (see Epilogue in GBCRS-1983). The snag with the archaeomagnetic method is that only sporadic geographic and temporal coverage is possible.

4.2 Archaeointensity Results

The experiments carried out in the laboratory lead directly to an estimate of the strength of the ancient field by comparing it with a known applied laboratory magnetic field. Basic techniques, mineralogical problems and reliability of results from many geographical regions are discussed in Chapter 3 of GBCRS-1983. About 65% of the data originates from Greece and surrounding regions. These results, which have good dating control, are of high quality and span the last 8000 years. In Figure 3 archaeointensities are presented as ancient field strengths normalized to the present axial dipole field value calculated for the collecting site.

Figure 3a illustrates Walton's results for Greece covering the last 2500 years in detail (see Figure 3-6 in GBCRS-1983). Figure 3b shows data compiled by Thomas (1983) including results from Bulgaria and Yugoslavia (Kovacheva, 1980) and some of Walton's results

392

(see Figure 3-17 in GBCRS-1983).

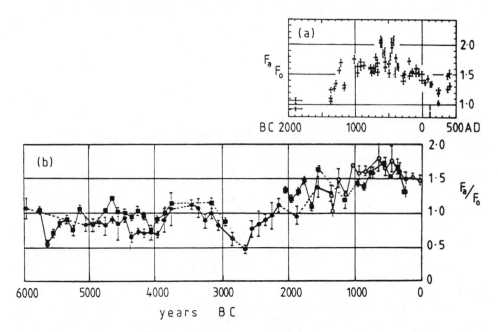

Fig. 3. Ratios of archaeointensities (F_a) to the present dipole field
intensity (F_0): (a) Walton's results for Greece (see GBCRS-1983);
(b) Thomas' (GBCRS-1983) compilation, including Bulgarian and
Yugoslavian results (black squares) from Kovacheva (1980),
Greek results (black circles) from Thomas (1983) and some of
Walton's results (open circles).

Walton's data (Figure 3a), exhibit two minima at about 250AD and 1400BC
separated by about 1650 years. Thomas' extended data set (Figure
3b) identifies an older minimum at about 2700BC, suggesting a possible
3000 year cycle. Both these estimates fall within the range of
periodicities identified from the lake sediment records (section 3.6).
Shorter oscillations, of duration around 500 years, are also evident
through the 2500 years covered by Walton's data.

4.3 Dipole or nondipole origin?

Archaeointensities are sometimes not presented directly as field values
normalized to the present dipole field as in Figure 3, but indirectly
as moments of the virtual geocentric dipole (VDM) which would give
rise to the measured ancient field. Use of procedures such as this
implies that the measured variations in ancient field strength actually
represent variations in the geomagnetic dipole moment. However, it
is absolutely necessary to compare results from a number of widely
separated geographical regions to identify dipole symmetry securely.

It is not possible to do this in the archaeomagnetic context because about 65% of the global data base is derived from Greece and the Middle East. If we proceed by analogy with lake sediment data (section 3), we are led to the conclusion that periods of a few thousands of years should originate from variations in the <u>nondipole</u> field rather than the dipole field. Again, periods of several hundreds of years should also have a nondipole origin by analogy with the historic geomagnetic record.

McElhinny and Senanayake (1982) (see section 3.16 GBCRS-1983) have attempted to remove the signal due to variations of the nondipole field by averaging results from around the world at 500 to 1000 year intervals. Their curve is shown in Figure 4. However, our understanding of the lake sediment secular variation record tells us that such intervals are <u>too short</u> to average out the signal from the nondipole field. Neither can we be sure that nondipole components have been averaged out spatially over the global surface because of the high density of data originating from a single geographic region. The two minima at 6500 and 10500 years BP in Figure 4 suggest the presence of a period of around 4000 years, which is typical of nondipole sources (section 3.5). Since the nondipole signal has not been entirely averaged out, it is pertinent to ask whether the small maximum at about 8500 years BP could mask the maximum of a longer half period. If so, a fair estimate of the long period would be roughly 12000 years.

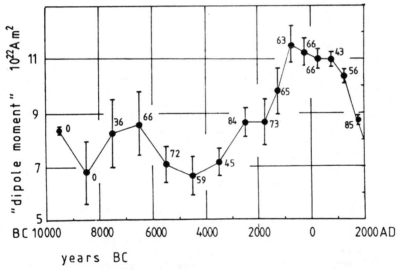

Fig. 4. 500 year interval averages of purported virtual dipole moments calculated from archaeointensity determinations (after McElhinny and Senanayake, 1982). Although data from various parts of the world are included, the data set is strongly biased to sites in Greece and the Middle East (percentages from this region are given adjacent to each point), so that the curve must retain a substantial contribution from the nondipole field.

5. ^{14}C ACTIVITY IN ATMOSPHERIC CO_2

5.1 Long term variations

^{14}C measurements made on wood samples, dated in calendar years by their tree rings, has revealed that systematic differences exist between radiocarbon years and calendar years (de Vries, 1958; Damon, Lerman and Long, 1978) and this has been attributed to a slow variation, through Holocene time, of specific ^{14}C activity.

A sine function provides a very good fit to $\Delta^{14}C$ values plotted as a function of time (e.g. see Figure 1 of Neffel, Oeschger and Suess, 1981). If extrapolated, the half period of 5800 years suggests a periodicity of about 11500 years, which is much longer than the periods originating from nondipole field sources deduced from spectral analyses of lake sediment palaeomagnetic data (section 3.6). The length of this period is similar to the rough estimate of the long period associated with archaeomagnetic intensities (section 4.3).

It is generally agreed that changes in atmospheric ^{14}C concentration on the longer time scale are due mainly to changes in the geomagnetic dipole moment (see Sonett, 1984), though longer runs of both radiocarbon and archaeomagnetic data are necessary for the relationship to be firmly established.

5.2 Short term variations

Until recently the fine structure (Suess wiggles) in the ^{14}C level was thought by most investigators to be noise. However Neftel, Oeschger and Suess (1981) and Sonett (1984) have shown that non random features do exist in the power spectrum. In particular, a 200 year periodicity extends over the entire 8500 year long tree ring record. Sonett (1984) has arrived at the same conclusion and also identifies the sporadic existence of periods of 1500 to 2000 years.

We have seen that periods of a few thousand of years duration are characteristic of the nondipole geomagnetic field (section 3.6). Now, should we expect variations in the strength of the nondipole field as a whole to influence the incoming cosmic ray flux, noting that higher order fields fall off as r^{-n-1} with distance from their source? Probably not. However we have seen that the drifting part of the nondipole field is essentially quadrupolar in character (section 1.2). Quadrupolar fields fall off as r^{-3}. Therefore could the persistent presence of a strong drifting quadrupole field maintain a small axial assymetric input of ^{14}C against atmospheric mixing? If the 2000 year period is in fact caused by a drifting quadrupole field, it would not be in phase at all longitudes: maximum ^{14}C generation would be concentrated at the strong positive foci of the quadrupole (e.g. north pole down in the northern hemisphere) and carried along with each focus as it drifted westwards. The effectiveness of this mechanism will depend on the strength of the quadrupole, and if this has varied on the time scale of the nondipole field, as we should expect, its influence on the radiocarbon concentration would be sporadic in agreement with Sonett's observation about the sporadic nature of the 1500 - 2000 year spectral peak. This hypothesis could be tested if additional records similar to the bristlecone pine ones were to become available. The object would then be to investigate the phase relationships of the periods around 1500 - 2000 years.

6. FUTURE RESEARCH

6.1   We need high quality, well dated, stacked type curves from a substantially increased number of geographical regions. IAGA Resolution No. 7 of the Canberra Assembly in 1979 recognized the need for such data. It stated:

IAGA noting that a comprehensive study along an east-west profile of geomagnetic secular variations recorded in sediments deposited since the last glaciation in lakes is likely to yield new data basic to our understanding of the origin of the geomagnetic field and that such a profile would be optimally located in the Eurasian and American continents, within a band of latitudes between 40° and 50°, recommends that National Bodies in the respective countries support the preparation of a cooperative project in which standardized techniques will be used.

So far, after nearly a decade, data are notably missing from China and the USSR, and these are absolutely essential if a close enough spacing is to be achieved to trace the path and evolution of each NDF focus as it drifts.

6.2   We need to obtain more reliable information about long term trends of the type curves if a successful attempt is to be made to identify any signature from the main dipole. The radiocarbon activity curve (section 5.1) suggests that we should be looking for a period of around 12000 years. Multiple coring with palaeomagnetically proven coring equipment will be necessary to eliminate long wavelength trends introduced inadvertently by the coring process.

6.3   A prime objective must be to gain a sufficiently good under-standing of the magnetization process to be able to recover variations in intensity of the palaeofield from the measured intensities of remanent magnetization. It is essential that any procedure developed for normalization should be thoroughly tested against archaeo-intensity data from the same region. Greece is the only place where it is feasible to carry out such a test.

6.4   Tree ring calibrations of the radiocarbon time scale should be carried out at geographic locations around mid-northern latitudes to test thoroughly the effectiveness of atmospheric mixing along the lines proposed for palaeomagnetic cores in the IAGA Resolution of 1979 (see 6.1).

6.5   Further efforts are necessary to extend the geographical coverage of high quality, well dated archaeointensity determinations so as to permit global analyses. For this an improved understanding of the effects of weathering and of mineralogical changes induced by heating in the laboratory will be required, because measurements will have to be made on materials whose properties are not ideal for archaeointensity determinations to obtain global coverage.

REFERENCES

Barton, C.E., 1982, Phil. Trans. Roy. Soc., London, A306, 203.

Barton, C.E., 1983, Geophys. Surveys, 5, 335.

Barton, C.E. and McElhinny, M.W., 1982, Geophys. J. Roy. astr. Soc., 68, 709.

Barraclough, D.R., 1974, Geophys. J. Roy. astr. Soc., 36, 497.

Caufield, D.E. and Berner, R.A., 1987, Geochim. Cosmochim. Acta, 51, 645.

Creer, K.M., 1981, Nature, 292, 208.

Creer, K.M., 1982, Hydrobiologia, 92, 587.

Creer, K.M., 1985, Geophys. Surveys, 7, 125.

Creer, K.M. and Tucholka, P., 1982a, Can. J. Earth Sci., 19, 1106.

Creer, K.M. and Tucholka, P., 1982b, Phil. Trans. Roy. Soc. London, A306, 87.

Creer, K.M. and Tucholka, P., 1982c, J. Geophys., 51, 188.

Creer, K.M., Valencio, D.A., Sinito, A.M., Tucholka, P. and Vilas, J.F., 1983, Geophys. J. Roy. astr. Soc., 74, 199.

Damon, P.E., Lerman, J.C. and Long, A., 1978, Ann. Rev. Earth planet. Sci., 6, 457.

de Vries, H., 1958, Nederl. Akad. Wetensch. Proc. Ser. B, 61, 94.

GBCRS-1983, 'Geomagnetism of Baked Clays and Recent Sediments', eds. Creer, K.M., Tucholka, P. and Barton, C.E., Elsevier, Amsterdam, 324.

Hyodo, M. and Yaskawa, K., 1985, Rock Magnetism and Palaeogeophysics, Tokyo, 12, 1-5.

King, R.F., 1955, Mon. Not. Roy. astr. Soc., 61, 57.

Kovacheva, M., 1980, Europe. Geophys. J. Roy. astr. Soc., 61, 57.

Levi, S. and Banerjee, S.K., 1976, Earth planet. Sci. Lett., 29, 219.

Mackereth, F.J.H., 1971, Earth planet. Sci. Lett., 12, 332.

Mcelhinny M.W. and Senanayake, W.E., 1982, J. geomag. geoelect., 34 39.

Neftel, A., Oeschger, H. and Suess, H.E., 1981, Earth planet. Sci. Lett., 56, 127.

Parkinson, W.D., 1986, 'Introduction to Geomagnetism', Scottish Academic Press, Edinburgh, 433.

Sonett, C.P., 1984, Rev. Geophys. Space Physics, 22, 239.

Thomas, R., 1983, Geophys. Surveys, 5, 381.

Tucker, P., 1981, Earth planet. Sci. Lett., 56, 298.

Turner, G.M. and Thompson, R., 1982, Geophys. J. Roy. astr. Soc., 70, 789.

Verosub, K.L. and Banerjee, S.K., 1977, Rev. Geophys. and Space Phys., 15, 145.

397

Verosub, K.L., Mehringer, P.J. and Waterstraat, P., 1986, <u>J. geophys</u> <u>Res</u>., 91, 3609.
Yukutake, T. and Tachinaka, H., 1968, <u>Res. Inst</u>., Tokyo, 46, 1027.
Yukutake, T. and Tachinaka, H., 1969, <u>Res. Inst</u>., Tokyo, 47, 65.

GEOMAGNETIC SECULAR VARIATION AS DETERMINED FROM PALEOMAGNETIC STUDIES OF LAKE SEDIMENTS AND ITS RELATION TO THE STUDY OF THE COSMIC RAY FLUX

Kenneth L. Verosub
Department of Geology
University of California
Davis, California 95616
U.S.A.

ABSTRACT. Recent advances in paleomagnetism have made it possible to obtain reproducible, high-resolution records of geomagnetic secular variation from lacustrine sediments. The changes in geomagnetic field directions recorded by these sediments provide new information about the orientation of the Earth's magnetic field, the relative importance of the dipole and non-dipole fields, and the rates at which the field directions can change. Although there are still some problems with the methodology, the lake sediments also appear to be able to provide a detailed record of changes in geomagnetic field intensity. This information will be very important in separating the effects of the geomagnetic field from those of solar variability with regard to the production of cosmogenic radioisotopes in the Earth's atmosphere.

1. INTRODUCTION

The production of cosmogenic radioisotopes in the Earth's atmosphere is controlled by the flux of galactic cosmic rays. This flux is modulated by geomagnetic secular variation and by solar variability. Recent progress relating to the paleomagnetic study of lacustrine sediments has significantly increased the contribution which paleomagnetism can make to our understanding of these pheonomena. In this paper, I describe in detail a high-resolution paleomagnetic study of one particular lake and show how the data from this and related studies provide direct information about changes in the direction and intensity of the geomagnetic field as well as indirect information relevant to other types of observations.

2. SOURCES OF GEOMAGNETIC SECULAR VARIATION DATA

There are five sources of information about changes in the direction and intensity of the Earth's magnetic field: historical records, lava flows, archaeological materials, speleothems (cave deposits), and

F. R. Stephenson and A. W. Wolfendale (eds.),
Secular Solar and Geomagnetic Variations in the Last 10,000 Years, 399–418.
© *1988 by Kluwer Academic Publishers.*

rapidly-deposited sediments. Historical records are inherently the most accurate because they are based on direct measurements of the magnetic field; however, they can provide information only about the most recent behavior of the field. Comprehensive directional data is available only for the last 200 years although sparse data exists for the 200 years before that. Intensity measurements have only been made for the last 150 years. Any understanding of geomagnetic field behavior on a time scale longer than the last few hundred years must involve paleomagnetic measurements.

In general, lava flows are believed to be the most reliable recorders of the direction and intensity of the field. But the occurrence of volcanoes is geographically limited, and even where lava flows are present, they can only provide a discontinuous record of the magnetic field. Archaeomagnetic features can provide individual field directions which are accurately-dated, but a continuous record must be assembled from many sites. Some speleothems apparently do preserve an accurate record of geomagnetic field behavior, but problems of resolution and dating limit their usefulness.

Originally, rapidly-deposited sediments, particularly lacustrine sediments, were thought to be ideal as sources of secular variation data, and indeed, some of the first paleomagnetic studies were designed to determine secular variation from lake sediments (Ising, 1942; Johnson et al., 1948). Slow deposition in quiet waters was thought to produce an accurate and continuous record of the geomagnetic field. Furthermore, because the time scale for changes in the field was of the order of a few hundred years, and typical paleomagnetic samples included about 2.5 cm of sediment, adequate resolution of 50 years or less per sample could be achieved with a sedimentation rate of 0.5 mm/year or more. Such rates are not unusual for lacustrine environments.

3. PALEOMAGNETIC STUDY OF RAPIDLY-DEPOSITED SEDIMENTS

3.1 First- and Second-Generation Studies

Many years passed before believable records of secular variation could be obtained from lake sediments. One of the first such records came from Lake Windermere in England (Mackereth, 1971; Creer et al., 1972). Other lakes in Great Britain soon yielded paleomagnetic records which showed the same general features as the Lake Windermere record (Thompson, 1975). However, in many cases, published curves from adjacent lakes, or even the same lake, showed almost no similarities (Creer et al., 1976; Vitorello and Van der Voo, 1977). One of the primary problems with these early studies was that they were usually based on a single core from a given lake, which meant that there was no means of assessing the variability of the paleomagnetic recording process. Other problems arose from the nature of the coring devices and the coring procedures. In addition, the chronology for the entire core was often based on a handful of radiocarbon dates and the assumption that the rate of sedimentation had been uniform. It has now been shown that local climatic fluctuations can produce significant changes in the rate of sedimentation and that uniform rates are the exception rather than the rule.

In recent years, paleomagnetists have overcome many of the problems which plagued earlier studies, and consistent, believable records of secular variation are becoming available for many areas of the world. Several factors have contributed to the success of these "second-generation" secular variation studies, not the least of which has been more careful attention to the coring technique. Development of the Mackereth piston corer (Mackereth, 1969) and of techniques for slowly driving core barrels into the sediment from small moored platforms (Cushing and Wright, 1965) have made it feasible to collect several cores up to 10 cm in diameter and up to 6 m long from a small area in a given lake. In some cases, the azimuthal orientation of the cores can be determined. Where reentry of a hole is deemed desirable, the use of casing has led to accurate determination of the depth of each core segment. In a neighboring hole, the depth control can then be used to obtain a sequence of cores which overlap those of the first hole.

The chronological framework of the studies has also been significantly improved. Anywhere from ten to forty radiocarbon dates may be used to resolve changes in the rate of sedimentation. Ancillary geochemical studies provide a firmer basis for corrections to the radiocarbon dates if they are needed. In some cases identifiable tephra layers or changes in the pollen spectra can be used to provide additional chronologic control.

More attention has also been given to determining the nature of the magnetic carriers and the processes by which the sediment has acquired its magnetization. These factors play an important role in understanding the uses and limitations of records of secular variation.

3.2 Fish Lake, Oregon: A Case Study

As noted above, second-generation paleomagnetic studies of lacustrine sediments are now yielding believable curves of secular variation. One such study was done on cores from Fish Lake, a small lake in Steens Mountain, Harney Co., Oregon (Verosub et al., 1986). Here, eleven 10-cm diameter cores were obtained from five separate holes distributed over an area of less then 50 sq. meters on the lake bottom. The coring was done by Peter Mehringer and his students from Washington State University. The cores were obtained from a moored raft using a chain-hoist operated piston corer, and the interval sampled by each core was controlled to provide multiple-overlap of the entire section. Individual core segments were 1-3 m long. A total of 20.9 m of sediment was recovered, representing complete recovery of the entire 8.9 m section of post-glacial lake sediments.

The paleomagnetic record was derived from a composite section composed of nine overlapping cores with a total length of 16 meters. Small plastic boxes (2.5 cm x 2.5 cm x 1.8 cm) were used to collect 455 samples, representing nearly continuous sampling of the 16 meters. Each box was fully oriented with respect to the axis of the core segment. Six tephra layers and other thin distinct bands were used to align the overlapping segments of different cores with an uncertainty of less than 0.3 cm (Figure 1). Age control was based on 18 radiocarbon dates from the lake itself as well as 19 other radiocarbon dates from two nearby lakes containing the same six tephra layers. The sampled sediments span the interval from 13,500 years B.P. to the present although the sedimentation rate was low during the first 3,500 years.

4. SECULAR VARIATION OF GEOMAGNETIC FIELD DIRECTIONS

4.1 Paleomagnetic Procedures and Results - Fish Lake

The samples were subjected to alternating field demagnetization. In this procedure, a sample is exposed to an alternating magnetic field whose intensity decreases uniformly to zero. The decreasing field randomizes the directions of those magnetic carriers which have a coercive force lower than the initial field intensity. Each step of the demagnetization involves an initial field with a higher intensity. The direction of the magnetization of the sample is measured after each step so that the directions associated with the highest steps correspond to the magnetization of the most stable magnetic carriers. The samples from Fish Lake were found to possess a strong and extremely stable remanence. Associated rock magnetic studies were used to demonstrate that the principal magnetic carrier was magnetite. This result is not surprising because most of the detritus coming into the lake is derived from Miocene basalt.

An example of the declination records from two overlapping cores is shown in Figure 2. The high quality of the paleomagnetic record can be seen in the low scatter and high degree of serial correlation within each core as well as in the excellent agreement between corresponding segments of different cores. The cores were not azimuthally oriented, but the large overlap and precise stratigraphic correlation between them made it possible to align the declination records and obtain a single, unoriented composite record. The complete azimuthal orientation was recovered by using one of the tephra layers found in the core. This tephra layer is associated with the eruption of Mt. Mazama and the formation of the Crater Lake caldera in Oregon 6,800 radiocarbon years ago (Bacon, 1983). Deposits from this event are found in outcrop and in lake sediments throughout the northwestern United States. By matching the declination in the core near this tephra layer with the declination of samples from outcrops, as measured by Duane Champion of the United

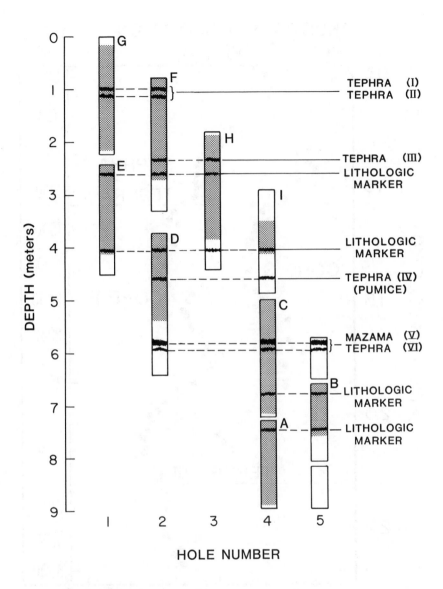

Fig. 1. Relationship of core segments from Fish Lake as determined
         from tephra layers and lithologic markers. The shaded
         area represents the core segments from which paleomagnetic
         samples were collected.

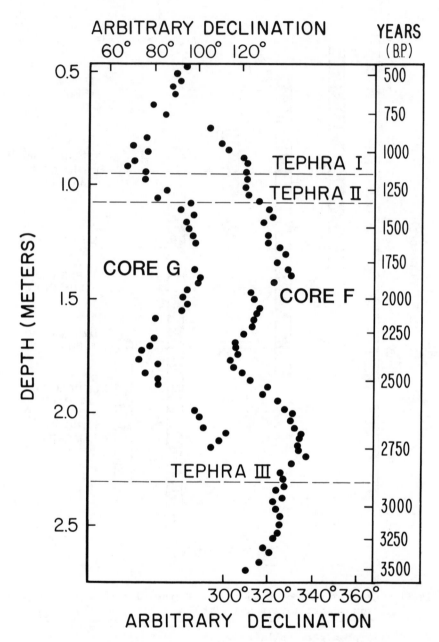

Fig. 2. Declination record of two cores from Fish Lake. The declinations are arbitrary because the cores were not azimuthally oriented.

States Geological Survey, the composite record could be properly oriented. Various consistency checks demonstrated that the cores had penetrated vertically and had not twisted during the coring procedure.

Comparison of the inclination at the Mazama tephra in the core with that measured from outcrop samples of Mazama pyroclastic deposits showed that an inclination error was present. A similar conclusion was reached by comparing inclinations at the top of the core with historical records. The age of the samples at the top of the core was determined from the location in the core of the decrease in Gramineae (grass) pollen resulting from the beginning of European settlement and cultivation in the area in 1877. The occurrence of a systematic inclination error in certain types of lake sediments has been known for several years (King, 1955), and there is a theoretical model which accounts for it (King and Rees, 1966).

After correction for the inclination error, the data were smoothed using a 7-point Gaussian weighting function (Verosub et al., 1986). The conventional smoothing function in this situation would have been a three-point running mean. However, such a function has the undesirable effect of introducing spurious frequencies into the data (Holloway, 1958). With the Gaussian weighting function, the primary contribution still arises from the three central points but the problem of spurious frequencies is avoided.

The resulting record of secular variation for the last 10,000 years is shown in Figure 3. The same data represented as virtual geomagnetic poles (VGPs) is shown in Figure 4. The VGP transformation assumes that each geomagnetic direction is produced by a geocentric dipole source. The VGP represents the point where the axis of the dipole source intersects the surface of the earth. Clearly, the idea that a geocentric dipole source produces an observed geomagnetic direction is not consistent with the fact that the geomagnetic field at any point on the earth results from the superposition of a dipole field and a non-dipole field. However, the representation of field directions as VGPs is a useful means of comparing data from different sites, especially those located at different latitudes, and the technique is very common even though it has no physical reality

4.2 Comparison with Other Records

Verosub and Mehringer (1984) have compared a portion of the record from Fish Lake with an archaeomagnetic record compiled by Sternberg (1983) from sites in the southwestern United States. The archaeomagnetic record spans the interval from 750 A.D. to 1450 A.D. and is based on 73 independently-dated archaeological features, mainly hearths. The VGP path, which is shown in the lower portion of Figure 5, can be compared to the appropriate portion of the Fish Lake record, which is shown in the upper portion of the same figure. Because the dating of the

406

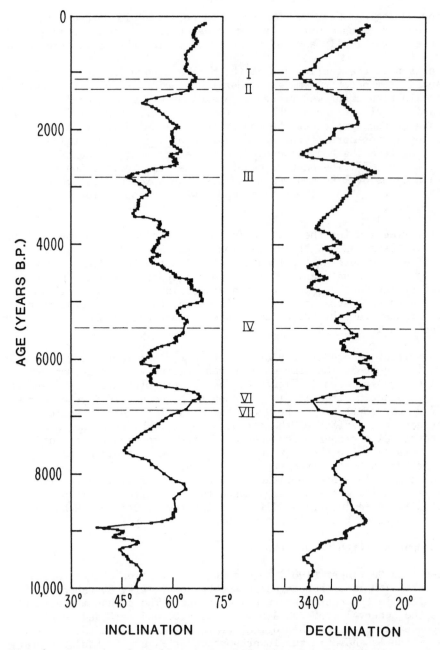

Fig. 3. Records of inclination and declination for the last
10,000 years as recorded at Fish Lake, Oregon. The Roman
numerals indicate the location of six tephra layers used
in correlating the individual core segments.

archaeological features was based on dendrochronology, the dates are given in calendar years, and the radiocarbon ages of the Fish Lake record must be converted to calendar ages. In the upper part of Figure 5, the arrows represent intervals of 100 radiocarbon years. The calendar age ranges from the radiocarbon calibration curves of Klein <u>et al</u>. (1982) are given for four points. For the remaining points, only the midpoint of the range is shown.

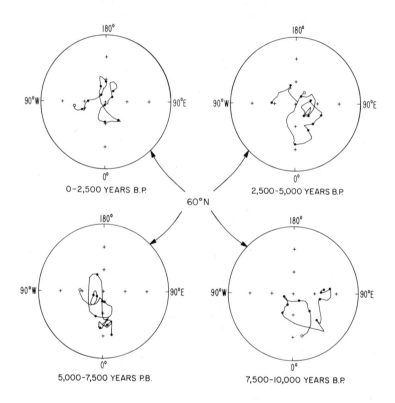

Fig. 4. Virtual geomagnetic poles for 2,500-year intervals corresponding to the paleomagnetic record from Fish Lake shown in Figure 3. The closed circles represent 250-year intervals. The youngest point of each curve is shown with an open circle.

408

The agreement between the two records is as good as or better than that between any other published pair of records, leaving little doubt that the two curves record the passage of the same non-dipole feature of the geomagnetic field (Verosub and Mehringer, 1984). The curves have similar shapes and the same sense of movement although the archaeomagnetic curve is translated toward the south and rotated slightly counterclockwise with respect to the Fish Lake curve. In three out of four cases the age range of a point shown in the upper part of Figure 5 includes the age of the corresponding archaeomagnetic point. The discrepancies between the two curves could be caused by a variety of factors including possible dating problems in both records, questions about the relationship between dates and the time of magnetization, and the 900 km distance between the two sites.

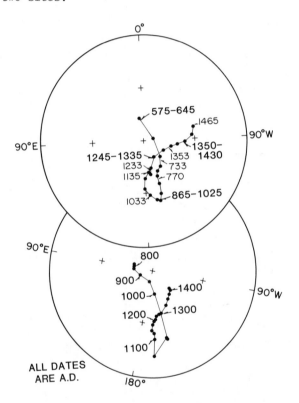

Fig. 5. Virtual geomagnetic poles from western North America for the time interval 750-1450 A.D. The upper circle represents data from Fish Lake, Oregon; the lower one represents archaeomagnetic data for Sternberg (1983) from the American Southwest. The bounding circle is at 70°N.

Another relevant record is an unpublished study by Champion (1980) of
directions associated with some dated Holocene lava flows in the western
United States (Figure 6). Given the fact that the directions from the
lava flows are not themselves internally consistent, there is reasonable
agreement between the Fish Lake record and that of the lava flows over
the last 4,500 years B.P. At about 6,000 years B.P., there is a cluster
of lava flow directions whose inclinations agree with the Fish Lake
record, but whose declinations differ by about 20°. This disagreement is
surprising because the time interval is so close to the 6,800 year B.P.
Mazama datum where, by construction, the two records coincide. All of
the lava flows which date near 6,000 years are from the same site so
there may be a systematic problem with that site. Beyond 10,000 years
B.P., the agreement is about the same as it is for the first 4,500 years
B.P.

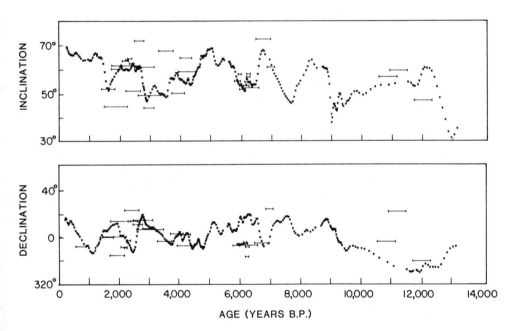

Fig. 6.   Comparison of the paleomagnetic record from Fish Lake with
          paleomagnetic directions determined by Champion (1980)
          from dated lava flows in the western United States. The
          bars indicate the uncertainties in the ages of the lava
          flows.

410

Based on the agreement between these records from different sources, it
appears that the record from Fish Lake does provide an accurate record
of the secular variation of geomagnetic field directions in western
North America during the last 10,000 years. Similar records are
available from lakes in east-central North America (Lund and Banerjee,
1985), Great Britain (Turner and Thompson, 1981), Iceland (Thompson and
Turner, 1985), Greece (Creer et al., 1981), Israel (Thompson et al.,
1985), Argentina (Creer et al., 1983), and Australia (Barton and
McElhinny, 1981; Constable and McElhinny, 1985). Several of the
second-generation European curves have been combined to produce a
composite "master curve" for that region (Turner and Thompson, 1981;
Creer and Tucholka, 1982a). Creer and Tucholka (1982b) have also

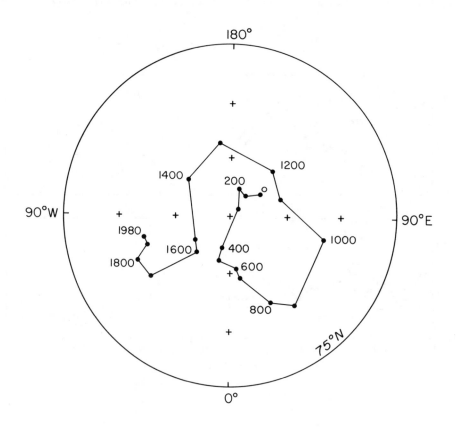

Fig. 7   Position of the geomagnetic pole at 100-year intervals in
         the Northern Hemisphere as determined from paleomagnetic
         data.   The curve is based on data from Champion (1980)
         and Merrill and McElhinny (1983).

produced a composite curve for east-central North America, based on both first- and second-generation studies.

4.3 Applications to the Study of the Cosmic Ray Flux

Although the geomagnetic field intensity is of primary importance in modulating the galactic cosmic ray flux, detailed curves of geomagnetic field directions can provide important insights into the modulation process. For example, the amplitudes of the swings in inclination and declination are directly related to the ratio of the non-dipole field to the dipole field. Until recently, it had been assumed that the modulation of the cosmic ray flux was controlled by the dipole field (Barton et al., 1979). However, recent calculations indicate the contribution of the non-dipole field should not be neglected (Shea et al., 1987). If this inference proves to be correct, the directional data will be important in determining when the non-dipole field was strong and how much it contributed to the production of cosmogenic isotopes.

The directional data can also be used to determine the actual location of the geomagnetic pole. Such information is important because the modulation process depends on the orientation of geomagnetic field with respect to the solar wind. As noted above, the virtual geomagnetic pole of a single direction has no physical reality. However, when the VGPs are averaged globally for a particular time interval, the effects of the non-dipole field at different sites cancel, and the mean VGP gives the orientation of the dipole field.

This approach was used by Champion (1980) who grouped the available data for the last 2,000 years into seven regions. He calculated an average VGP for each region at 100-year intervals and used these VGPs to determine the position of the geomagnetic pole. Merrill and McElhinny (1983, p.100) added data from an eighth region to produce the path shown in Figure 7. One interesting feature of this path is that it indicates that between about 1200 and 1500 A.D. the pole moved rapidly from a position near Siberia to a position near its present site in Greenland. This shift can also be inferred from the frequency of auroral sightings as recorded in Japan and China (Siscoe and Verosub, 1983). The agreement of the auroral record with the paleomagnetic record increases our confidence in the accuracy of the auroral record and implies that other inferences drawn from it are probably also valid.

The observed shift in the position of the geomagnetic pole raises the question of how rapidly the pole can change its position. The directional data from Fish Lake has provided one answer to this question. For each successive pair of samples, rates of angular change in VGP were computed from the directions and the interpolated age difference. Age differences of less than 10 years were excluded because of their large relative uncertainty. A histogram of the rates of change is shown in Figure 8. From the histogram it appears that the VGP

observed at a single site often changes as rapidly as 1°/decade but that
rates as high as 2.5°/decade are possible. These changes can be
compared to the shift observed between 1200 and 1500 A.D. The three
100-year intervals associated with this shift each correspond to a rate
of about 0.8°/decade. Admittedly this comparison must be done cautiously
because the histogram refers to changes in VGP while the observed shift
reflects changes in the actual geomagnetic pole. Nevertheless the
observed shift appears to fall within the normal range of rates of
change and does not appear to represent anomalous movement of the pole.

5. SECULAR VARIATION OF GEOMAGNETIC FIELD INTENSITY

5.1 Paleomagnetic Procedures and Results – Fish Lake

Determination of the geomagnetic field intensity (paleointensity) from
samples of lake sediments is not as simple as determination of their
paleomagnetic directions. The natural remanent magnetization (NRM) of a
sample is proportional to both the intensity of the geomagnetic field
and the concentration of magnetic carriers. In order to isolate the
contribution of the intensity of the field, some parameter which depends
only on the concentration of magnetic carriers is first measured. This

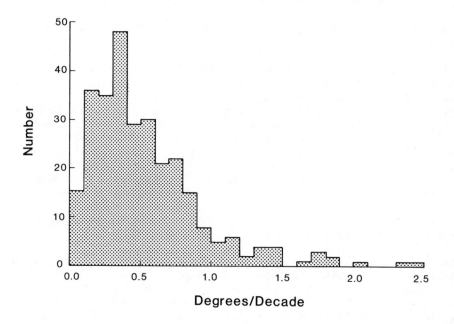

Fig. 8.   Histogram of the rates of change of the VGP for the
          last 13,500 years B.P. as determined from the paleo-
          magnetic record from Fish Lake.

parameter is then used to normalize the NRM values. The result is a measure of the relative intensity of the field, not its absolute value. Several different parameters have been suggested for the normalization. At the present time, the most commonly used parameter is the intensity of the anhysteretic remanent magnetization (ARM). ARM is the magnetization which a sample acquires when subjected to a decreasing alternating magnetic field in the presence of a constant applied field. In effect, the ARM procedure is similar to the alternating field demagnetization except that the goal is not to randomize the direction of the magnetic carriers but to magnetize them along the direction of the applied direct field. For a given applied field, the intensity of the ARM is proportional to the concentration of magnetic carriers.

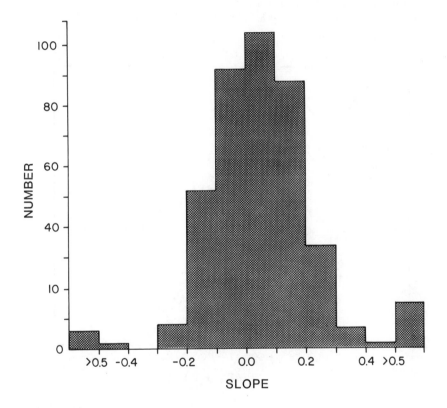

Fig. 9. Histogram of the slopes of a straight line fit to the NRM/ARM ratio versus the demagnetization level for the samples from Fish Lake. If the ARM procedure were activating the same magnetic carriers involved in the acquisition of the original NRM, the slopes would be zero.

414

In order to yield a valid paleointensity determination, the ARM process must affect the same magnetic carriers that were involved in the original magnetization of the sediment (Levi and Banerjee, 1976). To verify that this is the case, the behavior of the NRM during alternating field demagnetization is compared to the behavior of the ARM during alternating field demagnetization. If the NRM/ARM ratio is independent of the peak demagnetizing field, the paleointensity determination is considered reliable (Levi and Merrill, 1976).

The methodology of the NRM/ARM technique is still evolving, and the procedures are not completely standardized (King et al., 1983). In addition, there are clearly certain types of magnetic carriers for which the NRM/ARM technique is not appropriate (Amerigian, 1977), and there is some question about the feasibility of making paleointensity determinations on sediments at all (Turner and Thompson, 1979). However,

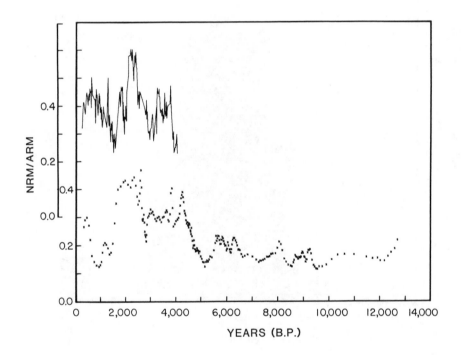

Fig. 10.   NRM/ARM ratios as a function of age for samples from
Fish Lake, Oregon (lower curve) and Le Boeuf Lake,
Pennsylvania (upper curve). The NRM/ARM ratio appears
to be an accurate measure of relative paleointensity
of the geomagnetic field.

at the present time NRM/ARM technique is the method of choice for such studies.

The technique was used to determine relative paleointensity values for the samples from Fish Lake. Because the magnetic carrier is fine-grained magnetite, the samples were good candidates for the NRM/ARM technique (Levi and Banerjee, 1976). After the demagnetization of the NRM, each sample was given an ARM which was then demagnetized at the same levels as the demagnetization of the NRM. The ratio of NRM to ARM was calculated for each demagnetization level, and a linear regression program was used to determine the slope of the curve of NRM/ARM versus demagnetization level. An ideal sample would have a slope of zero. A histogram of the slopes of the Fish Lake samples is shown in Figure 9. Most of the slopes conform to a normal distribution which is centered on 0.05. A few of the slopes are clearly outside the normal distribution.

In order to confine the study to those samples most appropriate for the NRM/ARM technique, samples for which the absolute value of the slope exceeded the arbitrarily chosen value of 0.2 were eliminated from the study. For the remaining samples, the NRM/ARM ratio at an intermediate demagnetization level (20 mT) was used. These data were then smoothed using a Gaussian smoothing function. The resulting curve is shown in Figure 10.

5.2 Comparison with Other Records

We can compare the record from Fish Lake with the record reported by King et al. (1983) for the last 4,000 years at Le Boeuf Lake in Pennsylvania (Figure 10). The two records are quite similar; the only significant difference between them is that the low in the Le Boeuf Lake record occurs at 1,500 years B.P., 500 years earlier than the low in the Fish Lake record. Such a discrepancy can probably be attributed to a problem with some of the radiocarbon dates from Le Boeuf Lake. The fact that there is such good agreement in the NRM/ARM records from two lakes over 3500 km apart indicates that they must be responding to the same input signal. The only parameter which is likely to be coherent over such a distance is the intensity of the magnetic field. This agreement is strong evidence for the validity of the NRM/ARM method and for the potential of lacustrine sediments in providing a reliable record of the secular variation of the geomagnetic field intensity.

The NRM/ARM record from Fish Lake implies that there has been considerable variation in the field intensity in western North America during the past 5,000 years B.P. Because only two points from the Fish Lake record overlap with the period of instrumentally determined intensity values, it is difficult to convert the relative paleointensity value to absolute values. Nevertheless, an NRM/ARM ratio of approximately 0.3 appears to correspond to the most recent geomagnetic field intensity. In that case, the Fish Lake record implies that 1,000

years ago in western North America the field was about 50% weaker than
its present value, but 2,000 years ago, it was almost 40% stronger.
Between about 2,500 and 4,000 years B.P., the field was fairly constant
and comparable to the present value although 4,000 years ago there was a
peak almost 20% stronger than the present value. Farther back in time,
from about 5,000 years B.P. to about 13,500 years B.P., the field
appears to have been fairly constant with only small fluctuations around
a level about 25% weaker than the present value.

5.3 Applications to the Study of the Cosmic Ray Flux.

As noted above, the intensity of the dipole field is generally believed
to be of primary importance in modulating the galactic cosmic ray flux.
Over the years, many researchers have used the available paleointensity
data to calculate the effects of changes in the intensity of the dipole
field on radiocarbon production (Bucha, 1969; Barbetti and Flude, 1979;
Barton et al., 1979). In some cases, the procedure has been reversed and
the data on radiocarbon abundances have been inverted to calculate
geomagnetic field intensities (Barton et al., 1979).

Whichever approach has been used, the interpretation of the results has
been limited by the resolution with which changes in the strength of the
dipole field could be determined. For the most part, such information
has been derived from paleointensity measurements of lava flows and
archaeological features. As with directional data, the measurements
must be averaged globally in order to eliminate the effects of the
non-dipole field. However, the difficulty in making reliable
paleointensity measurements, combined with the sparsity of suitable
materials older than about 2,000 years, has meant that the most of the
global-averaging must be done over time intervals of 500 or 1000 years.
The NRM/ARM technique can apparently provide relative paleointensity
information for a given site on a much finer time scale. Eventually, it
should be possible to combine data from a well-dated,
globally-distributed set of lakes to obtain a record of the strength of
the dipole field over the last 10,000 years with a time-resolution
comparable to that available for the orientation of the dipole (Figure
7). Such a record would be invaluable in any endeavor to separate the
effects of geomagnetic secular variation from those of solar variability
in determining the production of cosmogenic radioisotopes.

6. SUMMARY

Paleomagnetic results from samples from Fish Lake, Oregon, illustrate
how second-generation paleomagnetic studies of rapidly-deposited
sediments can yield high-resolution, reproducible records of the
geomagnetic secular variation. In particular, the changes in
geomagnetic field directions as recorded at Fish Lake provide new
information about the relative importance of dipole and non-dipole
components and about rates of change of the pole position. When

combined with similar records from other localities on several
continents, the Fish Lake record can provide information about the
orientation and location of the geomagnetic pole. All of these
parameters are relevant to the study of the cosmic ray flux. Even more
useful is information about the intensity of the geomagnetic field.
Although still in a developmental stage, the NRM/ARM technique appears
to be able to provide reliable relative paleointensity data. The record
from Fish Lake shows that there have been large fluctuations in
intensity during the past 5,000 years but the preceding 8,500 years
represented a period of fairly constant field intensity. Such
information is critical in any attempt to separate solar factors from
geomagnetic ones in the control of the galactic cosmic ray flux and the
production of cosmogenic radioisotopes in the Earth's atmosphere.

REFERENCES

Amerigian, C., 1977, Earth Planet Sci. Lett., 36, 434.

Bacon, C.R., 1983, J. Volcanol. Geotherm. Res., 18, 57.

Barbetti, M., and Flude, K., 1979, Nature, 279, 202.

Barton, C.E., and McElhinny, M.W., 1981, Geophys. J., 67, 465.

Barton, C.E., Merrill, R.T., and Barbetti, M., 1979, Phys. Earth
   Planet. Int., 20, 96.

Bucha, V., 1969, Nature, 224, 681.

Champion, D.E., 1980, Ph.D. Thesis, 314, Calif. Inst. Technol.,
   Pasadena.

Constable, C.G., and McElhinny, M.W., 1985, Geophys. J., 81, 103.

Creer, K.M., Gross, D.L., and Lineback, J.A., 1976, Geol. Soc. Amer.
   Bull., 87, 531.

Creer, K.M., Readman, R.W., and Papamarinopoulos, S., 1981, Geophys.
   J., 66, 193.

Creer, K.M., Thompson, R., Molyneux, L., and Mackereth, F.J.H.,
   1972, Earth Planet. Sci. Lett., 14, 155.

Creer, K.M., and Tucholka, P., 1982a, J. Geophys., 51, 188.

Creer, K.M., and Tucholka, P., 1982b, Can. J. Earth Sci., 19, 1106.

Creer, K.M., Valencio, D.A., Sinito, A.M., Tucholka, P., and Vilas,
   J.F.A., 1983, Geophys. J., 74, 223.

Cushing, E.J., and Wright, H.E. Jr., 1965, Ecology, 46, 380.

Holloway, J.L., 1958, Adv. Geophys., 4, 351.

Ising, F., 1942, Ark. Mat. Astron. Fys., 29, 1.

Johnson, E.A., Murphy, T., and Torreson, O.W., 1948, Terr. Magn.
   Atmos. Elec., 53, 349.

King, J.W., Banerjee, S.K., and Marvin, J., 1983, J. Geophys. Res.,
   88, 5911.

King, R.F., 1955, Mon. Not. R. Astron. Soc. Geophys. Suppl., 7, 115.

King, R.F., and Rees, A.I., 1966, J. Geophys. Res., 71, 561.

Koein, J., Lerman, J.C., Damon, P.E., and Ralph, E.K., 1982, Radiocarbon,, 24, 103.

Levi, S., and Bannerjee, S., 1976, Earth Planet. Sci. Lett., 29,
   219.

418

Levi, S., and Merrill, R.T., 1976, Earth Planet. Sci. Lett., 32, 171.

Lund., S.P., and Banerjee, S.K., 1985, J. Geophys. Res., 90, 803.

Mackereth, F.J.H., 1969, Limol. Oceanogr., 14, 145.

Mackereth, F.J.H., 1971, Earth Planet. Sci. Lett., 12, 332.

Merrill, R.T., and McElhinny, M.W., 1983, 'The Earth's Magnetic Field', Academic Press, London, 401.

Shea, M.A., Smart, D.F., and Lal, D., 1987, In, 'Proc. XIX IUGG General Assembly', 2, 438.

Siscoe, G.L., and Verosub, K.L., 1983, Geophys. Res. Lett., 10, 345.

Sternberg, R.S., 1983, In, 'Geomagnetism of Baked Clays and Recent Sediments', edited by K.M. Creer, P. Tucholka, and C.E. Barton, Elsevier, Amsterdam, 159.

Thompson, R., 1975, Geophys. J., 43, 847.

Thompson, R., Turner, G.M., 1983, Phys. Earth Planet. Int., 38, 250.

Thompson, R., Turner, G.M., Stiller, M., and Kaufman, A., 1985, Quat. Res., 23, 175.

Turner, G.M., and Thompson, R., 1979, Earth Planet. Sci., Lett., 42, 412.

Turner, G.M., and Thompson, R., 1981, Geophys. J., 65, 703.

Verosub, K.L., and Mehringer, P.J.Jr., 1984, Science, 224, 387.

Verosub, K.L., Mehringer, P.J. Jr., and Waterstraat, P., 1986, J. Geophys. Res., 91, 3609.

Vitorello, I., and Van der Voo, R., 1977, Quat. Res., 7, 398.

ARCHEOMAGNETIC RESULTS FROM UNITED STATES, 10,000 B.C. TO PRESENT,
AND WITH SOME DATA FROM MESOAMERICA

Robert L. DuBois
School of Geology and Geophysics
University of Oklahoma
Norman, Oklahoma 73019
U.S.A.

Detailed knowledge about the Earth's magnetic field for the past few
thousands of years can be obtained from archeomagnetic studies of
materials collected from archeological sites. High quality geomagnetic
data for various lengths of time have been recorded at magnetic
observatories located in many parts of the world. This source of data,
however, is mostly limited in record length to some fifty to three
hundred years. Exceptionally long records from this source are
available for Europe going back to approximately 1600 A.D. (Malin
and Bullard 1981). Since around 200 B.C. the Chinese have observed the
interaction of lodestone with the magnetic field and were probably the
first to use a "magnetic compass" for determining direction. It will
be most interesting if early geomagnetic information can be obtained
from some ancient Chinese records.

Secular variation of the geomagnetic field causes changes in
direction from a few degrees of amplitude over a period of ten to
a few tens of years to 10 to 30 degrees over a period of a few
hundred to a few thousand years. The archeomagnetic (AM) data for
southwestern U.S. suggests variations of about 10° with a time
period of 100 to 300 years, a variation of some 20° with a time
period of about 600 years and one with a period of about 1300 to 1500
years with a variation of 20 to 30°. These magnetic variations result
from changes in the internally derived field, possibly both dipole
and nondipole parts, coupled with a geologic contribution of local
to regional extent. Such geologic contributions are considered to
be related to magnetic anomalies in the crust and/or to differences
in conductivity of the mantle which distort the internally derived
field at or near the surface.

The results presented in this paper have been obtained from
magnetic measurements made on baked clay specimens collected from
ancient sites. The specimens mainly come from hearths or roasting
pits, but some have been collected from baked plaster walls, floors
or post-hole rims. The size of the hearths range from 15 cm up
to as much as one meter in diameter and from 15 to 40 cm
in depth. These features contain a baked layer a few millimeters
up to 10 cm or more in thickness composed of mostly clay to silt

F. R. Stephenson and A. W. Wolfendale (eds.),
Secular Solar and Geomagnetic Variations in the Last 10,000 Years, 419–436.
© 1988 by Kluwer Academic Publishers.

size particles but locally containing small rock fragments. The
hearths vary from a simple hole excavated in the ground in which a
fire or hot coals has been placed to features that have been especially
prepared and lined with a thick layer of material closely resembling
that which was used in the manufacture of pottery. The clay layer
bakes harder than the material surrounding the hearth. The firing of
modern artificially prepared hearths and the refiring of ancient
hearths has been studied using multithermal couples placed outward
from the fire-hearth interface at various intervals to some 50 cm. The
results of these experiments suggest that the temperature ranges from
600°C at the fire-surface of the hearth to normal ground temperatures
at 50 cm when firing the hearth continuously for twenty-four hours.
A steep thermal gradient exists across the fired rim of the hearth
such that 2 to 5 cm away from the fire interface the temperatures may
only reach 100°C or so. The maximum temperature reached in hearths
at any position is dependent upon the length of firing time, availability of air, conductivity, and thickness of insulating ash layer
accumulating in the hearth. The baking process of the hearth usually
produces a colour change tending towards reds or oranges. In ancient
hearths having an original blackish or brownish pre-fired colour, the
colour change will range from orange or red at the fire interface
outward to buffs, browns, grays and finally to the natural blackish
colours. This colour banding occurs in layers ranging from a few mm
in thickness up to several cm across. Magnetic measurements made on
specimens collected from artifically fired hearths suggest an increase
in magnetic intensity with firing and a consistent magnetic direction
acquired parallel to the existing field at the site. Unfired samples
collected from the hearth before firing have a direction in the same
general trend as the existing field but with considerable variation
between specimens. Whereas the colour change associated with the
baking would suggest a process leading to a chemical remanent
magnetization (CRM) being acquired by the material, some of the
magnetism is a partial thermal remanent magnetism (PTRM). Some remains
of a detrital remanent magnetism (DRM) acquired during pre-firing
time may still exist. The CRM results from both dehydration and
oxidation of iron associated with clay minerals or other existing
iron-bearing minerals. The observed colour may in some cases result
from the presence of original reddish coloured material still existent
after firing. The CRM acquired at the fire-hearth interface probably
changes directions with repeated high intensity firings. The magnetism
formed some distance back from the fire interface retains most of its
original direction acquired during the initial chemical process brought
on by the firing. At these distances, repeated firings would only
change the direction in that ferromagnetic fraction having unblocking
temperatures below that of the temperature gradient at a particular
position. Whereas most baking of post-hole rims and plaster areas
will have similar chemical and thermal histories as a hearth, cases
are found in which, due to abundant fuel supply and associated air-drift
conditions, firing temperatures in excess of 800°C have been reached.
Material in this case has been vitrified and the high firing
temperature is well evidenced.

Much of the baked clay collected is quite uniform and fine grained in character, but occasionally pebbles or rock fragments are present. Depending upon their magnetic properties, these inhomogeneities can distort the magnetic information obtained from the specimens if the firing temperature was not sufficiently high. To neutralize the magnetic effects of fragments, the firing temperature would have had to exceed the blocking temperature spectrum of the ferromagnetic minerals present in them. To study the magnetic effects of fragments in a specimen, both strongly and weakly magnetized fragments were introduced into a prepared hearth. Fragments having a high blocking temperature and a strong magnetism affected the measured direction.

To access whether the site existed in a distorted geomagnetic field or the feature being collected had sufficient magnetic moment to distort the local field, magnetometer surveys were run over the area. For most sites in the southwest no magnetic distortion of either the field itself or resulting from the feature being collected was found. For many of the sites collected during the early days of this project a magnetic gradiometer was used to check each individual feature collected. When these experiments continued to show a lack of distortion of the magnetic field by the hearth or burnt wall, this preliminary exercise was abandoned. In Mesoamerica the question of locally distorted fields is always a consideration because of the nearby volcanic units. Particularly when the hearth being sampled is close to an existing flow, it is possible that the hearth cooled in a distorted field or that the current field distortion would affect field orientation of specimens.

METHODS

Collections in the field were made by removing excess material from around a small column still afixed in place, then encasing the column in a plaster cube using a mold. The mold was levelled to vertical before introducing the plaster by using a two-axis cross-test level and by tilting the mold in a modeling clay seal at the base. The columns of baked clay are usually cylindrical in shape and about three centimeters in diameter and also in length. If there is any question of movement of the column during excavation or subsequent casting, the column is discarded. Horizontal reference directions are obtained by either using a Brunton measurement along a selected edge of the mold or by a specially constructed orienter which aligns the diagonal to magnetic north. Eight to twelve individually oriented columns constitute a set from an individual feature. To assure accuracy of orientation both the cross-test level and Brunton are checked and calibrated in the lab; untested instruments can be a source of error sometimes in excess of one degree. Plaster is tested in the laboratory to be magnetic free before it is used in the field. Testing of the plaster was a necessary procedure as some samples from around the world were found to be quite magnetic. Horizontal magnetic orientation measurements are recorded generally to the nearest half degree and in special sets to the nearest one-quarter degree. Once received from the field, the specimens are cleaned by brushing to remove possible surface magnetic contaminants.

Magnetic measurements of the specimens have been done using a static, spinner or cryogenic magnetometer with most of the measurements made using a spinner. The repeatability of the measurements has been good because the external dimensions of the cube could be used as reference for emplacing the specimen in the instrument. Generally, alpha 95 values of two-tenths of a degree or less for ten repeat measurements on the same cube were obtained. An alpha 95 value represents a radius of the cone of confidence about the mean direction at the 95% confidence level. Multi-hearth comparisons have suggested that sets collected from the same or adjacent hearths generally gave directions of magnetism with means differing less than one-half degree. All samples are magnetically "cleaned" using either an alternating field or thermal demagnetization process. A typical procedure is to take four cubes of a set of eight or twelve and stepwise AF demagnetize in 50 to 100 oersted P-P steps until the mean direction is consistent and the alpha 95 value has passed through a low and has been increasing for the past few steps. The remaining cubes of a set are then cleaned at several different AF field intensity values selected from the stepwise series. Considering the steep thermal gradient that exists across a hearth, optimum AF demagnetization for individual cubes from a hearth could vary depending upon their original position. The final data set for averaging to determine the mean direction may well contain cubes that have been demagnetized at different levels. Whereas some samples are adequately cleaned by 150 oersted demagnetizing fields, others require much larger fields of magnetic cleaning. This condition has been investigated using AF fields as high as 3200 oersted and routinely 1600 or 2400 oersted P-P fields are used. All samples must be magnetically cleaned of low stability components in order to give reliable information. Many samples contain a viscous remanent magnetism (VRM), and some, particularly from Mesoamerica, contain in addition an isothermal remanent magnetism (IRM). The final mean direction of a set of cubes can be arrived at by using a minimum alpha technique, doing a lost vector calculation, looking at a stable endpoint direction or applying a principal component analysis method. While the latter method is the most time consuming, it is probably the most useful. AF and thermal demagnetization associated with magnetization and re-magnetization experiments support the conclusion that the magnet-ization in these baked clays is a combination of a CRM, PTRM, VRM and in some cases an IRM.

Archeological dating of the samples is an important problem in archeomagnetic investigations and cannot be overemphasized. The dating of the sites reported here has been done by various techniques: (1) the archeological context in which the sample occurs and by which the archeologist interprets the age of the feature; (2) radiocarbon analysis is used in general and particularly with sites of older ages; (3) tree-ring dated features; (4) historical records. Some features will not have absolute chronological control, but only relative ages based on archeological context or stratigraphic position. At the

time of collection of some features the archeologist has had little
time to analyze his collections from the site and in fact may not have
yet completed his excavations. As a result the initial age assignment
may have to be updated later by the archeologist. Also the relationship
of the dated material, for example in radiocarbon or tree-ring dating,
to the hearth may be difficult to establish. As this project has pro-
gressed, an attempt has been made to provide the archeologist, as
requested, with some preliminary chronological information based on the
archeomagnetic results to help him in his work. One consequence of this
may be that the archeologist is using earlier archeomagnetic results
to determine dates given for subsequent features at this or other sites.

Presentation of data is by archeomagnetic apparent polar curves
(AAPC). A running mean averaging was employed in order to arrive
at a curve representative of the polar data. Averaging windows of
one degree were used with one-half degree overlaps between successive
windows. The AAPC representation is useful in that declination and
inclination values are combined in the pole calculation and need not
be individually corrected for differences in spatial position of sites.
The pole calculations assume a symmetrical global dipole field and
locate a dipole field that would have a magnetic flux direction at
the site similar to that of the specimen.

The data are grouped together by geographical area. The Chaco
Canyon set includes samples collected between latitudes 35.9°N to
36.2°N and between longitudes of 107.7°W and 108.1°W. The Santa Fe
data were collected within latitudes of 34.0°N to 37.0°N and longitudes
of 104.0°W to 107.3°W. The northern New Mexico set includes these
sets as well as other samples and encompassed an area of 34.0°N to
38.0°N and 104.0°W to 109.0°W longitude. The sampling area for the
south Arizona set is between latitudes of 30.0°N to 34.0°N and longitudes of 109.0°W to 115.0°W. The range for the southwest United
States was from 30.0°N to 39.0°N of latitude and from 102.0°W to 115.0°W
of longitude and therein includes samples from all of the above sets.
Samples for the central United States set were collected between
28.0°N to 43.0°N latitude and 83.0°W to 102.0°W longitude. The
Mesoamerica set consisted of samples collected from latitudes of
10.0°N to 25.0°N and longitudes of 85.0°W to 110.0°W.

RESULTS

To familiarize readers with archeomagnetic polar plots and to provide a
modern analog, Figure 1 presents the observatory derived data as given
in Malin & Bullard (1981) as inclination and declination values and
these are calculated as virtual geomagnetic poles for the London area.
The resulting curve has the same shape characteristics as would a
declination versus inclination plot but here is referenced to geographic coordinates and the axis of rotation of the Earth.

An AAPC for southwestern United States (DuBois, 1967) with data
points added, Figure 2, shows the general configuration of archeomagnetic pole position locations from approximately 200 A.D. to modern.
This curve is presented for reference to compare the subsequent data
for various times and geographical areas of the U.S. The general

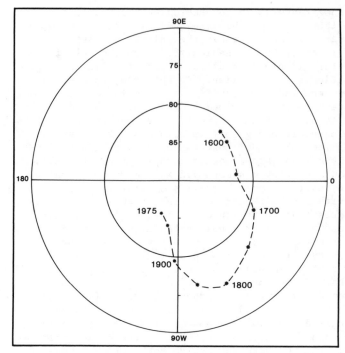

Fig. 1. Virtual geomagnetic poles for the London area calculated from observatory data (Malin & Bullard, 1981).

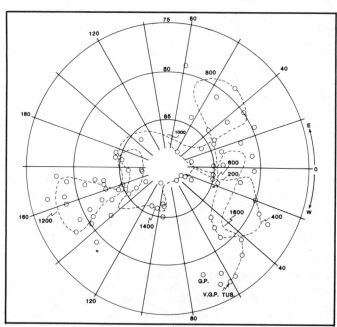

Fig. 2. A general configuration of an archeomagnetic apparent pole curve (AAPC) from approximately 200 A.D. to modern (DuBois, 1967, with data points added).

shape of the curve and position of the time marks will be modified
with additional and new data acquired since 1967. Whereas this curve
was developed from approximately 150 data points, its refined version
results from some ten-fold increase in sample sets. The curve is one
of general counter-clockwise shape moving forward in time. Three
clockwise loops occur at 500-600 A.D., 800-900 A.D., and at 1100-1200
A.D. and a counterclockwise loop occurs at 1400 A.D. Seventeenth to
twentieth century data seem to be forming another loop similar in
general position to the 500-600 A.D. loop. The polar curve circles
the axis of rotation of the Earth with the small loops extending outward
and plots at north latitudes greater than 75 degrees. Plots of
inclination versus time for this data set (DuBois, 1967) suggest a
maximum change of about 20 degrees, ranging in value from 40 to 60
degrees in magnitude. Plots of declination values suggest changes
of approximately 30 degrees, ranging from 15°E to approximately 15°W
declination.

The archeomagnetic results from some 2000 hearths or other baked
features are presented in the following figures. This large data set
is subdivided into groups based upon geographical regions of North
America. Within a region the groups are subdivided into sub-groups
based upon general age. A further subdivision of these sub-groups
is made using values of alpha 95. For various geographical regions
comparisons will be made of plots limited by different maximum values
of alpha 95.

AM data are first presented for a restricted geographical area,
Chaco Canyon. These data are then compared for specific time ranges
with data from the Sante Fe area and then the two data sets are combined
with other values from nearby areas to develop plots for a northern
New Mexico region. Separate plots are similarly developed for south
Arizona to compare archeomagnetic values with those from northern New
Mexico. Subsequently all of the data from the southwestern United
States are combined into a larger data set encompassing the entire
region. Data for central and east-central United States are presented
as are separate data sets for Mesoamerica.

The largest single data set presented comes from Chaco Canyon
in northwestern New Mexico and contains abundant materials ranging
in age from 1000 A.D. to the late 1400s and with additional samples
dating from the seventh century. Figures 3, 4, 5 and 6 are plots of
pole positions derived from samples ranging in age from 1150 A.D. to
approximately 1500 A.D. Figure 3 plots the pole positions of samples
that have alpha 95 values of 3.99 degrees or less. The outer circle
is 70°N latitude and the inner one is 80°N. GM is placed at 0° longitude.
Figures 4, 5 and 6 are plots of the same data set but using samples
with alpha 95 values of 2.99, 2.49 and 1.99 degrees, respectively.
Comparing Figures 3 and 4 suggests that using samples with alpha 95
values of 2.99 degrees or less excludes some of the sample sets
throughout the distribution but also excludes many of the outlying
points. Such a change in distribution is also noticed in Figures 5
and 6 for data sets with alpha 95 values of 2.49 and 1.99 degrees,
respectively. In the case of Figure 4, the total number of sample
sets has been reduced significantly such that a calculation of mean

426

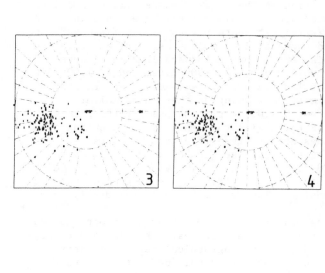

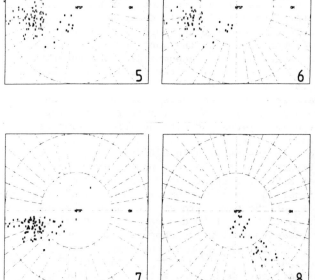

Fig. 3. Pole positions of samples from Chaco Canyon with ages 1150 A.D. to 1500 A.D. and having alpha 95 values of 3.99 degrees or less. Zeros represent ages of 1300 A.D. to 1500 A.D.

Fig. 4. Pole positions of samples from Chaco Canyon with ages of 1150 A.D. to 1500 A.D. and having alpha 95 values of 2.99 degrees or less. Zeros represent ages of 1300 A.D. to 1500 A.D.

Fig. 5. Pole positions of samples from Chaco Canyon with ages 1150 A.D. to 1500 A.D. and having alpha 95 values of 2.49 degrees or less. Zeros represent ages of 1300 A.D. to 1500 A.D.

Fig. 6. Pole positions of samples from Chaco Canyon with ages of 1150 A.D. to 1500 A.D. and having alpha 95 values of 1.99 degrees or less. Zeros represent ages of 1300 A.D. to 1500 A.D.

Fig. 7. Pole positions of samples from Chaco Canyon with ages of 800 A.D. to 1200 A.D. and having alpha 95 values of 2.99 degrees or less. Zeros represent ages of 1000 A.D. to 1200 A.D.

Fig. 8. Pole positions of samples from the Santa Fe area with ages of 1400 A.D. to modern. Zeros represent ages of 1600 A.D. to modern.

position for some locations of poles could not be reliably accomplished.
Statistical calculations are in progress to analyze these distributions,
but for this paper an alpha 95 value of 2.99 is used as an upper limit
for sets to be plotted or included in the calculations for AAPC for
the southwest. Pole positions derived from samples with ages more
modern than 1300 A.D. are plotted in these figures as zeros whereas
those with ages less than 1300 A.D. are plotted as Xs. Figure 7 is
a plot of pole positions derived from samples ranging in age from
800 to 1200 A.D. and with alpha 95 values less than 2.99 degrees.
An age of 1000 A.D. is used to separate the data into Xs and Os. The
distribution of points occurs in the same general areas as that of
those of younger age just discussed, but displaced to lower values
of westerly longitudes. Samples from Chaco Canyon with ages less
than 800 A.D. and that have alpha 95 values of 2.99 or less are few
in number and therefore are not plotted separately. They will,
however, be included both in the data set from northern New Mexico
and in that for the southwest.

To compare AM from a nearby, but separated area, samples collected
from the Santa Fe area are presented in Figures 8 through 10, using
the same precision parameters as discussed above. Data for samples
with ages ranging from 1400 A.D. to modern are plotted as Figure 8
with zeros used for sites with ages more modern than 1600 A.D. Data
sets from the Santa Fe area with ages similar to those plotted for
Chaco Canyon as Figures 4 and 7 are plotted for purposes of comparison
as Figures 9 and 10. The distribution and position of both of these
data sets are comparable and suggest that, for defining secular
variation, both can be combined into a single set.

AM results for the Chaco Canyon and Santa Fe areas are combined
in a data set for northern New Mexico which also includes results
from other outlying sites in this region. Figures 11, 12, 13 and
14 show the combined data. Running mean positions were calculated
by the methods previously described to provide a way to make a comparison between this and other data sets. They do not justify
separate presentation here as they are not significantly different
from those presented later for the S.W. The circles and Xs, as plotted
on these figures, have the same age connotations as those outlined
above and 450 A.D. is used to separate poles in the range 0 to 900
A.D.

To compare AM results from south Arizona with those of northern
New Mexico, Figures 15 and 16 show plots of the data for their respective ages. Again, poles are plotted using Xs and Os with the same
criteria as described above. Running mean positions were calculated
for both of these age ranges but the small number of data points
limits their interpretation. They are, however, consistent with those
from northern New Mexico. The spatial distribution of data points
for south Arizona are in agreement in position with those for northern
New Mexico, and therefore the two sets are combined as one for the
entire southwest region. The combined set contains results from sites
of both younger and older ages than those presented as Figures 15
and 16 and are included with their proper group.

428

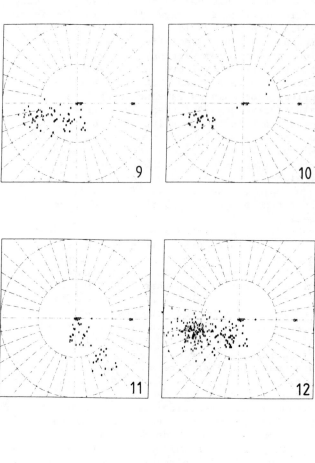

Fig. 9. Pole positions of samples from the Santa Fe area with ages of 1150 A.D. to 1500 A.D. having alpha 95 values of 2.99 degrees or less. Zeros represent ages of 1300 A.D. to 1500 A.D.

Fig. 10. Pole positions of samples from the Santa Fe area with ages of 800 A.D. to 1200 A.D. and having alpha 95 values of 2.99 degrees or less. Zeros represent ages of 1000 A.D. to 1200 A.D.

Fig. 11. Pole positions of samples from northern New Mexico with ages of 1400 to modern having alpha 95 values of 2.99 degrees or less. Zeros represent ages of 1600 A.D. to modern.

Fig. 12. Pole positions of samples from northern New Mexico with ages of 1150 A.D. to 1500 A.D. having alpha 95 values of 2.99 degrees or less. Zeros represent ages of 1300 A.D. to 1500 A.D.

Fig. 13. Pole positions of samples from northern New Mexico with ages of 800 A.D. to 1200 A.D. having alpha 95 values of 2.99 degrees or less. Zeros represent ages of 1000 to 1200 A.D.

Fig. 14. Pole positions of samples from northern New Mexico with ages of 0 A.D. to 900 A.D. having alpha 95 values of 2.99 degrees or less. Zeros represent ages of 450 A.D. to 900 A.D.

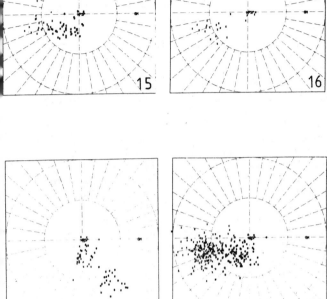

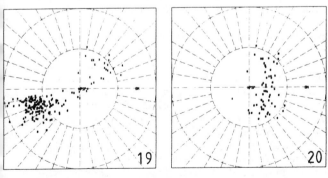

Fig. 15. Pole positions of samples from south Arizona with ages of 1150 A.D. to 1500 A.D. and having alpha 95 values of 2.99 degrees or less. Zeros represent ages of 1300 A.D. to 1500 A.D.

Fig. 16. Pole positions of samples from south Arizona with ages of 800 A.D. to 1200 A.D. and having alpha 95 values of 2.99 degrees or less. Zeros represent ages of 1000 A.D. to 1200 A.D.

Fig. 17. Pole positions of samples from south-west United States with ages of 1400 A.D. to modern and having alpha 95 values of 2.99 degrees or less. Zeros represent ages of 1000 to modern.

Fig. 18. Pole positions of samples from south-west United States with ages of 1150 A.D. to 1500 A.D. and having alpha 95 values of 2.99 degrees or less. Zeros represent ages of 1300 A.D. to 1500 A.D.

Fig. 19. Pole positions of samples from south-west United States with ages of 800 A.D. to 1200 A.D. and having alpha 95 values of 2.99 degrees or less. Zeros represent ages of 1000 A.D. to 1200 A.D.

Fig. 20. Pole positions of samples from southwest United States with ages of 0 A.D. to 900 A.D. and having alpha 95 values of 2.99 degrees or less. Zeros represent ages of 450 A.D. to 900 A.D.

AM results for the southwest region of the United States are presented in Figures 17 through 21 with the pole positions plotted as before. The same value of alpha 95, less than 2.99, is used as a selection factor for the data. Plots using a value of alpha 95 of less than 2.49 remove from consideration more of the outlying data points and could be a basis for calculating a mean position for the polar curve. Similarly, a data set with an alpha 95 value of less than 1.99 would exist for some time ranges, but this would not be true for most of the time ranges under consideration. The mean curve drawn for the southwest data set using the running mean method is presented in Figure 21 and has been developed using data with alpha 95 values less than 2.99 degrees. The age assignments are a preliminary result of reassessment in part based on data from Chaco Canyon (T. Windes, personal communication).

AM results from central U.S. are presented in Figures 22, 23 and 24 using the same selection criteria as above. The distribution of points in Figures 22 through 24 are similar to that in Figures 17 through 19. The similarity of data suggests that the AM method may not be able to resolve a difference in secular variation between these two areas. A statistical analysis in process may clarify this problem.

Results from archeological sites older than 2,000 years are presented in Figures 25 and 26, with Figure 25 presenting the data from 5000 B.C. to 0 A.D. and Figure 26 presenting that ranging in age from 8500 B.C. to 5000 B.C. The AM sets from these ages have, in general, more dispersion than that in sets used in the preceding series. To provide as much information as available, sets were used which had larger alpha 95 values, less than 5.99 degrees. Considering sets with larger dispersion and also including those with greater ranges in ages excludes the possibility of detailed interpretation of these figures. However, they do suggest that for at least the last 10,000 years secular variation of the geomagnetic field for North America can be represented by pole positions that are mostly contained within 70°N latitude. The data are distributed with regard to the axis of rotation in a manner similar to that for the Southwest as shown above for the past 2,000 years.

Archeomagnetic results from Mesoamerica are presented in Figures 27 through 31 using different age ranges for grouping the results. A selection value was used for alpha 95 of less than 3.99 degrees. This procedure was suggested by the long range of time represented by the data and the use of a larger value of alpha 95 includes more of the data points that may contribute to the problem of secular variation. The general sparsity of data from this area suggests this procedure. An AAPC is drawn for these data for graphical means, Figure 32, to give a very preliminary suggestion of the distribution of the data with regard to time. A more detailed anlaysis of this data set does not appear to be justified at this time considering the errors related to the following; (1) age assignment, (2) geologic contribution to present or past distorted local field, (3) local ground movement caused by recent tectonic activity or settling, (4) local magnetic disturbances causing sample misorientation.

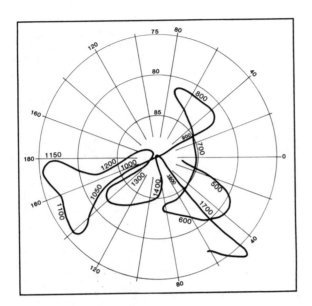

Fig. 21. Archeo-
magnetic Apparent
Pole Curve (AAPC)
based on data
from the southwest
United States
having alpha 95
values of 2.99
degrees or less.

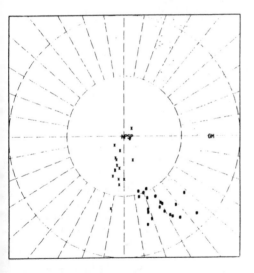

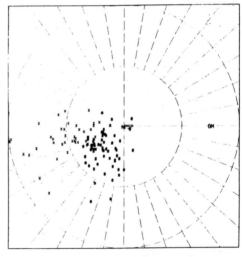

Fig. 22. Pole positions of samples
from the central United States
set with ages of 1400 A.D. to modern
and having alpha 95 values of 2.99
degrees or less. Zeros represent
ages of 1600 A.D. to modern.

Fig. 23. Pole positions of samples
from the central United States
set with ages of 1150 A.D. to
1500 A.D. and having alpha 95
values of 2.99 degrees or less.
Zeros represent ages of 1300
A.D. to 1500 A.D.

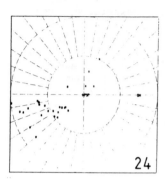

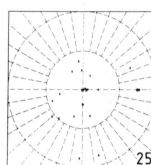

Fig. 24. Pole positions of samples from the central United States set with ages of 800 A.D. to 1200 A.D. and having alpha 95 values of 2.99 degrees or less.

Fig. 25. Pole positions of samples from ancient archeological sites with ages of 5000 B.C. to 0 A.D. and having alpha 95 values of 5.99 degrees or less.

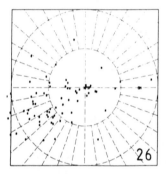

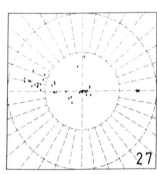

Fig. 26. Pole positions of samples from ancient archeological sites with ages of 8500 B.C. to 5000 B.C. and having alpha 95 values of 5.99 degrees or less.

Fig. 27. Pole positions of samples from Meso-america with ages of 1150 A.D. to modern and having alpha 95 values of 3.99 or less.

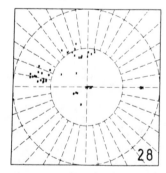

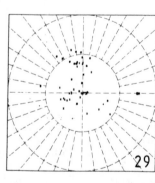

Fig. 28. Pole positions of samples from Meso-america with ages of 800 A.D. to 1300 A.D. and having alpha 95 values of 3.99 degrees or less. Zeros represent ages of 1000 A.D. to 1300 A.D.

Fig. 29. Pole positions of samples from Meso-america with ages of 400 A.D. to 900 A.D. and having alpha 95 values of 3.99 degrees or less. Zeros represent ages of 700 A.D. to 900 A.D.

Fig. 30 Pole positions of samples from Mesoamerica with ages of 200 B.C. to 500 A.D. and having alpha 95 values of 3.99 degrees or less. Zeros represent ages of 150 A.D. to 500 A.D.

Fig. 31. Pole positions of samples from Mesoamerica with ages of 1000 B.C. to 200 B.C. and having alpha 95 values of 3.99 degrees or less. Zeros represent ages of 600 B.C. to 200 B.C.

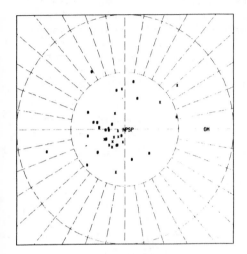 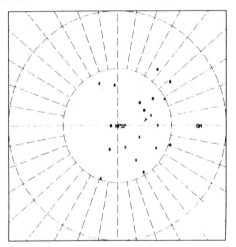

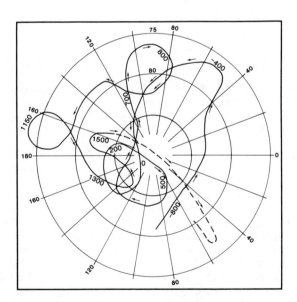

Fig. 32. Archeomagnetic Apparent Pole Curve (AAPC) based on data from Mesoamerica having alpha 95 values of 3.99 degrees or less.

434

Archeomagnetic paleointensity determinations using the Thellier,
E. and Thellier, O. (1959) double heating method for North America
are presented in Figure 33. This figure shows the ratio of the ancient
to modern geomagnetic field intensity at a given site as the vertical
axis and time along the horizontal axis. The data presented here are
from Lee (1975) and Hsue (1978), with open circles data from Mesoamerica
and solid circles data from southwestern United States. The values
plotted represent an average of results from different pieces of baked
clay from the same feature and were derived from a multipoint position
consideration of the Thellier curve. Some of the results have been
confirmed by using repeat experiments or reprocessed by the Shaw method.
The error bars shown represent variances in age assignment where
available and estimates of the error in the determination of the ratio
of ancient field intensity.

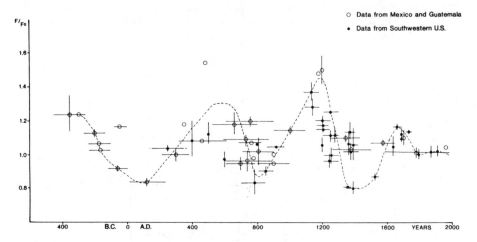

Fig. 33. A plot of the ratio of ancient to modern geomagnetic field
intensity versus time in years. Open circles represent
data from Mesoamerica and closed circles represent data
from southwestern   United States (Lee, 1975; Hsue, 1978).

DISCUSSION

The archeomagnetic apparent pole curve (AAPC) for the Southwest, shown
in Figure 21, is a compilation of all of the data obtained from the
combined Southwest area. The curve uses only data sets having alpha
95 values of 2.99 degrees or better and for which stability tests have
been completed using mostly the AF method, but in a few cases a thermal
cleaning. The curve suggests that secular variations can be described
by an irregular pole path that is essentially contained around the
axis of rotation and the region extending outward to the 75°N latitude
with the only exception being that of near modern time. The pole path
consists of four fairly well developed loops and a fifth partially
developed one. The loop of youngest age, designated A, is only partially

developed,   started about 1500 A.D. and extends to the present time.
This loop is preceded in age by a loop B which was developed around
1400 A.D. extending outward from the axis of rotation to its maximum
extension about that time. Preceding in age was a loop C loop extending
also out from the axis of rotation during 1000 and 1200 A.D. with its
maximum extension occurring somewhere near 1100 A.D. A loop D only
in part contains the axis of rotation position and developed during
the 700 to 900 A.D. period of time with the maximum of extension around
800 A.D. This loop is preceded by an E loop developed during 500 and
600 A.D. with a maximum extension of about 550 A.D. Spatially, loop
E occurs in a position near and overlapping parts of loop A. An F
loop is suggested by the Mesoamerican data occurring at a time preceding
that of the E loop with an age estimated to be around 200 A.D. These
loop-like forms of the AAPC can be open or closed with or without cross-
over points or enlarged ends (for example with C or D loops) and suggest
a feature of secular variation that lasts for 100 to 300 years. An
exception is the A loop which is still developing over a longer period.
The polar data plots shown lack short-term, low-amplitude, fine-scale
features which, if present, have been lost by the running mean method
of averaging data to produce these curves. Looking at modern observatory
records, as represented by pole positions, open or closed fine-scaled
loops are suggested by data from several of the observatories of the
world for times of their active recording. It will be interesting if
archeomagnetic methods can be refined so that such finer-scale
variations can be determined. For some time a figure of 1° per ten
years for apparent polar wandering along a curve has been used to
estimate, at the archeologist's request, alpha 95 values for a given sample
in terms of time. Such a procedure, of course, is only useful to compare
the relative dispersion of two samples, not as an estimate of age
assignment. Even then it is an oversimplification because the rate of
apparent polar wandering along such a curve depends upon the detail
plotted and the method used to develop the curve, because the curve
seems to be a fractal in that as the time resolution of the data is
increased, the more detail is observed. It will vary from 6 to 8 years
per degree up to as much as 20 considering the curves shown. One period
of rather rapid apparent polar wandering occurs as the pole moves from
loop D to loop C, a range of about 25 degrees in some 200 years. Loop
A, forming from 1500 years up to modern, seems to represent the lowest
rate of change of secular variation represented.

The AAPC for the Southwest is not uniformly distributed about
the axis of rotation. Only the D loop occurs on the far side away
from the sampling area of North America. All of the rest of the AAPC
occurs on the near-side region and is bounded by 75°N latitude and
0 to 180°W longitude.

Comparison of the individual regions as subdivided here for the
purposes of study suggests that the differences between regions is
so small that the AM method as applied here cannot resolve them.
Maybe with greater refinements and more samples small differences
between areas could be determined and fine-scale secular variations
noted. Considering VGPs calculated for modern field directions across
North America, small differences between regions are easily resolved

with each westward extension until west of 115°W longitude. Cox (1962)
in an analysis of VGP displacement angles from the geomagnetic pole
for the 1945 geomagnetic field suggested variances ranging about 4
to 7 degrees across southwestern U.S. (Note the angles calculated
by Cox are displacement angles between VGP and GP which are equivalent
to delta angles calculated by the author). Analysis in a different
way of the 1965 field for latitude 35°N suggested differences between
subsequent pole positions of about one and a half degrees in position
for each 5 degrees of longitude going westward, becoming about 1 degree
difference between positions at 95 and 100°W longitude and about one-
half degree difference in pole position where comparing 110 to 115°W
longitude. A similar analysis along a 270°W longitude with varying
degrees of latitude suggested small differences in pole position from
20°S latitude to 20°N latitude with increasing differences at higher
degrees of north and south latitude.

The scatter of poles as plotted, for example Figures 3 and 7 for
Chaco Canyon, about the AAPC can be explained by a normal statistical
distribution of points or by a physical variance. Such a variance
may be caused by differences in orientation or collecting methods used,
changes of the reference field across the area or differential inter-
or intrasite movements. One to three degrees of tilting, settling
or differential compaction could explain most of the scatter observed.
Such movements would produce scatter normal to or along the AAPC.

The estimates of ancient field intensity as shown in Figure 33
suggest that the field has changed in intensity from an upper value
of 140% to approximately 80% of the present field over periods of several
hundred years. The maximum field intensity in the late 1100s appears
to be in agreement with similar estimates made of the field in southern
Europe. A correlation is suggested between the times of maximum or
minimum field intensity with those of maximum extension of the loops
in the archeomagnetic pole curve. Such a correlation might favour
an interpretation that part of the field is producing the loop-like
extension of the archeomagnetic polar curve and at the same time causing
maximum/minimum values to occur in the intensity of the field.

REFERENCES

Cox, Allen, 1962, Journ. of Geomagnetism and Geoelectricity, 13, 101.
DuBois, Robert L., 1967, 'Space Magnetic Exploration and Technology
    Symposium', University of Nevada, Reno, Nevada.
Hsue, Tien Shaing, 1978, Thesis, University of Oklahoma, Norman, Ok.
Lee, Sheng-Shyong, 1975, Thesis, University of Oklahoma, Norman, Ok.
Malin, S.R.C., and Sir Edward Bullard, 1981, Philosophical
    Transactions of the Royal Society of London, A, 299, 357.
Thellier, E., and Thellier, O., 1959, Academy of Sciences of the USSR
    Bulletin, Geophysics Series, 7-12, 29.

GEOMAGNETIC INTENSITY MEASUREMENTS FROM NORTHERN GREECE AND THEIR
COMPARISON WITH OTHER DATA FROM CENTRAL AND N. EASTERN EUROPE FOR THE
PERIOD 0 - 2000 yr AD

S. P. Papamarinopoulos
Department of Geology
Geophysical Laboratory
University of Patras
Rio-Patras, Greece

ABSTRACT. The geomagnetic intensity variation in northern Greece
has been recorded in Byzantine vases. Minute cylindrical 3 x 3mm^2
samples have been drilled from fragments of vases and measured with
a cryogenic magnetometer. These mini-cores have been treated using
a modified version of the Thellier method. The data show a broad
agreement with the Bulgarian, Ukrainian and Czechoslovakian variation.
The geomagnetic intensity ratio F_A/F_D, which expresses the ancient
field intensity relative to the present day field intensity, exhibits
values as follows: 1.20 between 300-400, 1.49 between 400-700, 1.51
at about 1100, 0.89 at about 1400, 1.07 at about 1500 and 0.94 at
about 1650 yr AD. The data have been corrected for the anisotropy
and cooling rate effects.

INTRODUCTION

With the help of a battery-powered hand-held portable diamond drill,
very small pottery 3 x 3mm^2 cores have been extracted from dated
fragments of Byzantine vases of known archaeological age. These ages
were obtained with subjective archaeological criteria based on the
style, type, size, shape and painted patterns of the vases. The
fragments were obtained from the museum collection of the 9th
Byzantine Regional Archaeological office of Antiquities of
Thessaloniki.
    Archaeological criteria, although subjective, for some historic
periods of ancient Greek history, are superior to other direct or
indirect dating methods, such as thermoluminescence of carbon 14.
In the classical period, several centuries BC, archaeologists can
date with an accuracy of ± 25 years. However, in the protochristian
Byzantine and post-Byzantine periods these criteria are applied with
an induced uncertainty in the range of ± a few centuries. These
uncertainties produce large error bars in the time axis of the plots.
In addition, recently fabricated vases may produce further noise in
a particular curve. Certainly, thermoluminescence and magnetic tests
can pin-point such a vase, but this is rather a rare case.

F. R. Stephenson and A. W. Wolfendale (eds.),
Secular Solar and Geomagnetic Variations in the Last 10,000 Years, 437–442.
© 1988 by Kluwer Academic Publishers.

The best way to obtain a reliable curve for this period is to obtain samples from various collections. This procedure allows reduction in the time uncertainties and confusion errors in dating and also enables removal of possible fabrications. The addition of samples coming from ceramic kilns or fired walls in churches of known age is another source of data.

DETERMINATION OF THE GEOMAGNETIC INTENSITY

The equation which governs the determination of the Earth's ancient field intensity is:

$$F_A = (M_A/M_L) \cdot F_L \tag{1}$$

The parameters contained in this equation are defined as follows:
F_A : The geomagnetic field intensity in ancient times.
F_L : The magnetic intensity of the artificially applied field in the laboratory.
M_A : The thermoremanent moment in ancient times.
M_L : The acquired magnetic moment after re-heating in a known laboratory field.
Since the appearance of the technique for the determination of the ancient geomagnetic intensity, (Thellier and Thellier, 1959), equation (1) has been modified considerably when the intrinsic anisotropy effect has been recognized in pots (Rogers et al., 1979).
The new formula is expressed by equation (2).

$$F_A \cdot \cos\theta = C \cdot F_V^L/(1 + f) \tag{2}$$

The parameters in this equation are defined as follows:
F_A : The intensity of the magnetic field in ancient times.
θ : The angle which is defined by the sample's ancient magnetization and the vertical.
F_V^L : The oven vertical magnetic field at the laboratory.
f : The negative slope of the line obtained by plotting the parameter M_V^L (the vertical component of the sample's moment) after remagnetization versus the parameter M_V^O (the moment remaining after subsequent demagnetization).

A CORRECTION PARAMETER FOR THE INTRINSIC ANISOTROPY EFFECT IN THE THERMOREMANENCE

Rogers et al. (1979) have found that the effect is dominant in vases. The thermoremanence produced by a given field depends on the direction of the field with respect to an easy plane of magnetization. The thermoremanence produced by a given field parallel to the easy plane is higher than that due to the same field when it is perpendicular to the easy plane. The ratio between the two may be as high as 2, whereas 1.5 is typical. In fact the quantity C expresses a combined effect which is discussed analytically by Aitken et al. (1981, 1983).

Experience over 1000 measurements on various types of vases has shown that the effect of anisotropy is sometimes ±5% and more frequently ±2% for cores drilled from the same area of a pot (Aitken et al. 1983). All Byzantine data shown in Figure 1 have been corrected for the anisotropy effect, by theoretical calculations and/or by measuring the effect itself.

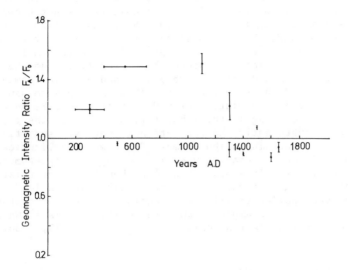

Fig. 1.   The ratio F_A/F_D, where F_A is the ancient
         field intensity after multiple corrections and
         F_D the present day field intensity in Northern
         Greece, plotted versus time in calendrical years
         AD. Mean and S. Dev. values of F_A/F_D are shown
         (vertical error bars). Horizontal error bars
         exist for samples for which the archaeological
         dating contains the length in centuries of un-
         certainties (after Papamarinopoulos, 1987).

Equation 2 needs to be modified when other types of anisotropy are taken into account. For instance, shape anisotropy - which is due to strongly magnetized samples through internal demagnetizing fields - may be of importance both as magnetic susceptibility anisotropy and/or thermoremanence anisotropy. Magnetic refraction is another phenomenon which certainly distorts the thermoremanence signal, when a core consists of layers which may have different magnetic susceptibilities from each other.

THE COOLING RATE EFFECT

Considerable theoretical and experimental work has been carried out
by Dodson and McClelland-Brown (1980) over the effect of the cooling
rate in the determination of the ancient field intensity from rocks.
However, in vases the work is limited. Fox and Aitken (1980) and
Aitken et al. (1983) examined the possible effect with the evaluation
of experimental results for more than 50 samples. The cooling rate
in ancient times and in different levels of cultures is not known.
If one assumes that the ancient people let the kilns cool for one
or two days, one wonders what the possible effect could be for 5 minute
coolings, which were applied for these particular minute samples in
the specific Thellier variant introduced by Walton (1977) at the
Oxford laboratory. Fox and Aitken (1980) obtained a ratio between
the apparent field and the true field of 1.11 with sample to sample
scatter having a standard deviation of ± 0.02. It should be noted
that for artificial cooling in the laboratory over 4 hrs, the ratio
is 1.08 ± 0.02. Consequently the correction is not very sensitive
to the real cooling rate in ancient times. For regular cylindrical
or cubic samples which are not measured in a cryogenic magnetometer,
the cooling time is usually about 4 hours and hence for the related
results the cooling rate correction is not significant. A 10%
reduction has been applied to all the Byzantine samples due to the
cooling rate effect. An example of such reduction appears in Walton's
results obtained from ancient Athenian vases, Walton (1979, 1984).

PARTICULAR UNEXPLORED PROBLEMS

Serious problems which have not been studied thoroughly in the case
of vases and the determination of the ancient geomagnetic field
intensity are as follows:
i.   The magnetic properties of the vases under study.
ii.   The magnetic properties of the magnetic grains locked in the
      baked clay structure.
iii. The making and the baking of the vases in ancient cultures.
iv.   Contamination of the vases from metallic iron tools or other
      elements which may exist in the vicinity of the kilns.
v.   The effect of an anomalous magnetic field due to an extensive
      magnetized metalliferous deposit in the area of the operation
      of the kiln. This effect could be either positive or negative
      depending on which side of the ore body the kiln was built.
    Distorting magnetic fields of the type discussed earlier may
induce a significant error in ancient geomagnetic intensity determinations in spite of the satisfaction of the requirements of the
Thellier technique. Locally deforming magnetic fields or strong demagnetizing fields within the sample may cause an otherwise unexplained
scatter in archaeomagnetic curves.

COMPARISON OF THE GEOMAGNETIC DATA BETWEEN CENTRAL AND SOUTH EASTERN
EUROPE

The variations of F_A/F_D for normalized published curves is examined
briefly for the period 0-2000 yr AD. For the time interval 0-500 yr
all curves show an initial decreasing trend at 0yr and over the
following centuries F_A/F_D values range between 1.3 and 1.4 in
Czechoslovakia, 1.4 and 1.2 in Ukraine. In Bulgaria over the same
time interval the values of F_A/F_D fluctuate at the level of 1.2. A
similar level exists in Greece. At about 500 yr all curves show an
increase reaching peak values at about 800 yr: 1.65 in Czechoslovakia
at about 950 yr, 1.70 in Ukraine and 1.85 at about 1050 yr in Bulgaria.
At about 400-700 yr AD in Greece the peak value of F_A/F_D is 1.51,
significantly lower than the other peaks. This may be explained from
the fact that the latter curve is incomplete. From the peak point
and after, all curves show a continuous decreasing trend which is shown
dramatically in the Czechoslovakian and Greek geomagnetic curves.
F_A/F_D reaches levels in the centuries 1400 and onwards into recent
years with values 1.1, 1.0 and 0.95. The Ukrainian, and Bulgarian
curves follow this general decreasing trend and they reach the same
levels as the previous curves through a fluctuating mode.
      An interruption in this general trend is found only in the
Bulgarian curve at 1650 yr with an F_A/F_D value of 1.5. This value at
that time exists only in the Bulgarian record and it is unexplained.
These data have been taken from Bucha (1967), Rusakov and Zagny (1973),
Kovacheva (1983) and Papamarinopoulos (1987).

CONCLUDING REMARKS

The same general trend seems to exist in all examined geomagnetic
curves in Central and South Eastern Europe. In spite of the differences which undoubtedly exist from curve to curve within the intensity
variations, one cannot conclude that the data illustrate a dipole
field only on the basis of these similarities.
      Normalized data from other parts of the world are needed. The
period 0-2000 yr AD may provide the answer in intensity studies,
because in many cultures fired objects exist around the world, whereas
for the BC period the ancient cultures are limited in numbers and
in geographic localities. Therefore the studied time interval should
be a target for many laboratories, if we really wish to conclude about
the nature of the geomagnetic field at least for this period.

ACKNOWLEDGEMENTS

The author is indebted to the 9th Byzantine regional Office of
Antiquities and in particular to Mrs. E. Nikolaidou and Mrs. C.
Tsioumi for collecting the sherds, the Ministry of Culture for giving
permission for both the study and export of the sherds abroad, the
Research Laboratory for Archaeology and History of Art of Oxford and in

particular Professor M. Aitken FRS, for offering the facilities of
the laboratory, for undertaking the measurements and Ms   Gill Spencer
of the same laboratory for undertaking the measurements.

REFERENCES

Aitken, M.J., Alcock, P.A., Bussel, G.D. and Shaw, C.J., 1981,
   Archaeometry, 23, 53.
Aitken, M.J., Alcock, F.A., Bussel, G.D. and Shaw, C.J., 1983,
   'Geomagnetism of Baked Clays and Recent Sediments' (eds. Creer, K.M.,
   Tucholka, P. and Barton, C.E.), Amsterdam, 122.
Bucha, V., 1967, Archaeometry, 10, 12.
Dodson, M.H., and McClelland-Brown, E., 1980, J. Geophys. Res., 85,
   2625.
Fox, J.M.W. and Aitken, M.J., 1980, Nature, 283, 462.
Kovacheva, M., 1983, 'Geomagnetism of Baked Clays and Recent Sediments',
   (eds. Creer, K.M., Tucholka, P. and Barton, C.E.), Amsterdam, 106.
Papamarinopoulos, S., 1987, J. Geomag. Geoelectr., 39, 261.
Rogers, J., Fox, J.M.W. and Aitken, M.J., 1979, Nature, 277, 644.
Rusakov, O.M. and Zagny, G.F., 1973, Archaeometry, 15, 275.
Thellier, E. and Thellier, O., 1959, Ann. Geophys., 15, 285.
Walton, D., 1977, Archaeometry, 19, 192.
Walton, D., 1979, Nature, 277, 643.
Walton, D., 1984, Nature, 310, 741.

THE SUN AND COSMIC RAYS

A.W. Wolfendale,
Department of Physics,
University of Durham,
Durham, U.K.

ABSTRACT. A brief summary is given of the relevance of the sun to
cosmic rays. It is pointed out that the sun generates low energy
particles and is therefore a useful test-source, at least for the
injection of cosmic rays which are then accelerated to higher energies
by other mechanisms - such as supernova remnant shocks. The
importance of the modulation of Galactic cosmic rays by the solar wind
is stressed, as is the relevance of the derived 'interstellar cosmic
ray spectrum' to the origin of the Galactic particles by way of
studies of diffuse cosmic gamma rays.
    Turning to the thorny question of solar neutrinos, the present
status of the observations is examined, with particular reference
to the backgrounds. An analysis is made of the apparent time-
variability of the neutrino signal and the relevance of the recently
detected neutrino burst from the supernova SN 1987a in the Large
Magellanic Cloud. It is concluded that variability is not yet proven;
however, the low signal level is still a cause for concern, and may
yet require a dramatic explanation such as finite neutrino mass and
neutrino oscillation.

1. INTRODUCTION

The role of the Sun in cosmic ray studies is an interesting one. At
low energies, below a few GeV, it is an occasional source of cosmic
rays and at these and higher energies its solar wind modulates those
cosmic rays coming from further afield, the modulation being both in
terms of effects on the number arriving at earth and on their energy.
Insofar as the origin of the bulk of the cosmic radiation is uncertain,
careful searches have been made for anisotropies in arrival directions
which can be attributable to Galactic phenomena and thereby give clues
as to the particle sources. The effect of solar modulation is such
that one must go to energies of at least 3×10^{11} eV before these effects
are negligible and the resulting anisotropy, which amounts to an
amplitude of the first harmonic of only $\delta I/I \sim 5 \times 10^{-4}$, can be
attributed to galactic phenomena. Although this energy, 3×10^{11} eV,

F. R. Stephenson and A. W. Wolfendale (eds.),
Secular Solar and Geomagnetic Variations in the Last 10,000 Years, 443–453.
© *1988 by Kluwer Academic Publishers.*

is very small compared with the maximum energy detected to date, 10^{20}eV, it is not negligible – the energy spectrum of cosmic rays falls rapidly with energy ($j(E) \propto E^{-2.7}$ approximately, to $\sim 10^{13}$eV) and the median energy is only about 10^{10}eV so that solar modulation has a significant effect on the total cosmic ray energy flux arriving at the earth. Knowledge of solar effects is therefore important for the cosmic ray physicist.

A case in point concerns the comparatively new subject of γ-ray astronomy. Here, measurements have been made of the flux of γ-rays of energy up to about 5 x 10^9eV from satellite – borne detectors. Most of the γ-rays are thought to be produced by the interaction of cosmic ray particles, protons and electrons, with the gas in the interstellar medium and since for certain regions of the Galaxy at least the gas densities are known, this means that the cosmic ray particle intensities can be explored in remote regions of the Galaxy. Here then is a powerful technique for studying the origin problem. A difficulty arises, however, in that the primary protons generating the detected γ-rays are largely in the range 10^9 - 10^{10}eV, i.e. in the region where solar modulation effects are quite significant. Part of the current argument concerning the likelihood, or otherwise, of large scale 'cosmic ray gradients' (see e.g. Bhat et al, 1986a), stems from the uncertainty in the modulation correction which is necessary for a derivation of the local cosmic ray datum spectrum. In fact, the argument is often turned round and an interstellar proton spectrum is assumed, by extrapolating back from the region above 10^{11}eV where modulation effects are small, and features of the solar wind are implied from the comparison of this spectrum with the observed one. The present author believes that this is a very dubious procedure.

The physicist interested in much higher energy cosmic rays than those generated in solar flares has other interests in solar cosmic rays too. One interest concerns the possibility that stars like the sun act as injectors of particles which are later accelerated by other mechanisms, e.g. by shocks of various kinds in the ISM. In parenthesis we note that taking sun-like stars alone, the energy budget in cosmic rays is about 10^{-5} of that required to produce the detected 0.5 eV cm^{-3} in galactic cosmic rays (the lack of observation of solar protons above 10GeV or so in the period for which measurements have been made – about 40 years – is of course, no great obstacle in itself since occasional flares might generate much higher energies). Low energy particles are themselves important insofar as many acceleration mechanisms which are efficient alone have difficulty in the early stages where particles are subject to ionization loss (see, for example, Ginzburg and Syrovatsky, 1964). Solar-type stars probably contribute a small fraction of the injected cosmic rays, with stars of other types such as Wolf-Rayet stars, OB stars and novae (Giler et al., 1987) probably being important too. The fact that the measured mass composition of cosmic rays is not too different from that of the sun – and indeed of the general ISM – points to the importance of studying solar-type stars.

Solar phenomena are also of considerable relevance to the general question of the acceleration of Galactic cosmic rays. Satellite

studies of energetic particles in the solar wind have lent support to
contemporary theories of the shock acceleration of particles (e.g. Wild
et al., 1963, Murphy et al., 1987 and references therein), and in turn
the hypothesis that some CR at least are accelerated by supernova shocks
in the ISM. Blandford and Cowie (1982) have made specific suggestions
for the energy gain by cosmic rays in the presence of shocks and Bhat
et al. (1985) have claimed that γ-ray observations of the nearby Loop
I SNR can be explained quantitatively by this mechanism. Simon et al.
(1985) and Giler et al. (1987) have pointed out an additional mechanism
for acceleration involving very weak shocks (from SNR and other
energetic phenomena) in the general ISM.

The question of the variability of solar effects is an interesting
one. In addition to solar flare-generated cosmic rays there is the
well known anticorrelation of cosmic ray intensity with sunspot number.
First 'discovered' by Forbush (1957) this topic continues to be of
interest through attempts to form a precise understanding of the
effects of the time-variable solar wind on Galactic cosmic rays. An
extended array of neutron-monitors over the earth's surface augmented
by spacecraft observations provide the cosmic ray data, the monitors
responding to neutron recoils from low energy protons incident on the
top of the atmosphere.

Most interesting, perhaps, and the main subject of this work, is
the problem of neutrino generation in the sun. Here is a topic that
embraces a number of areas of modern science, from the detailed nuclear
reactions responsible for neutrino generation in the solar core,
through the use of sophisticated radiochemical techniques to detect
the neutrino, to a detailed knowledge of cosmic ray background effects
expected deep underground, these backgrounds relating to neutrinos
generated in the upper levels of the atmosphere and to protons
produced near to the detection apparatus by muons which have penetrated
the overlying rock. In what follows, an examination of the present
status of this experiment will be given, particular emphasis being
placed on the apparent variability of the neutrino signal.

2. THE SEARCH FOR SOLAR NEUTRINOS

2.1 Solar Neutrino generation

It is generally accepted that solar energy is generated by nuclear fusion
reactions in the solar core. The initial suggestion appears to have
been made by Russell (1919) and Eddington (1920), detailed reactions
having been put forward by Bethe (1939) and more recently by Bahcall
(1978), Bahcall et al. (1982) and others.

The basic reactions are given in Table 1 and the energy spectra
in Figure 1.

2.2 Solar Neutrino Detection

Following suggestions by earlier workers, including Potecorvo (1946),
R. Davis and collaborators have, since the late 1960's (e.g. Davis

Table 1 The proton-proton chain for the 'standard model' (after
Bahcall et al., 1982). The numbers in parenthesis are the
percentages of terminations of the chain in which the
various reactions are calculated to occur.

	Reaction	% of terminations	Maximum neutrino energy (MeV)
	$p+p \rightarrow {}^2D + e^+ + \nu$	(99.75)	0.420
	or		
	$p+e^-+p \rightarrow {}^2H + \nu$	(0.25)	1.44 (monoenergetic)
	${}^2H + p \rightarrow {}^3He + \gamma$	(100)	
I	${}^3He + {}^3He \rightarrow {}^4He + 2\nu$	(86)	
	or		
II	${}^3He + {}^4He \rightarrow {}^7Be + \gamma$		
	${}^7Be + e^- \rightarrow {}^7Li + \nu$	(14)	9.861 (90%), 0.383 (10%) (both monoenergetic)
	${}^7Li + p \rightarrow 2\,{}^4He$		
	or		
III	${}^7Be + p \rightarrow {}^8B + \gamma$		
	${}^8B \rightarrow {}^8Be^* + e^+ + \nu$	(0.015)	14.06
	${}^8Be^* \rightarrow 2\,{}^4He$		

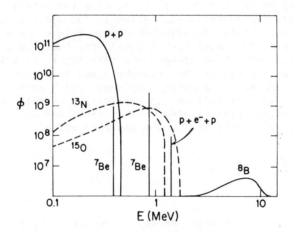

Fig. 1. The energy spectrum of solar neutrinos on the 'standard
model', (after Bahcall et al. 1982).
The fluxes are given in number $cm^{-2}s^{-1}MeV^{-1}$ for
continuum sources and number $cm^{-2}s^{-1}$ for line sources.
Insofar as there is some doubt as to the accuracy of the
standard model these fluxes are subject to some uncertainty;
similarly, the cross-sections for neutrino reactions
are not known very accurately and in turn there is uncertainty in the predicted rates of neutrino interactions.

et al., 1968), been using the following reaction

$$\nu_e + {}^{37}_{17}C\ell \rightarrow {}^{37}_{18}Ar + e^-$$

where ν_e is the electron neutrino. The threshold for the reaction is 0.81MeV so that only neutrinos from the decays above this energy are detectable. Insofar as phase space arguments dictate that the cross section is proportional to the square of the energy in excess of the threshold, the neutrinos from ^8B (with $E_{max} = 14.1$MeV) are expected to be the dominant source of the detected counts. It is interesting to note that the flux of these neutrinos is proportional to the 24th power of the solar core temperature (Gough 1984) and this means that there is the possibility of making an extraordinarily accurate detection of the core temperature if such neutrinos can be detected.

The apparatus used by Davis comprises a large cylindrical tank containing 100,000 gallons of $C_2C\ell_4$, the tank being located in a mine at Lead, South Dakota, at a depth of 4400 metres water equivalent. The $^{37}_{18}Ar$ atoms so generated are swept out by periodic purges with helium and 'counted' (K-capture X-rays) with a very low background proportional counter, pulse shape analysis helping to distinguish against background. The half life of $^{37}_{18}Ar$ is 35 days so that purges are carried out at intervals of the order of this period.

The resulting mean production rate for the data so far is plotted as a function of time in Figure 2. Also plotted are the estimated background and the theoretically-expected rate. It is immediately apparent that all is not well.

2.3 The Expected Background

Before considering the significance of the result, it is imperative to examine the predicted background – it is necessary to demonstrate that there have been any solar neutrinos detected at all. That the signal is close to the estimated background is not the fault of the experimenters – when the depth at which to operate was chosen, the predicted rate of detection of neutrinos was more than an order of magnitude higher than eventually observed (and nearly an order higher than finally predicted!).

The background in the experiment is due largely to protons which cause the reaction $p({}^{37}C\ell, {}^{37}Ar)n$, the protons coming from neutrons generated by the electromagnetic interaction of fast muons with nuclei in and near the underground tank. Its magnitude can be estimated from a knowledge of the cross section of muons for neutron-production and the energy spectrum of the muons at the depth in question but a more accurate method is to actually measure the background at shallower depths, using a smaller quantity of $C_2C\ell_4$, where the whole signal is background. A check on the effect of the form of the energy spectrum of the muons is afforded by determining the counting rate as a function of depth at the shallow depths (Davis et al., 1968; Davis, 1972; Wolfendale et al., 1972). This was done (Figure 2) and the predicted background was normalised to observation at a shallow depth where the statistical precision is largest. The extent to which the background

448

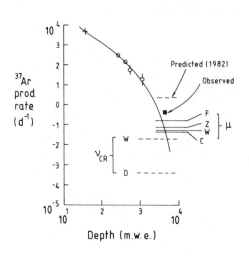

Fig. 2. Rate of production of ^{37}Ar in 10^5 gallons of C_2Cl_4. The full line is the predicted background from muon interactions (Wolfendale et al., 1972)., the circles are the background measurements - see text. The solid square is the observed rate at the depth of operation of the main experiment. It appears that solar neutrinos have been detected, although the rate is below the prediction of the standard model (denoted Predicted; Bahcall et al., 1982). 'μ' and 'ν_{CR}' represent the estimates of background from these causes 4400 m.w.e.

Key: F, Fireman et al., (1979); Z, Zatsepin et al. (1981); W, Wolfendale et al. (1972); C, Cassiday (1973); D, Domogatsky et al. (1977).

fits prediction at the shallower depths is apparent in the Figure and seen to be good. Thus, the predicted background should be trustworthy at the depth of operation. (Other, independent, estimates of the background were made by Cassiday, 1973; Fireman et al., 1979 and Zatsepin et al., 1981. The overall average is close to our value of 0.06 atoms day^{-1}).

There is another background that can be considered here: ^{37}Ar produced by the absorption of ν_μ and ν_e generated by cosmic ray interactions in the earth's atmosphere. Two estimates of this background have been made, as shown in Figure 2. Although they are disparate, both are very small, and much less than the muon background. It is likely that the lower estimate - being more recent - is more accurate.

The conclusion to be drawn is that solar neutrinos almost certainly have been detected and we proceed to an examination of the reasons for the paucity of the flux and, more particularly, the time sequence of the measurements, which will be described.

2.4 Reasons for the Low Rate

Many reasons have been put forward for the 'solar neutrino paradox' as it has come to be called. Although the solar neutrino output is very sensitive to core temperatures (see Section 2.2) the solar luminosity is fixed, by observation, and rather dramatic changes to the core composition are necessary in order to retain the luminosity but reduce the solar neutrino flux (Hoyle, 1975, has however made such a suggestion - as have others!).

A rather popular theory requires neutrino - oscillations (e.g. Close, 1980), a theory that appeared to receive support from the claim by Lyubimov et al. (1980) to have determined the rest mass of the electron neutrino ($14 < m_\nu < 46eV$), a finite rest mass being a prerequisite for oscillations. However, more recent work has depressed the upper limit to the electron neutrino mass and this theory is losing ground. Other theories have included the suggestion (Fowler, 1972) that there is a resonance in the (^{3}He, ^{3}He) reaction which reduces the fraction of reactions leading to the side branch from which the ^{8}B nuclei are generated, but the resonance has been shown not to exist. Magnetic monopoles located at the center of the sun have been invoked (Trefil et al., 1983) and WIMPS - dark matter candidates - which affect the core temperature appropriately without reducing the solar luminosity (e.g. Steigman et al. (1978), Faulkner et al. (1986)).

None of the preceding explanations is satisfying and explanation must await more experimentation. The forthcoming experiments using gallium and, separately, $C_2C\ell_4$ at the Baksan underground laboratory (see Chudakov, 1987) and the European Gallex (gallium) experiment (see Kirsten, 1986), all of which should be operational in a few years time, are awaited.

2.5 The Time Sequence of the Detected Counts

Almost as interesting as the low flux of detected solar neutrinos (indeed, probably more interesting, if confirmed) is the time sequence of the results. Figure 3 shows the latest published summary.

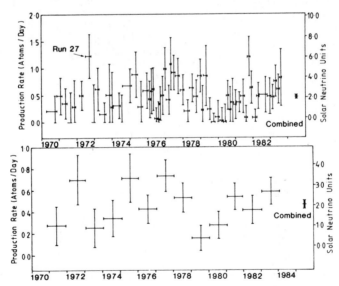

Fig. 3. Summary of ^{37}Ar production rates in the Brookhaven solar neutrino experiment (Davis, private communication, 1985).

Even without any detailed statistical analysis it can be seen that the rate appears not to be constant but to have excursions over a year or two. Insofar as the time taken for heat generated in the solar core to reach the surface is the Kelvin-Helmholtz time, $T_{K-H} \sim 10^7 y$, it would have been expected that changes in neutrino emission would occur over similar large time scales, if at all.

Inevitably, apparent time variations of the order of years have led to searches for correlations with solar surface -effects, viz. sunspots and solar flares - and, as is often the case with sunspots, correlations have been claimed! Thus, Sakurai (1984) claims a near-2 year periodicity (which appears to be visible by inspection of the data of Figure 3) corresponding to the sometimes-seen 2.2y periodic variation of sunspots which overlies the 11-year sunspot cycle (Currie 1973; Sakurai 1979). Figure 4a shows the Davis results rebinned in 4-monthly intervals - an apparent periodicity clearly seen. In a related vein, Lal and Subramanian (1985) plot the solar neutrino count rate versus the number of spots per month with the result shown in Figure 4b. There is thus circumstantial evidence for a correlation of solar ν-rate and sunspot number and a cause is clearly needed for a connection between a core phenomenon (ν-rate) and a clearly surface phenomenon (sunspots). (However, see Ching Shuk Chi, 1986, for a discussion of the statistical significance of the variations).

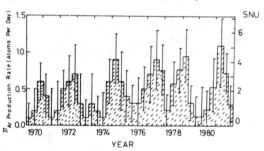

Fig. 4a. Apparent periodic variation of the four-monthly mean solar neutrino flux (Sakurai, 1984).

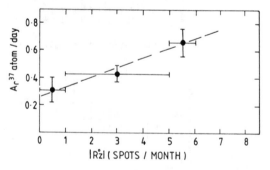

Fig. 4b. Correlation between ^{37}Ar production rate and $|R_2|$, the number of new sunspots per month (Lal and Subramanian, 1985).

One connection could be, in principle, by way of the background, viz if the cosmic ray background were higher than expected and were modulated by solar phenomena. However, we have shown that the mean background is well founded (at least the major contributor is) and furthermore the modulation at the appropriate proton energies is very small indeed. Another possibility worthy of study concerns neutrinos generated in the solar atmosphere from solar protons interacting on the far side of the earth. Erofeeva et al. (1983) have calculated the expected number of neutrinos of energy above 10eV to be expected from a solar flare containing 10^{34} protons ('such solar flares occur several times per solar cycle'). Using our estimate of the flux of atmospheric neutrinos above the same energy (Osborne et al., 1965) the solar contribution gives an increase of neutrino background of only about 0.5% when averaged over a 35-day exposure period. In view of the fact that the expected neutrino background is at least a factor ten below the signal, here again it appears impossible to put forward a background modulation as the cause.

Returning to the possibility that the basic solar neutrino rate is modulated, there has been the suggestion that g-mode oscillations in the sun may cause mixing in the solar core. These modes, which have their maximum amplitude in the core, have been detected (by Delache and Scherrer, 1983) and are the subject of much contemporary research.

3. NEUTRINOS FROM SUPERNOVAE

Apart from its intrinsic interest, the recent detection of neutrinos from the supernova 1987a in the Large Magellanic Cloud (preprints from the Kamiokande, IMB and Baksan experiments) may have relevance to the solar neutrino problem. It has been estimated (Bahcall, 1987) that the SN should have led to just one extra ^{37}Ar atom in the Davis experiment - an undetectable 'excess'. However, similar SN occurring in our own galaxy could give rise to significant excesses. For example, a similar SN near the Galactic Centre would give rise to an extra 30-40 ^{37}Ar atoms corresponding, approximately, to a doubling of the 'normal' rate. Thus, Run 27 (Figure 3) might have been due to such an event, the SN not being recognised as such in any other wavelength band because of obscuration (Pallister and Wolfendale, 1974, had, in fact made this suggestion).

It is conceivable that some of the other upward excursions in Figure 3 are also due to hitherto undetected SN although the rate of such SN would need to be much greater than the canonical 1 per century Type II (~ 1 per 30 years for all SN). However, it is not impossible that there is a class of rather frequent SN which give copious neutrino emission without marked optical signatures.

Even the allocation of 'Run 27' to the presence of a Galactic SN would reduce the chance of there being non-random variations in the solar neutrino flux; the presence of others would reduce the chance further still. Thus, although the mean solar neutrino flux is still low the evidence for embarrassing short term variations has probably been weakened.

4. CONCLUSIONS

The importance of studies of solar cosmic rays and other solar phenomena to the origin of cosmic rays is still considerable; the sun is a useful test-source for studying acceleration models, at least for the injectors of cosmic rays, the particles then being accelerated in other regions (e.g. SNR). Furthermore, modulation of Galactic cosmic rays by the solar wind is important, not only in its own right but because of the need to know the magnitude of the cosmic ray spectrum outside the solar cavity so that cosmic gamma ray data may be interpreted.

Concerning solar neutrinos (the 'Davis experiment'), the situation with the low mean rate is still obscure. However, the recently discovered neutrino burst from the supernova in the Large Magellanic Cloud (SN 1987a) raises the possibility that one or two of the upward excursions in the observed 'solar' neutrino rate are due to un-recognised supernovae in the Galaxy. If so, the case for variations in output of solar neutrinos - never strong statistically - is further reduced. The conclusion to be drawn from the present work with respect to solar variation is thus that the solar neutrino results do not necessarily require such variations over the lifetime of the Davis experiment ($\sim$ 20 years).

ACKNOWLEDGEMENTS

The author wishes to thank the following for helpful comments or correspondnce: D. Evans, D.O. Gough, K. Sakurai, G.M. Simnett, A. Subramanian and E.C.M. Young

REFERENCES

Bahcall, J.N., 1978, Rev. Mod. Phys., 50, 881.
Bahcall, J.N. et al., 1982, Rev. Mod. Phys., 54, 767.
Bahcall, J.N., Dar, A. and Piran, A., 1987, Nature, 326, 135.
Bethe, H., 1939, Phys. rev., 55, 434.
Bhat, C.L. et al., 1985, Nature, 314, 515.
                    1986a, J. Phys. G., 12, 1087.
Blandford, R.D. and Cowie, L.L., 1982, Astrophys. J., 260, 625.
Cassiday, G.L., 1973, 'Proc. 13th Int. Cosmic Ray Conf.', Denver, 3, 1958.
Ching Shuk Chi, E., 1986, University of Hong Kong Report (E.C.M. Young, Physics Department).
Chudakov, A.E. et al., 1987, 'Proc. 20th Int. Cosmic Ray Conf.', Moscow (in press).
Close, F.E., 1980, Nature, 284, 507.
Currie, R.G., 1973, Astrophys. Space Sci., 20, 509.
Davis, R., Harmer, D.S. and Hoffman, K.C., 1968, Phys. Rev. Lett., 20, 1205.
Delache, P. and Scherrer, P.H., 1983, Nature, 306, 651.

Domogatsky, G.V. and Eramzhyan, R.A., 1977, Izr. An. SSSR, Ser. Fiz., 41, 1969.

Eddington, A.S., 1926, 'The Internal Constitution of the Stars', (Cambridge Univ. Press).

Erofeeva, I.N. et al., 1983, 'Proc. 18th Int. Cosmic Ray Conf.', 7, 104.

Faulkner, J., Gough, D.O. and Vahia, M.N., 1986, Nature, 321, 226.

Fireman, E.L., 1979, 'Proc. 16th Int. Cosmic Ray Conf.', Kyoto, 13, 389.

Forbush, S.E., 1957, 'Proc. Nat. Acad. Sci.', 43, 28.

Fowler, W.A., 1972, Nature, 238, 24.

Giler, M., Wdowczyk, J. and Wolfendale, A.W., 1987, 'Proc. 20th Int. Cosmic Ray Conf.', Moscow (in press).

Ginzburg, V.L. and Syrovatsky, S.I., 1964, 'The Origin of Cosmic Rays', (Pergamon Press, London).

Gough, D.O., 1984, Adv. Space Res., 4, 85.

Hoyle, F., 1975, Astrophys. J., 197, L127.

Kirsten, T., 1986, 'Neutrino 86', Sendai, Japan.

Lal, S. and Subramanian, A., 1985, 'Proc. 18th Int. Cosmic Ray. Conf.', La Jolla, 5, 426.

Lyumibov, V.A. et al., 1980, Phys. Lett., 94B, 266.

Markov, M.A., 1964, Phys. Lett., 10, 122.

Murphy, R.J., Dermer, C.D. and Ramaty, R., 1987, Astrophys. J., 63, 721.

Osborne, J.L., Said, S.S., and Wolfendale, A.W., 1965, Proc. Phys. Soc., 86, 93.

Pallister, W.P. and Wolfendale, A.W., 1974, 'Neutrinos 1974', Am. Inst. Phys., 273.

Pontecorvo, B., 1946, Natn. Res. Coun. Canad. Rep. P.D. 205.

Rowley, J.K. et al., 1980, Proc. Conf. Ancient Sun', 45, (R.O. Pepin, J.A. Eddy and R.B. Merrill, Editors).

Sakurai, K., 1979, Nature, 278, 146; 1984 Space Scienc Rev., 38, 243.

Steigman, G. et al., 1978, Astrophys. J., 83, 1050.

Trefil, J.S., Kelly, H.P. and Rood, R.T., 1983, Nature, 203, 111.

Simon, M., Heinrich, W. and Mathis, K.D., 1985, 'Proc. Int. Cosmic Ray. Conf.', La Jolla, 3, 230.

Wild, J.P., Smerd, S.F. and Weiss, A.A., 1963, Ann. Rev. Astr. Ap., 1, 291.

Wolfendale, A.W., Young, E.C.M. and Davis, R., 1972, Nature, 238, 130.

Zatsepin, G.T., Kopylov, A.V. and Shirokova, E.K., 1981, Sov. J. Nuc. Phys., 33, 200.

TERRESTRIAL VARIATIONS WITHIN GIVEN ENERGY, MASS AND MOMENTUM BUDGETS;
PALEOCLIMATE, SEA LEVEL, PALEOMAGNETISM, DIFFERENTIAL ROTATION AND
GEODYNAMICS

Nils-Axel Mörner
Paleogeophysics & Geodynamics,
Geological Institute, Stockholm University
S-10691 Stockholm
Sweden

ABSTRACT. It has been claimed that certain terrestrial variables
record solar variability. This need not be true, however. The proposed
correlation between atmospheric 14C production and paleoclimate is
shown to be untenable. It is demonstrated that the variations in
climate, sea level and paleomagnetism can be fully understood in terms
of variations within given budgets of energy, mass and momentum. It
is the redistribution of heat and mass and the interchange of momentum
that primarily drive the terrestrial variables. The relations are
tested both among pre-instrumental paleo-records and among present-day
instrumental records. Even the main changes in seismic and volcanic
activity can be analysed in this context. The same applies for certain
long-term records. Despite complexity and interaction between multiple
variables, it all leads back to a "simple" matter of mass, energy
and momentum.

1. INTRODUCTION

Short-term paleoclimatic changes, atmospheric 14C-production variations
and Earth's geomagnetic field changes (as recorded by paleomagnetism
and archeomagnetism) are potential parameters for the search for
and recording of possible secular solar variations. New findings
have shown, however, that terrestrial variables are likely to be driven
by internal changes within a given energy, mass and momentum budget.
This opens up three scenarios.

1.1 Scenario 1 : Solar variability (Fig. 1A)

It has been claimed (e.g. Eddy, 1976, 1977a, 1977b) that there is
a close correlation between the atmospheric 14C-production variations
and the climatic changes on the Earth's surface. Under the assumption
that the short-term paleoclimatic changes were of global extent,
"verifications" of this correlation were exemplified by picking out
certain proposed warm and cold events and placing them at the peaks
and troughs in the 14C-production curve of Suess (1970). The paleo-

F. R. Stephenson and A. W. Wolfendale (eds.),
Secular Solar and Geomagnetic Variations in the Last 10,000 Years, 455–478.
© 1988 by Kluwer Academic Publishers.

456

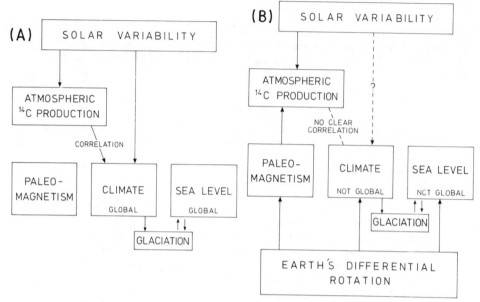

Fig. 1.   Origin of some terrestrial variables.   A: the external
origin (e.g. Eddy, 1977a).   The proposed correlation
between atmospheric 14C-production and paleoclimate is not
tenable, nor is the assumed globality of the short-term
changes in paleoclimate and sea level (see B).   B: The
internal origin (this paper, Mörner, 1984a, 1987b).   The
recorded variations are fully explained within a fixed
budget of energy, mass and momentum.

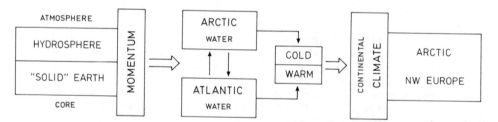

Fig. 2.   The interplay of cold Arctic water and warm Atlantic water
via the Gulf Stream have had a dominant controlling influence
on the continental climatic-environmental changes in
Scandinavia (and the rest of northwestern Europe, too).
These changes are driven by the interchange of angular momentum between the hydrosphere and the "solid" Earth (cf.
Fig. 4).

climatic data used were selected solely because of their fit, and
only represent a small (often unusual) percentage of our total paleoclimatic observational records. This is, of course, not good enough
for the formulation of such an important theory as a solar-terrestrial
linkage of short-term climatic changes. The background for picking
out certain paleoclimatic events is to be found in the previous belief
that the paleoclimatic signals had global extensions. Similarly,
the sea level oscillations were thought to be the direct function
of climatically controlled increases and decreases in the world's
glacial volume. This would imply that sea level oscillations, paleoclimatic changes and atmospheric 14C-production variations all
represented effects of the solar variability. This scenario is
illustrated in Figure 1A.

Against this scenario one must hold (1) the selection of not
representative, sometimes even directly odd, paleoclimatic "events",
(2) the absence of any correlation between atmospheric 14C-production
variations and paleoclimatic indices (Mörner 1973a; Stuiver, 1980;
Mörner, 1984a, 1984c), (3) the non-globality of short-term paleoclimatic changes and shifts (Mörner, 1984a, 1984b), and (4) the non-
globality of sea level changes (Mörner, 1976, 1987a, 1987b).

The atmospheric 14C-production is sensitive as much to the Sun's
solar wind changes as to the Earth's own geomagnetic field variations
(Mörner, 1978a, 1978b, 1984c). Both processes (the "external" and
the "internal") affect the incoming cosmic rays and hence the atmospheric 14C-production (Mörner, 1984c, Fig. 6).

1.2 Scenario 2 : Internal changes (Fig. 1B)

Modern eustatic sea level analyses indicate that oceanic water masses
and sea level change both with gravitational deformations of the
geoid surface (mass) and/or with rotational circulation changes
(momentum). Major climatic changes and shifts in the order of decades
and centuries are found to be regionally induced (not globally induced
as previously generally assumed). Their duration is about 50-150
years and the amplitude may vary from a few parts of a centigrade
up to several centigrades. This indicates that we are dealing with
energy redistributions over the globe. The only agent capable of
doing this and sustaining the signals for such a period of time is
the hydrosphere. Furthermore, these changes form frequency-changing
cycles calling for a non-constant terrestrial feed-back mechanism.
Interchange of momentum between the "solid" Earth and the hydrosphere
is advocated. This would give rise to a "pulsating" Gulf Stream
activity just as recorded by multiple data. The last centuries' records
of the Earth's rate of rotation (LOD) cannot be explained in terms
of atmospheric variations, nor does the core seem to be a source for
them except for periods of recorded geomagnetic disturbances (1836,
1910, 1970). The last decades' LOD-records include several short
LOD-increases. These correlate with El Nino (ENSO) events. In 1983
an especially distinct non-tidal increase of the LOD was recorded.
It indicates that angular momentum was transferred from the "solid"
Earth to the hydrosphere. The El Nino (ENSO) events have strong effects

on the coastal upwelling and by this on the atmospheric CO_2 content.
Pre-industrial CO_2 fluctuations in the order of 30 ppm in 100-150 years
are recorded. They are interpreted as records of significant changes
in coastal upwelling. The Earth must have experienced drastic hydrospheric events of the type "Super El Nino" or "Super ENSO" event
which affected eustatic sea level (mass and momentum), the distribution
of (hot or cold) sea surface water (energy) which affected the
neighbouring continental climate, and the coastal upwelling which
controlled the biological productivity and the atmospheric CO_2 content.
Similarly, the main trends in seismic and volcanic activity changes
seem explainable in terms of angular momentum stress changes on the
plate motions, subduction rates and asthenospheric flows. Despite
the complexity and interaction between multiple variables, it all leads
back to a "simple" question of mass, energy and momentum.

This implies that recorded changes in sea level, paleoclimate
paleomagnetism and geodynamics can all be understood in terms of
redistributions of heat (energy) and mass and the interchange of
momentum within a given mass-energy-momentum budget. This is
illustrated in Fig. 1B. It is this scenario that will be described
and discussed in this paper.

1.3 Scenario 3 : A combination

Admittedly we cannot exclude that scenario 2 is combined with simultaneous solar variations. Future research will reveal whether
there are any significant solar variability effects on the terrestrial
variables. At present, the important thing is to demonstrate that
the terrestrial variables are driven primarily by internal forces
that must not be misinterpreted in terms of solar variability as
is sometimes suggested (e.g. Eddy, 1977a).

2. SEA LEVEL

In 1969, I was able to separate the isostatic (i.e. crustal) and
eustatic (i.e. oceanic) factors behind the Scandinavian sea level
records, and to establish a detailed eustatic curve (Mörner, 1969).
At that time we all believed that eustasy was global and sensitively
depicted the glacial melting/freezing on the globe. I therefore
compared my northwest European eustatic curve with available paleoclimatic records in high-sedimentation-rate deep-sea cores from the
Atlantic (Mörner, 1973a). The result was striking; there was a very
close correlation between eustatic transgression maxima and warm
peaks and between eustatic regression minima and cold peaks. This
was interpreted in terms of global climatic changes. An interesting
fact emerged, however, viz. that these changes formed non-constant,
frequency-changing, cycles (Mörner, 1973a, 1973b). Already this
spoke for an origin in a terrestrial feed-back mechanism rather than
an extra-terrestrial beat (like a solar cycle beat).

In 1971, I noted that the global sea level data could not be
expressed in one eustatic curve (as generally assumed) and argued
that the ocean sea level distribution must have changed with time

in some presently unknown way (Mörner, 1971a, 1971b). With the
publication of satellite geoid maps (e.g. Gaposchkin, 1973), it became
obvious to me that the unknown factor was geoid changes with time.
    In 1976, I presented the new theory of geoid changes with time
(Mörner, 1976). This implies that eustasy is not a global phenomena,
but rather a regional phenomena (Mörner, 1976, 1980, 1981a, 1983,
1987b). The geoid is an equipotential surface that will deform with
any changes in the attraction of rotation potentials. We could no
longer expect even eustatic sea level to change in a similar way
(and certainly not in an identical way) over the globe; rather much
of the changes should be compensational. Many data indicating such
eustatic differentiation over the globe have been presented (Mörner
1976, 1980, 1981a, 1981b, 1983, 1987a,c; Newman et al., 1980, 1981;
Pirazzoli, 1977; Martin et al., 1986; Nunn, 1986, and others).
    However, still closer examinations of the sea level records
revealed that besides the non-correlations, there also existed some
correlations. Taira (1981) insisted that the Japanese eustatic and
climatic records were very similar to my Scandinavian records.
Similarly, Colquhoun et al. (e.g. 1983) claimed that the South
Carolina records (sea level and climate) exhibited large similarities
with our northwest European data. This needed some special explanation
(Mörner, 1984a).

3. CLIMATE

So we turned to the climatic changes. Besides the long-term changes
of Milankowitch cyclicity and the very short-term changes of sun-spot
cyclicity, there are some major climatic changes and shifts. Southern
Scandinavia was known to have an unusually detailed record of these
changes. Much of this material has been used as international
standard. We therefore devoted the Second Nordic Symposium on Climatic
Changes and related Problems in Stockholm 1983 especially to these
changes and their regional and global validity (Mörner and Karlén,
1984). This analysis revealed two fundamental characteristics of
these climatic changes and shifts (Mörner, 1984b): (1) that they
all had a duration in the order of 50-150 years, and (2) that they
had a non-global validity, i.e. that they were regionally induced.
    This meant that we were here dealing with climatic changes that
represented the redistribution of energy over the globe by means
of processes that had a capacity of sustaining the signals for some
50-150 years.
    This gave a clear indication of the origin (Mörner, 1984b).
We were not dealing with the gain and loss of incoming solar radiation.
Nor were we dealing with atmospheric processes, because they do not
have such a long "memory" (being rapidly balanced by compensational
changes in the rate of rotation of the "solid" Earth).
    The hydrosphere, on the other hand, is the ideal and logical
medium for the redistribution of energy over the globe and for sustaining these signals for some 50-150 years (Mörner, 1984b). This
would imply that the oceanic circulation must have undergone significant

460

fluctuations with time; usually in the form of some sort of pulsation or cyclicity.

We therefore turn to the hydrosphere, its variability (momentum), relation to continental climate (energy) and sea level (mass).

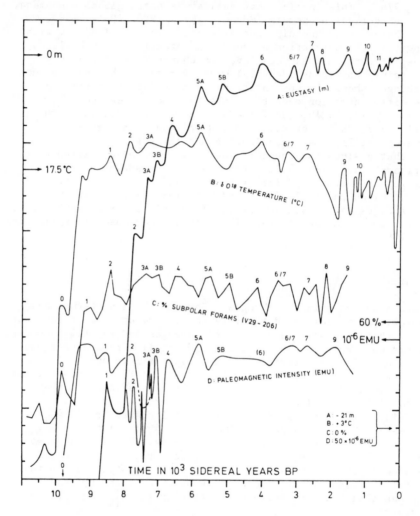

Fig. 3. Holocene short-term fluctuations (16 of them) in eustasy (A) in northwestern Europe, in continental temperature (B) on the island of Gotland in the Baltic, in percent subpolar forams (C) in the Denmark Strait between Iceland and Green-land sensitively registering the Gulf Stream pulsation, and in paleomagnetic intensity (D) in sediment cores from Southern Scandinavia. From Mörner (1984a, Fig. 6). Vertical scale in meters (A), centigrades (B), 10% (C) and 10x10⁻⁶ EMU (D). Geographical locations in Mörner, 1984a, Fig. 11.

4. OCEANOGRAPHY

At around 13,000 BP, there was a rapid immigration and spread of
a new fauna and flora (e.g. boreal molluscs) around Scandinavia and
in the North Sea. This initiated a rapid deglaciation of the
Fennoscandian ice cap, and a significant continental climatic amelior-
ation. A similar pulse came at around 10,000 BP. Both events must
represent significant intensifications of the warm water (heat) carried
by the Gulf Stream to northwestern Europe. Reversed effects are
noted at around 11,000 BP, 2500 BP and during the so-called Little
Ice Ages when pulses of southward expansions of Arctic water are
recorded.

   This exhibits the strong sensitivity of Scandinavian climate
to southward migration of Arctic water (coolings) and northward
migration of Atlantic water and intensifications of the Gulf Stream
(warmings). This is illustrated in Fig. 2.

   These redistributions of heat and mass must inevitably affect
the Earth's momentum budget and differential rotation. In order
to test this, we need to go into further details, however.

   First we recall that the northwest European eustatic sea level
changes show remarkable correlations with paleoclimatic changes in
six Atlantic deep-sea cores (Mörner 1973a) and that these changes
record frequency-changing cycles (Mörner, 1973b) calling for a non-
constant feed-back mechanism. Ocean circulation changes and interchange
of momentum between the hydrosphere and the "solid" Earth would generate
such correlations and such cycles.

   In Fig. 3 (Mörner. 1984a) we compare the eustatic changes in
northwestern Europe (A), the absolute temperature changes (B) in
Southern Scandinavia, the percentage subpolar forams (C) in a
deep-sea core from the Denmark Strait sensitively recording the
activity of the Gulf Stream, and the paleomagnetic intensity variations
(D) in South Scandinavian sediment cores (Mörner 1984a). It records
16 short-term oscillations during the Holocene which are expressed
almost fully in all four different records; curve C records the
pulsation of the Gulf Stream (mass, energy and momentum), curve B
the continental climate (energy) and curve A the regional sea level
(mass and momentum). Curve D may indicate that the core/mantle coupling
was simultaneously affected by differential rotation (momentum).

   This leads to the proposition of a mutual causal mechanism as
expressed in Fig. 4 and to the formulation of a "rotational-
gravitational-oceanographic" origin of short-term climatic changes
(Mörner, 1984a, 1984b).

   This model does not only explain the Atlantic data; it also
explains why there could be a correlation between Japanese and Atlantic-
European records. This is evident from the Earth's main oceanographic
circulation pattern (Fig. 5). If differential rotation and interchange
of momentum between the "solid" Earth and the hydrosphere are involved,
one would expect similarities in the behaviour of the Gulf Stream
in the Atlantic and the Kuroshio Current in the Pacific. This is
just what Taira's data suggest (Taira, 1981).

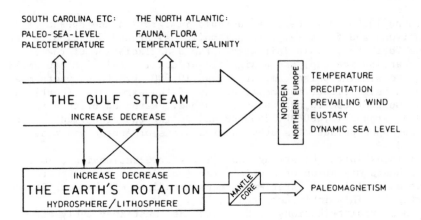

Fig. 4.   Driving causational mechanism for the mutual North Atlantic
          changes in eustasy (mass), climate (energy), Gulf Stream/
          Arctic water balance (momentum) and paleomagnetic intensity
          (core/mantle dislocation).   From Mörner (1984a, Fig. 12).

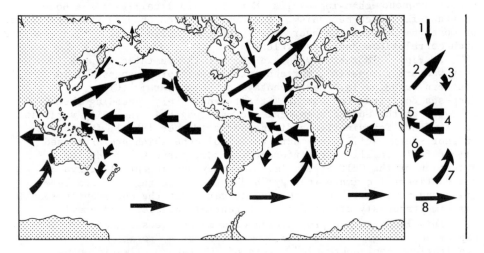

Fig. 5.   Main global oceanographic circulation patterns.   Interchange
          of momentum in the hydrosphere would lead to similar trends
          of the Kuroshio Current in the Pacific and the Gulf Stream
          in the Atlantic.   It would lead to quite different changes in
          the northern and southern hemispheres, however.   All this is
          consistent with observational data.   Black areas denote areas
          of coastal upwelling; i.e. the locations of main marine
          biological productivity and by this also the areas that play
          a significant role in the regulation of the atmospheric
          CO_2 content (cf. Fig. 14).

The totally different circulation pattern in the northern and southern hemispheres (Fig. 5) should - with the same mechanism - give rise to totally different climatic and eustatic records. This is what our observational data indicate.

In Fig. 5, we have also marked the major areas of coastal upwelling. These are places of exceptional biological productivity. Changes in the pattern and/or intensity of circulation will affect the upwelling and by this the biological productivity (Mörner, 1987b).

5. DIFFERENTIAL ROTATION

The Earth is a multi-layered system (Fig. 6) where any change in angular momentum of one layer must be compensated by another layer or sublayer (i.e. within the system) to keep the total momentum constant.

The short-term changes in the Earth's rate of rotation (i.e. the "solid" Earth) or length-of-the-day (LOD) of annual or shorter duration are compensated within the atmosphere (Barnes et al., 1983; Rosen and Salstein, 1983; Eubanks et al., 1983), or, rather, viceversa. The longer-term changes are thought to be compensated by the core. In the absence of observational data and means of quantifications, the role of the hydrosphere was previously assumed to be small. Our data shows that this cannot be tenable, however.

The Earth's rate of rotation, or rather the lengthening-of-the-day (LOD), has been continually measured since the beginning of the 17th century; with large uncertainties prior to 1650, poor resolution

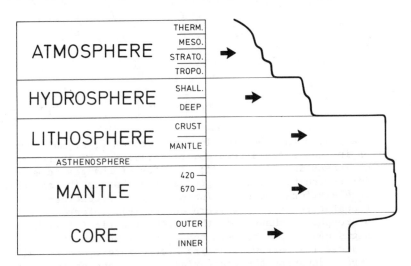

Fig. 6.   The Earth is a multi-layered system that experiences a differential rotation with interchange of angular momentum between the different layers and sublayers in order to keep the total momentum constant.

prior to 1780 and increasingly better resolution during the last 200
years (Stephenson and Morrison, 1984). The amplitude and rates of
the non-tidal changes in the LOD are given in Fig. 7 (B, C). We compare
these changes with the general changes in continental temperature
in northwestern Europe (Fig. 7E). The climatic shifts exhibit a striking

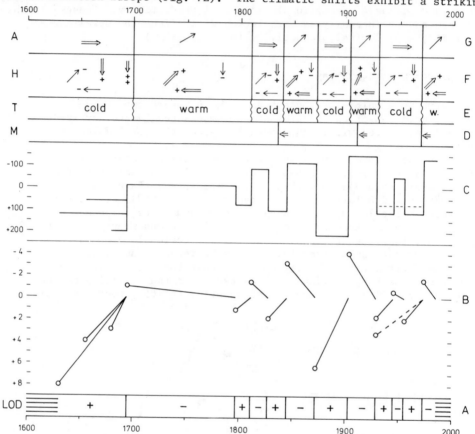

Fig. 7. Changes in the LOD from 1630 to 1985 (A-C; A: main periods
of increases and decreases; B: amplitudes of increases and
decreases in milliseconds; C: rates of changes in ms^{-3}yr),
corresponding levels of geomagnetic changes (D) and core
decelerations, mean cold/warm changes in continental
temperature in northwestern europe (E), inferred causational
hydrospheric changes in the North Atlantic (F) with plus-
signs denoting increases and minus-signs denoting decreases,
and vertical arrows denoting Arctic water, horizontal
arrows denoting the equatorial current and oblique arrows
denoting the Gulf Stream (thick arrows=increases, thin
arrows=decreases), and corresponding atmospheric circulation
pattern (G) with arrows denoting position and direction of
main westerly jetstreams (cf. Fig. 13).

correlation with the changes in LOD. This was also noted by Lambeck and Casenave (1976) who concluded: (1) "there is a close similarity between variations in the LOD and trends in various climatic indices", and (2) "surface pressure and sea level data represent only about 10 per cent" of the excitation necessary to explain the LOD variations (with the remaining 90 per cent unexplained). Courtillot et al. (1978) showed that the Earth's geomagnetic field was disturbed at about 1836, 1910 and 1970 in close relation to the changes from increasing to decreasing LOD (at 1840, 1903 and 1972 in Fig. 7D). This indicates that angular momentum was transferred from the core (slowing down) to the mantle (speeding up = decreasing LOD) at these three time levels.

In the search for "the missing excitation" (according to Lambeck and Casenave, 1976), I find it natural to turn also - (NB. not only) - to the hydrosphere. From data above, we know that large oceanic mass and energy redistributions occurred in the North Atlantic during the last 13,000 years.

In the North Atlantic, a speeding up of the "solid" Earth (decreasing LOD) and loss of momentum (slowing-down) of the hydrosphere would lead to an intensified equatorial eastward current, a northward migration of the water masses, i.e. the Gulf Stream, and a restriction of Arctic surface water influx. This would result in a warming of northwestern Europe and fits perfectly well for all major periods of LOD decrease (mantle speeding-up); viz. 1845-1872, 1903-1929 and after 1972 (also for the generally warm but poorly defined period 1695-1797). At the same time the main atmospheric jetstream would decrease and move northwards.

A slowing-down of the "solid" Earth (increasing LOD) and a gain of momentum (speeding-up) of the hydrosphere would lead to a decrease in the equatorial easterly current, a southward migration of the Gulf Stream and a southward penetration of Arctic surface water. At the same time the atmospheric westerly jetstream would increase and move southwards so that it would tend to block (i.e. retard) both the equatorial current and the Gulf Stream and may even lead to the loss of hot equatorial surface water to the southern hemisphere. All this would result in a cooling in northwestern Europe. This fits perfectly well with the periods of LOD-increases (mantle slowing-down); viz. prior to 1695 (representing the 2nd Little Ice Age), 1827-1845 and 1929-1972.

Originally (Mörner, 1987b), I had some problems of finding conclusive climatic expressions of the drastic (6.42 ms) increase in the LOD at 1872-1903. The long water temperature record from Den Helder in the Netherlands (van Malde, 1984), however, documents a cooling and low temperature period (Fig. 8) that fully agrees with the LOD-increase. During this period most tide gauges (but not the one at Den Helder) record a low (de Ronde, 1983; van Malde, 1984; Mörner, 1973c).

These hydrospheric and atmospheric effects are expressed in Fig. 7 (F, G). It seems quite clear that the transfer of angular momentum to the hydrosphere offers a simple explanation to, at least a major part of, "the missing excitation".

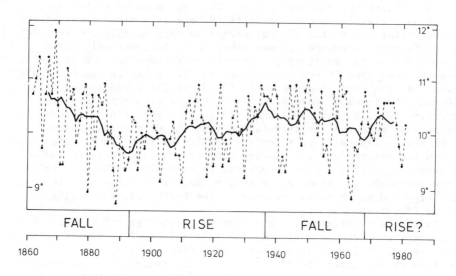

Fig. 8. Water temperature (in centigrades) at Den Helder (redrawn from Van Malde, 1984); annual mean temperature (dotted line) and 10 years running mean (thick line).

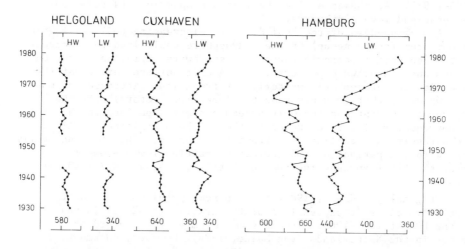

Fig. 9. Tidal records from Germany (redrawn from Siefert, 1984). The high (HW) and low (LW) water records (in meters above the zero level) are given for Helgoland off the coast, Cuxhaven at the coast and Hamburg up in the Elbe river estuary. All three stations show a change at around 1970. "Something must have happened outside the estuaries" according to Siefert (1984). Apparently this is a sensitive registration of the 1972-1973 change in rotation (Fig. 10).

We probably have to do with a combination of interchange of angular
momentum with the atmosphere (a minor part), the hydrosphere (a substantial, maybe even dominant, part) and the core (at the three periods
of geomagnetic disturbances).

It seems significant that the northwest European main eustatic
trend (Mörner, 1973c) records a low level prior to 1839, a distinct
rise (1839-1930), a decreased rise 1930-1950 and a stability (or even
fall) 1950-1970. The similarity with observed continental climate
and inferred hydrospheric circulation (Fig. 5) is striking. Tide gauge
data provides some additional details. The Dutch records   (de Ronde,
1983) record a low in the period 1885-1905. The Swinemünde tide gauge
(Mörner, 1973c) records a slow fall 1830-1855, a rapid rise 1855-1875,
a stability (very slow rise) 1875-1901, a rapid rise 1901-1922, a
stability (very slow rise) 1922-1946, and a rapid rise from 1946 to the
end of the record in 1955. This provides a fairly remarkable fit
with the LOD-record; rises during LOD-increases and stabilities or
falls during LOD-decreases. The 1972-change (Fig. 10) is well expressed

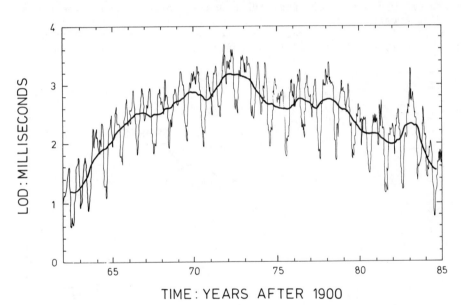

TIME: YEARS AFTER 1900

Fig. 10.   Changes in the LOD from 1962 to 1985 showing partly
            annual-seasonal fluctuations compensated within the
            atmosphere (thin line) and partly the main non-tidal
            variations (thick line) which besides a general increase
            up to 1972-73 and a general decrease afterwards, exhibits
            several minor fluctuations. These fluctuations represent
            increases of 1-2 years' duration. These are all directly
            related to El Nino (ENSO) years. The corresponding loss
            of momentum of the "solid" Earth must have been transferred
            to the hydrosphere.

in the German tidal records (Fig. 9). Siefert (1984) concluded that
"we can evaluate distinct changes in water levels since 1970, recorded
both in HW and LW" and that "something must have happened outside
the estuaries".

We now turn to the most recent records. Fig. 10 gives the LOD
variations from 1962 to 1985. Besides the annual and seasonal variations that are totally compensated within the atmosphere, there is
a secular, non-tidal, trend; partly the main rise (slowing-down) up
to 1972 and fall (speeding-up) afterwards, and partly a number of
minor fluctuations. The minor increases in the LOD (i.e. slowing-downs
of the "solid" Earth) are especially distinct. They last for 1 year
or, at the most, 2 years. Also, they all have a close relation to
El Nino events. These fluctuations are not compensated within the
atmosphere (that is clear) and are too short to be compensated by
the core. Both their amplitude and durations are ideal for a compen-
sation within the hydrosphere, however. We therefore propose that
these fluctuations (i.e. the LOD increases) represent momentum variation
in the hydrosphere and that they are closely linked to causal mechanism
of El Nino events.

In Fig. 11, we give a detailed curve of the non-tidal LOD changes
from 1977 to 1984. At around 1983, there is an increase of 0.4 ms

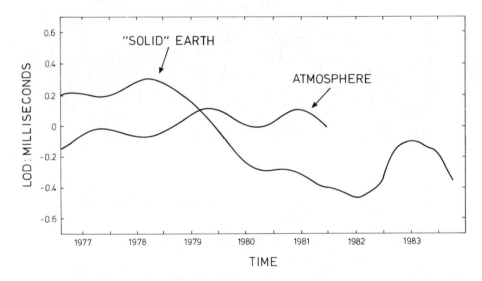

Fig. 11. Non-tidal changes in the LOD (of the "solid" Earth) from
1976.5 to 1984, and corresponding atmospheric momentum
changes from 1976.5 to 1981.5. We learn that the atmosphere has nothing to do with the non-tidal variations in
the LOD. We note that the large 0.4 ms increase around
1983 fits perfectly well with the occurrence of the last
(very large) El Nino (ENSO) event, and take this as firm
evidence that angular momentum was transferred from the
"solid" Earth to the hydrosphere at this time.

(i.e. a slowing-down of the "solid" Earth). This coincides perfectly well with the last El Nino event. This must represent a significant transfer of angular momentum from the "solid" Earth to the hydrosphere. Therefore, the eastward migration of hot surface water (raising the sea level in the east and lowering it in the west) in the Pacific should mainly, or at least partly, be an effect of this differential rotation (cf. Fig. 13).

6. EL NINO AND "SUPER EL NINO" EVENTS

The El Nino/Southern Oscillation (ENSO) events occur semi-regularly. They have drastic meteorological, oceanographic and biological effects (e.g. Cane, 1983; Barber and Chevez, 1983; Philander, 1983; Newell and Hsiung, 1984; Rasmussen and Wallace, 1983; WCDP, 1985). The atmospheric circulation is upset. The westerly jetstreams are intensified and moved equatorwards (Fig. 12). Hot sea surface water in the Pacific is transported eastwards. Sea level rises along the equatorial American west coast and falls in the West Pacific (Fig. 13). At the same time the thermocline moves down in the east and up in the west. The coastal upwelling outside South America is stopped or reduced. This has drastic biological consequences (e.g. Barber

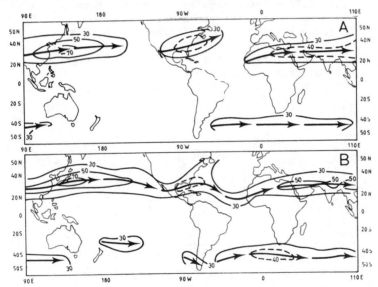

Fig. 12. Westerly jetstreams in normal years (A) and during the 1982/83 El Nino year (B). Wind velocity figure in m/s (redrawn from WCDP, 1985). We note that the North Atlantic jetstream in El Nino years was intensified and displaced considerably to the south but with a tendency also of a northwestern direction (i.e. of a splitting into two branches; cf. Fig. 7F).

470

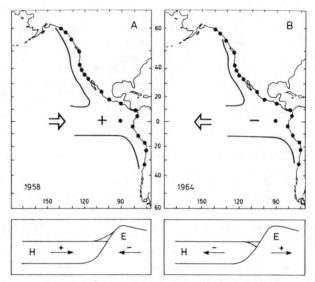

Fig. 13. Sea level changes along the American west coast in the ENSO
          year 1958 (A) and in the non-ENSO year 1964 (B). The mass
          redistributions are here claimed not only to be caused by
          wind forces (as generally assumed) but rather mainly by
          changes in hydrospheric angular momentum (cf. Fig. 11).
          Corresponding models of "solid" Earth deceleration and
          hydrospheric acceleration in 1958 (A), and of "solid"
          Earth acceleration and hydrospheric deceleration in 1964
          (B) are given in the lower part.

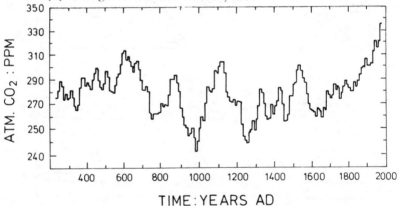

Fig. 14. Atmospheric CO_2 variations during the last 1800 years
          (redrawn from Bojkov, 1983, Fig. 5). Several pre-
          industrial 30 ppm changes are recorded during intervals
          of 100-150 years. These changes are taken to represent
          considerable changes in coastal up-welling and hence
          provide independent evidence of global changes of the type
          "Super El Nino - ENSO" events.

and Chevez, 1983) and also leads to a decrease in the global atmospheric CO_2 content (Newell and Hsiung, 1984). The Earth's angular momentum decreases (Eubanks et al., 1984, 1986).

Enfield and Allen (1980) made a detailed study of the sea level changes all along the American west coast. These data are here used to demonstrate the effect on sea level of ENSO and non-ENSO years. Fig. 13 gives the sea level rise in 1958, an ENSO year, and the sea level fall in 1964, a non-ENSO year. It also provides simple models of the corresponding interchanges of momentum between the "solid" Earth and the hydrosphere. This figure together with the evidence in Fig. 11 of a non-tidal deceleration of the "solid" Earth are taken as convincing indications of a transfer of angular momentum from the "solid" Earth to the hydrosphere during ENSO years. This seems to provide a solid base for a similar interpretation of the major, 50-150 years long, events found to characterize the climatic-eustatic changes during, at least, the last 13,000 years.

We have, above, already discussed the 50-150 years events when it concerns eustasy and climate, and their possible origin in the gain and loss of angular momentum in the hydrosphere (by differential rotation and interchange of momentum with the "solid" Earth). The effect of present-day ENSO events on the atmospheric CO_2 content (Newell and Hsiung, 1984) opens another analogy. Stuiver et al. (in Bojkov, 1983) have presented a record of the atmospheric CO_2 variations during the last 1800 years (Fig. 14). It records several high-amplitude fluctuations of about 30 ppm in 100-150 years in pre-industrial time (Fig. 14). These fluctuations seem only explainable in terms of large-scale changes in the coastal upwelling. This fits perfectly well with the theory of significant changes in the hydrospheric angular momentum here presented.

In principle, these 50-150 years long changes in the hydrospheric momentum with corresponding effects on eustatic sea level, heat distribution (climate) and coastal upwelling represent some sort of large-scale analogies to the El Nino (ENSO) events (Mörner, 1984a). We may, therefore, term them "Super El Nino" or "Super ENSO" events (Mörner, 1985).

7. SEISMIC AND VOLCANIC ACTIVITY VARIATIONS

Because of the large difference in mass between the mantle and the lithosphere, they should have experienced differential rotation with respect to each other, if they had not been coupled together via the asthenosphere. The asthenosphere, however, is a partially melted, low-viscosity (10^{18}-10^{20} Poise) layer. Therefore, large stress variations are likely to be created in the asthenosphere - and transferred into the lithosphere - in connection with accelerations and decelerations of the mantle-lithosphere system (i.e. LOD-variations).

It has previously been claimed (e.g. Anderson, 1974) that there is a causal connection between the changes in the Earth's rate of rotation at around 1900 AD and a general increase in global volcanism and seismicity at about the same time. A closer examination of the last centuries' records (Mörner, 1986a, 1986b) reveals partly the existence of certain times when seismicity, tectonics and volcanism seem to change their activity (increasing/decreasing) in a global east-west pattern that suggests the effects of accelerations and decelerations of plate motions and subduction in the longitudinal dimension, and partly the existence of certain times of increases and decreases in seismic activity in a global north-south pattern suggesting the effects of corresponding plate motions and expansions/ contractions in the latitudinal dimension. This means the identification of a very high sensitivity of plate motions and subduction processes as well as of asthenospheric flows and lithospheric stress to variations in the Earth's rate of rotation and corresponding interchanges in momentum between its different layers and sublayers.

This question (further discussed in Mörner, 1986a, 1986b) is here mentioned because it provides another and independent expression of the effects of the Earth's differential rotation.

8. LONG-TERM, 10^3-10^6 YEARS, RECORDS

The present model can also be applied on long-term records; the Holocene 10^3 yrs records, the Milankowitch parameters 10^4-10^5 yrs records, and the long-term 10^6 yrs records. Only a short review of these data will here be presented for the sake of comparisons with the short-term records primarily discussed.

During the Last Ice Age, mass was taken from the oceans (falling in the order of 100 m) and redistributed to mid and high latitudes in the form of contintental ice caps (in North America north of Lat. 40°N, and in Europe north of Lat. 53°N). This led to a new mass distribution (with consequences on the gravity potential and the rotation) and to a shortening of the Earth's radius inevitably increasing the angular momentum. With the melting of the ice and the rise of the ocean level, this process was reversed with corresponding adjustments of the gravity potential and the angular momentum. The general sea level rise up until about 6000 BP (by about 100 m at a mean rate of about 10 mm/year) must have caused a general deceleration allowing the Gulf Stream to move northwards (heating NW Europe and the Arctic) and the westerly jet streams to move northwards, too. This is well recorded in the general Holocene warming of the mid and high latitudes in the European-Arctic sector. In Mid Holocene time, and probably as a function of the end of the general sea level rise, the climatic regiment changed in many areas; viz. from drier to wetter in northern Europe, from wetter to drier in Africa and India, from warm and dry to cold and wet in Japan, etc. This seems to represent the displacements of the westerly jet streams in response to angular momentum changes (earlier deceleration period over).

The global spread of Holocene eustatic sea level data between about 8000 BP and the present (Mörner, 1971a) record a cyclic geoid deformation pattern (Mörner, 1976, 1978a). This cyclic pattern exhibits close similarities with the archaeomagnetic cycle of Bucha (1970)

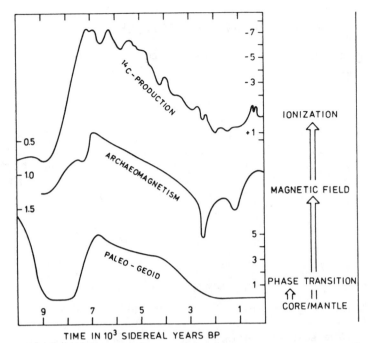

Fig. 15.   Holocene cyclic long-term changes in paleogeoid (sea level)
configuration, archeomagnetism and atmospheric 14C-
production (from Mörner, 1984a). The proposed mutual
origin (core/mantle dislocation) and causal connection are
illustrated to the right.

and the main atmospheric 14C-cycle of Suess (1970) extended backwards
via Swedish varve data (Mörner, 1979, 1980b). This calls for a mutual
causation in a core/mantle dislocation due to differential rotation
as illustrated in Fig. 15 (Mörner, 1978a, 1984a).
     The so-called Milankowitch astronomical variables (eccentricity,
tilt and precession) have during the last decades been used extensively
to explain the Earth's climatic changes on the 10^4-10^5 yrs time base.
The effective factor has been claimed to be the insolation variations
(e.g. Berger, 1978). In opposite to the general belief of an
insolation forcing, I have several times (Mörner, 1978a, 1980a, 1981c,
1984a, 1984c, 1986c) claimed that the Milankowitch variables must
have considerable effects on internal processes like the geoid configuration, differential rotation, geomagnetism, heat flow, etc.,
and that these variables may have had a significant, maybe even dominant,
influence on paleoclimate. This is illustrated in Fig. 16. The
identification of the typical Milankowitch cyclicities in sedimentary
records and sea level changes from periods of non glacial environment
can hardly be understood in other terms than in forcing via internal
variables.
     The deep-sea records include several so-called deep-sea hiatuses
(i.e. erosion and/or non-deposition surfaces). These have been inter-

474

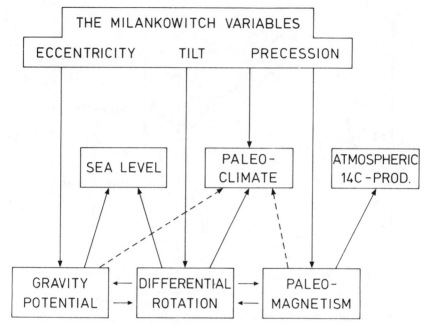

Fig. 16. The so-called Milankowitch variables have multiple effects. The effects on various internal processes must not be forgotten. They are known to drive significant sea level changes (for example, during non-glacial periods). They are likely to produce significant, maybe even dominant, paleoclimatic changes. Similarly, long-term atmospheric 14C-production variations are likely to be set up via the geomagnetic field variations (both the eccentricity and and tilt cycles have been claimed to be present in paleomagnetic data: e.g. Wollin et al., 1977; Kent and Opdyke, 1976).

preted in terms of sea level changes (Rona, 1973; Moore et al., 1978; Thiede, 1981; Barron and Keller, 1982). There is no physical reason for such a causal correlation, however. This has led me to propose a different mechanism (Mörner, 1985b). The deep-sea hiatus must represent major changes in the bottom currents. The most effective way of changing the bottom currents is transfer of angular momentum between the "solid" Earth and the hydrosphere, i.e. rotational changes. This is illustrated in Fig. 17. Differential rotation due to the interchange of angular momentum between the hydrosphere (H) and the "solid" Earth (L) is capable of (1) changing the sediment/water friction leading to the formation of deep-sea hiatuses, (2) affecting the interchange of water with the Arctic Basin leading to changed oxygen isotope composition and apparent coolings and warmings, (3) affecting the vertical mixing leading to a variety of biological-chemical effects, and (4) deforming the sea level by geoidal eustatic and dynamic sea surface changes. This provides a simple explanation for the recorded

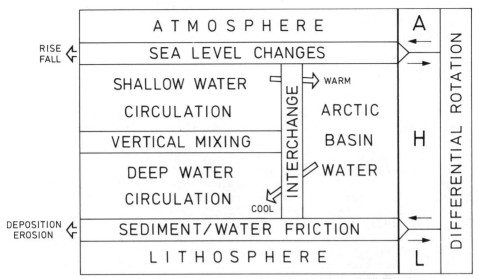

Fig. 17. Differential rotation and interchange of momentum between
the "solid" Earth (L) and the hydrosphere (H) offers a
simple and logical explanation to the occurrence of deep-
sea hiatus and their proposed correlation with sea level
changes and shelf unconformities, as well as with oxygen
isotope variations and some biological and geochemical
changes. Variations in the exchange between the Atlantic
and Arctic basins driven by angular momentum changes seem
finally to provide a solution to this intriguing problem
(cf. Mörner, 1985).

changes and their mutual correlation. Again, we find differential
rotation in the center.

9. CONCLUSIONS

The short-term terrestrial variations in paleoclimate, sea level and
paleomagnetism can be satisfactorily explained within terrestrial
budgets of fixed energy, mass and momentum by the redistribution of
heat and mass over the globe and by the interchange of momentum due
to differential rotation. The model was developed for the major short-
term changes and shifts during the last 20,000 years (Mörner, 1984a).
It offers an explanation not only to the paleoclimatic and sea level
changes, but also to the paleomagnetic changes, the circulation changes,
the pre-industrial CO_2 variations, the biological productivity variations
(and energy sources) and the changes in coastal upwelling. The
"missing excitation" of the LOD variations is explained. The LOD
variations seem closely - and primarily - related to hydrospheric
circulation changes. The differential rotation concept provides a

logical explanation for some variations (longitudinal and latitudinal) in seismic and volcanic activity, as well as to a variety of long-term variables.

10. REFERENCES

Anderson, L.D., 1974, Science, 49.
Barber, R.T., and Chavez, F.P., 1983, Science, 222, 1203.
Barber, R.H.T., Hide, R., White, A.A., and Wilson, C.A., 1983, Proc. Roy. Soc. London, Ser. A., 387, 31.
Barron, J.A. and Keller, G., 1982, Geology, 10, 577.
Berger, A.L., 1978, Quaternary Res., 9, 139.
Bojkov, R.D., 1983, WMO Proj. Res. Monitor. Atm. CO_2, Rep. 10.
Bucha, V., 1970, In: 'Radiocarbon variations and Absolute Chronology', I. Olsson, ed., Almquist and Wiksell, 501.
Cane, M.-A, 1983, Science, 222, 1189.
Colquhoun, D.J., Brooks, M.J., Brown, J.G. and Stone, P.A., 1983, IUGG, Hamburg 1983, Abstracts ICL, 89.
Courtillot, V., Ducruix, J., and Le Mouel, J.-L, 1978, C.R. Acad. Sc. Paris, 287:D, 1095.
Eddy, J.A., 1976, Science, 192, 1189.
Eddy, J.A., 1977a, Climatic Change 1, 173.
Eddy, J.A., 1977b, Scientific American, vol. 235, No. 5, 80.
Enfield, D.B. and Allen, J.S., 1980, J. Phys. Oceanogr., 10, 557.
Eubanks, T.M., Steppe, J.A., Dickey, J.O., and Callahan, P.S., 1983, Jet. Prop. Lab. (Pasadena), Geod. Geoph. Prepr. No. 102, 1.
Eubanks, T.M., Dickey, J.O., and Steppe, J.A., 1984, Jet. Prop. Lab. (Pasadena), Geod. Geoph. Prepr, No. 111, 11.
Eubanks, T.M., Steppe, J.A., and Dickey, J.O., 1986, JPL Geodet. Geophys. Prepr. No. 143, 1.
Gaposchkin, E.M. 1973, In: '1973 Smithsonian Standard Earth (III)', (E.M. Gaposchkin, ed.), Smithsonian Astr. Observ., Spec. Rep. 353, 85.
Kent, D.V. and Opdyke, N.D. 1976, Nature 266, 156.
Lambeck, K. and Cazenave, A., 1976, Geophys. J.R. Astr. Soc., 46, 555.
van Malde, J., 1984, Rijkswaterstaat, Nota WW-WH 84.08, 20.
Martin, L., Flexor, J.-M., Blitzkov, D. and Sugio, K., 1986, Proc. Fifth Coral Reef Congr., Tahiti 1985, 3, 85.
Moore, J.R, T.C., van Andel, Tj.H., Sancetta, C. and Psias, N., 1978, Micropaleontol. 24, 114.
Mörner, N.-A, 1969, Swedish Geol. Surv., C-640, 1.
Mörner, N.-A, 1971a, Geol. en Mijnbouw, 50, 699.
Mörner, N.-A, 1971b, Geol. Soc. America Bull., 82, 787.
Mörner, N.-A, 1973a, Moreas, 2, 33.
Mörner, N.-A, 1973b, J. Interdiscipl. Cycl Res., 4, 189.
Mörner, N.-A, 1973c, Palaeogeogr. Palaeoclim. Palaeoecol., 14, 1.
Mörner, N.-A, 1976, J. Geol., 84, 123.
Mörner, N.-A, 1978a, In: 'Evolution of Planetary Atmospheres and Climatology of the Earth', CNES Colloques Intern., Toulouse, 221.

Mörner, N.-A, 1978b, Danish Meteorol, Inst., Climatol. papers, 4,
    45.
Mörner, N.-A, 1979, GeoJournal, 3, 287.
Mörner, N.-A, 1980a, In: 'Earth Rheology, Isostasy and Eustasy',
    (N.-A Mörner ed), John Wiley & Sons, 535.
Mörner, N.-A, 1980b, Palaeogeogr. Palaeoclim. Palaeoecol., 29, 289.
Mörner, N.-A, 1981a, IAHS Publ., 131, 277.
Mörner, N.-A, 1981b, Ann. Geophysique (CNES, Toulouse), 37, (1),
    69.
Mörner, N.-A, 1981c, Geol. Rundschau, 70, 691.
Mörner, N.-A, 1983, In: 'Mega-Morphology' (R. Gardner and H. Scooging,
    eds.), 73, Oxford Univ. Press.
Mörner, N.-A, 1984a, In: 'Climatic Changes on a Yearly to Millenial
    Basis', (N.-A Mörner and W. Karlén, eds.), 483, 507, Reidel Publ.
    Co.
Mörner, N.-A, 1984b, In: 'Climatic Changes on a Yearly to Millenial
    Basis', (N.-A, Mörner and W. Karlén, eds.)., 637, Reidel Publ.
    Co.
Mörner, N.-A, 1984c, Geogr. Ann. 66A, 1.
Mörner, N.-A, 1985a, IAMAP/IAPSO Joint Ass., Honolulu, Abstracts,
    31.
Mörner, N.-A, 1985b, In: 'Sea Level Changes – An Integrated Approach',
    Houston.
Mörner, N.-A, 1986a, In: 'The Origin of Arcs' (F.C. Wezel, ed.),
    79. Elsevier Sci. Publ.
Mörner, N.-A, 1986b, Revista CIAF, 11, 205, Bogota.
Mörner, N.-A, 1986c, Bull. INQUA Neotectonics Comm., 9, 8-13.
Mörner, N.-A, 1987a, In: 'Sea level changes' (M.J. Tooley and I.
    Shennan, eds.), 333, Blackwell.
Mörner, N.-A, 1987c, In: 'Late Quaternary Sea-Level Changes' (Y.
    Qin. & S. Zhao, eds.), 26, China Ocean Press.
Mörner, N.-A, 1987b, In: 'Long Term changes in marine Fish Populations'
    (T. Wyatt, ed.), in press. Vigo (Spain).
Mörner, N.-A, and Karlén, W., (eds.), 1984, 'Climatic Changes on
    a Yearly to Millenial Basis', Reidel Publ. Co., 677pp.
Newell, R.E. and Hsiung, J., 1984, In: 'Climatic Changes on a Yearly
    to Millenial Basis' (N.-A, Mörner and W. Karlén, eds.), 533,
    Reidel Publ. Co.
Newman, W., Marcus, L., Pardi, R., Paccione, J., and Tomacek, S.,
    1980, In: 'Earth Rheology, Isostasy and Eustasy' (N.-A. Mörner,
    ed.), 555, John Wiley & Sons.
Newman, W., Marcus, L., and Pardi, R., 1981, IAHS Publ. 131, 263.
Nunn, P.D., 1986, Geol. Soc. America Bull., 97, 999.
Philander, S.G.H., 1983, Nature, 302, 295.
Pirazzoli, P.A., 1977, Zeitschr. Geomorph., 21, 284.
Rasmusson, E.M. and Wallace, J.M., 1983, Science, 222, 1195.
Rona, A., 1973, Nature 244, 25.

de Ronde, J.G., 1983, In: 'Seismicity and Seismic Risk in the Offshore North Sea Area', (A.R. Ritsema and A. Gürpinar, eds.), 131, Reidel Publ. Co.

Rosen, R.D. and Salstein, D.A., 1983, J. Geophys. Res., 88, 5451.

Stephenson, F.R. and Morrison, L.V., 1984, Phil. Trans. Soc. Lond., A313, 47.

Stuiver, M., 1980, Nature, 286, 868.

Suess, H.E., 1970, In: 'Radiocarbon variations and Absolute Chronology' (I. Olsson, ed.), 595, Almquist and Wiksell.

Taira, K., 1981, Palaeogeogr. Palaeoclim. Palaeoecol., 36, 75.

Thiede, J., 1981, Science, 211, 1422.

WCDP, 1985, 'Climatic System Monitoring (CSM) of the World Climate data Program (WCDP)', WMO Geneve, 52.

Wollin, G., Ryan, W.B.F., Ericsson, D.B. and Foster, J.H., 1977. Geophys. Res. Letters, 4, 267.

GEOPHYSICAL OBSERVATIONS OVER THE LAST FEW MILLENNIA AND THEIR
IMPLICATIONS

S.K. Runcorn
School of Physics
University of Newcastle upon Tyne
Newcastle upon Tyne NE1 7RU
U.K.

Another field where secular changes over this period of time are
very important is the variations in the length of the day which are
closely related to the secular variation of the Earth's magnetic field.
One of the problems in understanding the changes of the geomagnetic
field is to obtain a continuous record, whether from historical
observations or from the palaeomagnetic record. I think this will be
a theme which comes into this conference a great deal.

In discussing the secular changes in the rate of rotation in the
Earth it is useful to mention the very interesting history of this
subject. After Newton published Principia 300 years ago (July 6, 1687),
it resulted in a great programme of research for applied mathematicians,
taken up particularly in France, to check whether what we physicists
call perturbations and what the astronomers call inequalities in the
motions of the Moon and planets could be wholly explained by Newtonian
gravitation. Halley had drawn attention to an acceleration in the
Moon's longitude of about 10" arc/century^2 obtained from ancient
eclipses. Laplace, supposing that this secular acceleration of the
Moon in longitude was part of a very long period perturbation in
the Moon's motion, explained by Newtonian gravitation, showed that
the Sun affected the eccentricity of the Moon's orbit. Thus in
the early 19th century it appeared that this rather important in-
equality in the Moon's motion could be explained by Newtonian mechanics
along with all its other well known inequalities. Laplace won a
prize for this work. Then Adams in the 1840s revised the calculations
of Laplace, showing they were incomplete, and that part, about 4.8"
arc/century^2, could not be explained by Newtonian gravitation.

In fact earlier, Kant had given the explanation. Interested
in the philosophy of physics, he drew attention to the importance,
which had not been prominent in Newtonian physics, of conservation
of energy. He asked the question "Where does the energy come from
which is obviously being dissipated in the ocean tides?" and suggested
that it came from the rotational kinetic energy of the Earth. Thus
this secular acceleration of the Moon was soon attributed to the
slowing down of the Earth's rotation by tidal friction because
the astronomers' unit of time was then tied to the rotation of the

F. R. Stephenson and A. W. Wolfendale (eds.),
Secular Solar and Geomagnetic Variations in the Last 10,000 Years, 479–488.
© 1988 by Kluwer Academic Publishers.

Earth. A lengthening of the second would give an apparent acceleration of the Moon's motion.

The accurate observations of the motion of the Moon had shown
that, in addition to this effect, there was another discrepancy between
the observed Moon's longitude and that calculated. This amounted
to about 15" arc over the period from 1650 to 1900: astronomers thought
that there was a sinusoidal term in the Moon's motion of unknown
origin. They naturally followed the familiar procedures (I am sure
we shall have examples in this conference) of fitting a sine wave
to it, determining the amplitude (of about 10" arc) and period (250
years) and the phase: they named it the "great empirical term": it
had some obscure origin which would later be discovered. They even
put it in the tables of the Moon's motion in the Nautical Almanac.
Just as everybody had accepted that the great empirical term was
real, Nature took a hand and the curve of the discrepancy of the
Moon's longitude suddenly departed from prediction. The origin
of this anomaly was not settled until 1939 when Spencer Jones published
an important paper in which he showed that this same discrepancy
in the mean motion of the Moon is also seen in the mean motions of
the Sun and Mercury and Venus, when allowance is made for their
different mean motions on the celestial sphere. Thus Spencer Jones
drew the conclusion that the discrepancy curve was not due to a defect
in the gravitational theory of the moon's motion but had a common
origin in the variation of the rotation of the Earth or changes in
the unit of time. In particular, he pointed out that the sudden
slope about 1900 AD implied an increase in the length of the day
of about 3 ms.

A recent representation of the data by Morrison and Stephenson
(1981) using many lunar occultations, shows changes in the length
of the day of the order of milli-seconds but the data is smoothed
and the change around 1900 curves over some years: even so, very
sharply compared to most solid earth phenomena. By differentiation,
they determine the torques which must be applied to the Earth's mantle
to cause these changes in the rate of rotation of the Earth, although
early workers thought that changes in the moment of inertia of the
Earth might account for these changes. It was soon recognised that
they were insufficient by many orders of magnitude: de Sitter said
that to change the length of the day by these amounts, the Himalayan
Mountains would have to be levelled. Morrison and Stephenson show
on the same graph the tidal frictional torque responsible for the
secular deceleration; it is an order of magnitude smaller than those
required to account for the fluctuations in the Earth's rotation.
The lunar and the solar tidal torques (about 2/5 as large) increase
the length of the day by about 2 ms/century. Hence in order to explain
the irregular changes in the length of the day, torques lasting
for rather short times that may be accelerating or decelerating and
much larger than the steady decelerating torques must be postulated.
I am suspicious that in the process of smoothing, the sharpness of
some of these changes is removed but in any case the torques required
can be approximated by impulsive torques of either sign occurring
every few decades.

Data on the geomagnetic field in historical times goes back
a little longer than the length of day data to William Gilbert. Gilbert,
by a model experiment, taking a sphere of magnetized lodestone, showed
that the variation of the angle of magnetic inclination or dip with
latitude was very similar to that of the Earth. So Gilbert concluded
in de Magnete in 1600 that the Earth had a magnetic field like that
of a uniformly magnetized sphere. Sydney Chapman once said that
this was the only successful model experiment every done in geophysics!
But at Gresham College London in 1635, Gelibrand discovered
the secular variation of the Earth's field, first in declination. In
due course, surveys over the globe were done and magnetic maps were
produced; one of the first by Halley shows lines of equal declination
over the Atlantic. One of the big advances was made by the Carnegie
Institution of Washington who constructed a non-magnetic ship. In
the early part of this century rather accurate maps were made, particularly over the oceans, and from these in the immediate post war
years Vestine (1953) discovered the westward drift. He showed that
the isoporic foci, that is, the localities where one component of
the field is changing most rapidly, move in the course of decades
generally to the west. There was a strong centre of secular change
off Africa, the vertical component decreasing by over 100 γ per year
($1\gamma = 1m\mu T$). (A gamma γ has been a useful unit because, even in
the last century, observatory instruments were accurate to about
this value).

There are two older maps for 1780 and 1885 by L.A. Bauer, first
Director of the Carnegie Institution, which show so clearly, in the
course of a hundred years, the westward drift of the geomagnetic
field, see Figure 1. He plotted Isapoclinics (Bauer, 1895) curves
of equal difference between the observed angle of magnetic dip and
that corresponding to a dipole aligned along the axis of rotation
(or the uniformly magnetized sphere of Gilbert). In the course of
a century, the pattern remains similar but both the line of zero
difference and the points of greatest difference move to the west.
Many other demonstrations from the analysis of historical data have
shown this phenomena. Understanding of the secular variation did
not get far until this century but models with moving magnets inside
the Earth were used to represent the data. When Alfven's ideas on
magnetohydrodynamics became accepted, and when the core had been
shown by seismologists to be liquid, it was realised that the westward
drift was due to relative motion of the core with respect to the
mantle at about 0.2° per year, i.e. velocities at the core mantle
boundary of $\sim$ 1 cm/s. To derive the pattern of velocities at the
core surface is now a very profound part of the theory of the secular
variation. But from the point of view of understanding the irregular
changes in the length of the day, it is sufficient to think about
this relative rotation of the core with respect to the mantle that
the westward drift demonstrates. The mantle of the Earth is a ferro-
magnesian silicate, which in the outer parts has a very low conductivity but, due to the rise of temperature with depth, it is electrically conducting, due to the process of semiconduction. The conductivity
σ is given by terms like $\sigma = \sigma_0 exp - (E/2\ kT)$ where E is an excitation

energy and T is a temperature, k being Boltzmann's constant. Like so
many processes in solid state physics, the conductivity curve

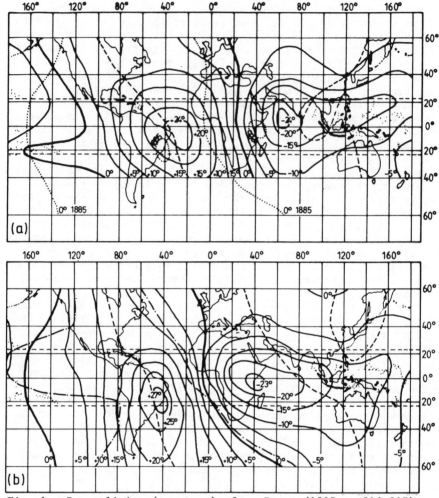

Fig. 1.   Isapoclinics (see text) after Bauer (1895, pp316-317):
          (a) for 1780, (b) for 1885. Comparison between the two
          plots clearly shows the westward drift of the Earth's
          magnetic field.

is S shaped and at depths 700-1000 km σ rises rapidly to flatten
off in the lower mantle. Semiconduction has enabled us to understand
results which had been gradually obtained by geomagnetic analysis
by Shuster, Chapman and then Lahiri and Price (1939). They showed
that to explain the relationship between the external and internal
sources of the daily variations and the magnetic storm variations
(separated by the method of spherical harmonic analysis), i.e. an
external driving varying field and these induced currents, a rapid

rise of conductivity at this depth of about 700 km had to be postulated. The secular variation observed at the surface contains periods of tens of years and this sets a limit to the mean conductivity of the mantle. Thus there is no doubt that the conductivity cannot continue according to the law Lahiri and Price found. The electrical conductivity cannot exceed about 1 $ohm^{-1}cm^{-1}$. This conductivity in the lower mantle is sufficient to give electromagnetic coupling between the core and mantle but there are many uncertainties in the theory. It was recognised early that variable electromagnetic torques between the core and the mantle could explain the variations in the length of the day quantitatively. It is therefore necessary to distinguish clearly between the two changes in the rotation of the mantle: the tidal deceleration in which angular momentum of the Earth is lost to the orbit of the Moon – and through the solar torque to the Earth's orbit – resulting in the Moon retreating from the Earth at a rate of about 5 cm/year; and the irregular fluctuation in the length of the day which involves interchange of angular momentum between the core and the mantle. Thus it was realised that the two phenomena, the westward drift and variations in the length of the day should be correlated. The first person to try to demonstrate this was Vestine (1953). He realised that a simple way of looking at the westward drift which could be used on early data was to use the representation of the quadrupole component common in the geomagnetic literature as a dipole displaced from the Earth's centre. Thus at present the dipole is inclined by about 11° to the axis and to best fit the quadrupole part of the field the dipole must be displaced towards the Pacific. Thus the latitude and longitude of the position of the best fitting dipole was plotted back to 1830 – Gauss' first spherical harmonic analysis. There is a motion in latitude as well as in longitude which Vestine thought was a good measure of the westward drift. He demonstrated the change in the westward drift velocity over these years and showed a rough correlation with changes in the length of the day. Ball et al. (1968) revised the data and plotted the rate of change of longitude of the eccentric dipole against the length of the day and the correlation is reasonably good. Thus we can conclude that the irregular fluctuations in the length of the day are caused by electromagnetic torques in the lower mantle reflecting changing magnetic fields generated in the core.

Considering the amplitude of these curves, it becomes clear that in the interchange of angular momentum between the core and the mantle, only the outer part of the core is being involved at least in these rather short period changes, in fact only the outer 200 km of the core. I think that this is very significant and it has a bearing on some very recent work by Stephenson and Morrison (1984).

In the early discussion of the tidal deceleration of the Earth's rotation it was evident to Spencer Jones and to Munk and MacDonald (1960) that there appeared to be a difference in the tidal frictional torque derived from the modern observatory data back to about 1650 and that derived from ancient eclipses. Spencer Jones showed in

1939 how one could separate out that part of the change in the
rotation of the Earth which is due to tidal friction and that part
which is due to internal processes within the Earth.

The irregular fluctuations, the interchange of angular momentum
between the core and the mantle, do not affect the total angular
momentum of the Earth but the effects of the lunar and the solar
tides do. The slowing down of the Earth's rotation by the lunar
tidal component transfers angular momentum to the orbit of the moon;
thus the secular deceleration from the old observations was the
difference between an apparent part due to the secular lengthening
of the unit of time used in the reductions and a real acceleration
of the Moon in its orbit. Spencer Jones had shown that the same
discrepancies appear in the motion of the Sun and Mercury and Venus.
Those are apparent and only reflect the changes in the rotation of
the Earth. Thus, the angular momentum being transferred to the Moon's
orbit can be derived from the astronomical observations and it is
possible to determine the tidal frictional torque in spite of the
large irregular changes. There is a discrepancy between the determinations of the lunar tidal frictional torque from the modern
observations and that formerly derived from the ancient eclipses
where the effect of the irregular fluctuations is unimportant.
Stephenson and Morrison's achievement in their study of the ancient
eclipse records is to remove this discrepancy which, at one time,
appeared to suggest that the lunar tidal frictional torque had changed
by a factor of 2 over the last 2,000 years.

Stephenson and Morrison, in settling this important issue, show
that from the ancient eclipse observations in addition to the tidal
slowing down of the Earth, there is, what has come to be unfortunately
called a secular acceleration of the Earth. Stephenson and Morrison
have taken out the tidal term and leave a non-tidal change to be
explained. It might be better to represent it by a sine curve of
about 2000 year period than a constant acceleration or a change
in acceleration around 1000 A.D.

I suggest the following interpretation: the non-tidal term that
Stephenson and Morrison have now established may be a very long
period core mantle coupling. If so, it may be reasonable that if
there is a period of 1000-2000 years in the strength of the geomagnetic
field, which is indicated by the archaeomagnetic data, and is required
to understand the C14 discrepancy, such a coupling would penetrate
down and involve the whole core. The dates assigned by the Egyptol-
ogists and the C14 dates showed a discrepancy; subsequently the C14
dates were calibrated in the famous tree ring curve by Suess (1970)
but the explanation in terms of a changing magnetic field has not
been adequately explained. It is clear that there are very long
periods of a few thousand year variations of the strength and direction
of the geomagnetic field and, therefore, the idea that there was
a long term variation in the Earth's rotation in addition to the
much shorter irregular fluctuations, is reasonable. It can be
reasonably deduced that, if this is the explanation of the non-tidal
term, about two thousand years ago the core was rotating to the east.
The mean rotation rate of core and mantle must be the same but it

seems that the varying torques which result from changes in the geomagnetic field being applied to the mantle can result statistically
in the velocity being at one time east and another time west.

I now turn to review the evidence that, looking at the astronomical data, the changes in the length of the day can establish
themselves in very short periods of time, e.g. a year. Geophysicists
have become accustomed to think of the geomagnetic field as having
periods perhaps between 50 and 10,000 years The isoporic foci and
their associated regions of chance- nests of oval curves on the
magnetic maps shows this well. This caused scientists to look for
sudden changes in the geomagnetic field. I suggested a long time
ago that to explain these sudden changes in the rate of the Earth's
rotation the torque on the mantle had to be a Dirac delta function
rather than a step. I suggested that the velocity of Alfven waves
in the core made it possible that a disturbance in the core would
propagate through it in a time of a few months to a year. This
speculation has only recently been shown to be realistic. Courtillot
and Le Mouel (1984) have discovered in the observatory records changes
with this kind of sharpness. Using annual means from data from many
observatories in Europe they discovered that in 1970 the second
derivative of the 3 components of the geomagnetic field changed
abruptly. They also think that there was a similar change around
1910. In taking annual means the data is somewhat contaminated by
external components of the Earth's field but spherical harmonic analyses
on world wide data show the jerk to be of internal origin. They
found that they could represent the data up to that point rather
well by a quadratic term in time but then if they did not change
the coefficient of the T^2 term, they found that the departures of
the observations from the quadratic began to increase after 1969
to 1970. At any one observatory the annual means are much contaminated
with the effect of magnetic storms and stormy days. When observations
over the whole globe are used, the difference before and after 1970
of the second derivative in the north horizontal component, the east
horizontal component and the vertical component show a distribution
over the globe; the quantity changes sign over nodal lines of global
extent: it is a global phenomena. Malin and Hodder's (1982) spherical
harmonic analysis shows that certain low degree harmonics predominate.

I come finally to the question of the polar motion. From laser
ranging of satellites and from VLBI the accuracy in determining the
changes in the length of the day and the motion of the pole has
increased dramatically. This wonderful advance in technique has
had the unfortunate result of making geophysicists think that perhaps
the historical data do   not need further study or are too inaccurate
to be worth using. I wish to state the obvious: the historical record
must be used to study these secular geophysical effects. The motion
of the pole has been observed by an international programme since
about 1900 but there is much noise in the data. The motion of the
pole observed can be separated into the annual motion and the 14
month Chandler wobble. The pole moves over an area roughly 10m
by 10m (Figure 2).

486

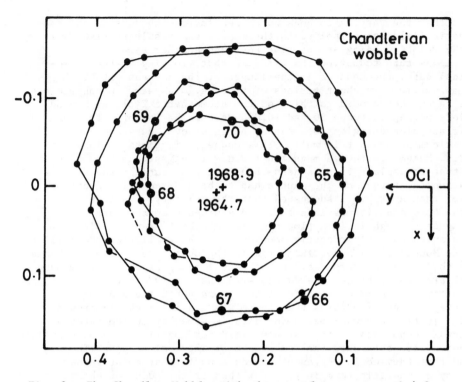

Fig. 2.   The Chandler Wobble with the annual term removed (after
Guinot, 1972).   The units of both axes are "arc.

        Recently, by making use of the very much more accurate data on the
changes in the length of the day, Hide (1984) has conclusively demonstrated that there are very short term, small but now accurately
measured changes in the rotation of the Earth which are explained as an
exchange of angular momentum between the mantle and the atmosphere.   He
has gone further and suggested that changes in the angular momentum
in the atmosphere play a role in exciting and maintaining the Chandler
motion but I admit to having difficulties mainly as regards orders
of magnitude.

        I suggested many years ago that the core might be responsible
for exciting the Chandler wobble.   My argument, which I think is
not widely accepted, is as follows (see Figure 3).   If the disturbance
in the core magnetic field which produces a torque on the mantle
due to the semiconducting lower mantle   is a very short
term phenomenon, it can be represented as a Dirac delta function.
If the torque has a component along the axis of rotation, it changes
the length of the day, and, of course, the westward drift.   If, however,
the magnetic field change produces a torque about an axis in the
equatorial plane it changes the pole position.   The alternative theory
of the excitation of the Chandler wobble, which was once popular,
was that the magnetic field could be represented by a step function,

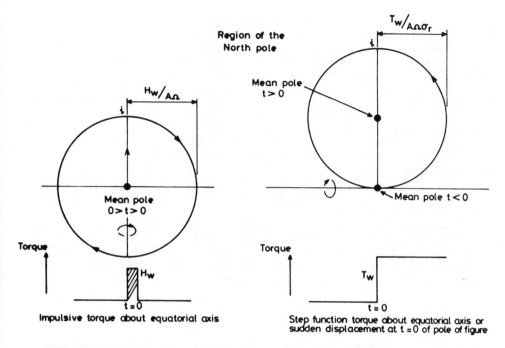

Fig. 3. Possible causes of excitation of the Chandler Wobble.

or that a large earthquake could change the position of the axis
of maximum moment of inertia: the effect of the two is the same.
I pointed out that these two alternative theories of the excitation
of the Chandler wobble should result in a quite different behaviour
pattern of the polar motion. The Chandler wobble comes about when
the axis of maximum moment of inertia and the axis of instantaneous
rotation do not coincide. Between the two is the axis of angular
momentum and because the Earth has a small ellipticity, the axis
of spin and the axis of angular momentum are nearly coincident. If
there is no Chandler motion then these three axes are coincident
and an impulsive torque is applied to the mantle; the axis of angular
momentum is moved away from the axis of maximum moment of inertia
(often called the axis of figure) and the axis of instantaneous
rotation (which is the pole determined by astronomical observations),
being very close to it, describes a circle with the axis of figure
as the centre. In the more general case of there already being a
polar motion, the effect of an impulse is to change the radius of
the polar motion curve but not its centre which remains the axis
of figure. However, if an earthquake were to excite the Chandler
wobble by changing the moment of inertia tensor, supposing the three
axes are coincident, then the axis of maximum moment of inertia will
move to some new position on the Earth's surface and then the
Chandler wobble will begin with the axis of instantaneous rotation
moving round on a circle with a new centre. In the more general

case of a polar motion already occurring, an additional excitation of this type will move the centre of the circular motion. The question arises whether the polar motion data allows one to distinguish between these theories of the excitation. An important contribution has been made by Guinot (1982) who took out the atmospheric 12 monthly motion and examined the Chandler wobble behaviour over some recent years when the data was beginning to be good. Plotting fortnightly mean poles he showed that from 1965 it described a path reasonably close to a circle and then in the spring of 1967 the polar motion deviated from this circular path and began another set of closed curves, nearly circles. Guinot calculated the centre before and after this departure from the circular paths. He found that the centre has not changed although the radius has. Thus there appears to be evidence in favour of the impulsive torque origin of the Chandler wobble excitation.

We have now very much more powerful techniques for accurately monitoring changes in the length of the day and the polar motion but we must await with some impatience the occurrence of such "jumps" and "jerks" as we have seen through the noise of the older data. Then some of the ideas outlined can be given a proper test.

REFERENCES

Ball, R.H., Kahle, A.B. and Vestine, E.H., 1968, Trans. Amer. Geophys. Union, 59, 152.
Bauer, L.A., 1895, Amer. Jl. Science, 50, 314.
Courtillot, V. and Le Mouel, J.L., 1984, Nature, 311, 709.
Guinot, B., 1972, In 'Rotation of the Earth' (eds. P. Melchior and S. Yumi), Reidel, 46.
Guinot, B., 1982, Geophys. Jl. R. Astr. Soc., 71, 295.
Hide, R., 1984, Phil. Trans. R. Soc., A, 313, 107.
Lahiri, B.N. and Price, A.T., 1939, Phil. Trans. R. Soc., A, 237, 509.
Malin, S.R. and Hodder, B.M., 1982, Nature, 296, 726.
Morrison, 1.V. and Stephenson, F.R., 1981, In 'Reference Coordinate Systems for Earth Dynamics', (eds. E.M. Gaposchkin and B. Kolaczek), Reidel, 181.
Munk, W.H. and MacDonald, G.J.F., 1960, 'The Rotation of the Earth', Cambridge Univ. Press, 323.
Spencer, Jones, H., 1939, Mon. Not. R. Astr. Soc., 99, 541.
Stephenson, F.R. and Morrison, L.V., 1984, Phil. Trans. R. Soc., A., 313, 47.
Suess, H.E., 1970, In 'Radiocarbon Variations and Absolute Chronology', Almquist and Wiksell, Stockholm, 595.
Vestine, E.H., 1953, Jl. Geophys. Res., 58, 127.

ASTRONOMY : EVOLUTION OF EVOLUTION

Professor Sir William McCrea,
University of Sussex.
U.K.

MILESTONES   ON   THE   WAY

1. STAR : CONSTITUTION
2. STAR : ENERGY-SOURCE
3. INTERSTELLAR MATTER : EVOLUTION OF THE GALAXY
4. UNIVERSE OF GALAXIES
5. EXPANDING UNIVERSE
6. HOT BIG BANG (LEMAITRE)
7. NUCLEO-SYNTHESIS : BACKGROUND RADIATION (GAMOW)
8. STEADY-STATE COSMOLOGY : B^2FH NUCLEO-SYNTHESIS
9. REVISED DISTANCES : HUBBLE TIME
10. MICRO-WAVE BACKGROUND DISCOVERED
11. SIMPLE STEADY STATE ABANDONED
12. SIMPLE BIG BANG TAKEN SERIOUSLY : HISTORY OF UNIVERSE
13. EARLY UNIVERSE : DARK MATTER
14. VERY EARLY UNIVERSE : INFLATION
15. ?

The Workshop is a study of some variations with time of certain features
of the Sun and Earth over an interval for which significant empirical
evidence appears to be available. So it may be said to have bearing
upon the evolution of these bodies - a rather specialized aspect
amenable to special expertise. The organizers have invited me to
go to the opposite extreme by trying to offer a brief survey of the
overall trend in the study of evolution in the astronomical Universe.

The aim of the study is to produce the biography of that Universe.

There are difficulties:-

We have to discover what we have to discuss as we go along,
Needed measurements may be lacking,
Needed physics may be lacking,
There is a shortage of 'archival' material.

489

F. R. Stephenson and A. W. Wolfendale (eds.),
Secular Solar and Geomagnetic Variations in the Last 10,000 Years, 489–498.
© 1988 by Kluwer Academic Publishers.

It is to be remarked that evolution must include origins. For example the *formation* of, say, the Sun is an incident in the *evolution* of the Galaxy.

The scientific study of evolution in astronomy began effectively early in the present century. When we see how much knowledge of astronomy and physics this was found to require, we can understand why it could not have happened any earlier.

Stars and the Galaxy

Remembering that the Sun is a star, it is true that all the natural *light* we normally see by day or night is starlight. So all the *matter* we see is in stars (apart from a small amount of other matter illumined by starlight). So our study begins with *stars*.

In the early 1900s astronomers deemed all stars to constitute *the stellar system*. Thus in 1914 A.S. Eddington published *Stellar motions and the structure of the Universe*. This was a pioneering study of this system's kinematics and dynamics.

Now a *star* is a body of hot gaseous plasma held together by self-gravitation. Again it was Eddington who in 1926 published *The internal constitution of the stars* which, more than any other single work, laid the foundations of modern astrophysics. From our immediate standpoint the great discovery was that the physical state throughout a star could be determined without knowledge of the source of the star's energy of radiation. For the work demonstrated that, whatever the energy source, were it completely switched off in any typical star that star would show little change for something like 10^7 years. This meant that astrophysicists had to discover an energy source that would operate under the conditions made known by Eddington's work. Previously they had supposed that they could not know the conditions before they knew the source, and since they could not know the source before they knew the conditions in which it was required to operate, there had seemed to be no way to get started on the problem.

But even knowing the conditions, it was some years before enough was known about the (nuclear) physics that goes on in those conditions. By the year 1938/39 such knowledge was adequate to enable H. Bethe and C.F. von Weizsäcker, independently, to identify the nuclear reactions most likely to be significant.

Thermonuclear energy-generation became recognised as the general source of a star's radiation. Therefore in due course it became possible to infer *the entire life history* of any *single star* from its start on the main-sequence (ZAMS) to its ultimate fate as a planetary nebula, supernova, white dwarf, neutron star (pulsar), or perhaps black hole. Also the somewhat more complex life-history of a close *binary star* has come to be worked out.

Besides stars the galaxy contains *interstellar* matter (ISM). In a remarkable Bakerian Lecture to The Royal Society in 1926, Eddington pioneered the study of this too. It is now known to comprise atoms, ions, molecules and dust-grains - much of it forming 'giant molecular clouds' (GMC). It contributes a few per cent to the total mass of the visible Galaxy.

Star-formations

The famous Hertzsprung-Russell diagram for the stars is effectively the graph of absolute luminosity against surface temperature. Starting around 1946 astronomers studied such diagrams for the stars of one particular stellar cluster at a time, and interpreted them in the light of the knowledge of stellar evolution just mentioned. They were able to infer:

All stars of any one cluster are of *closely the same age*.

Some clusters were formed recently - astronomically speaking, say within the past few 10^7 years - *from ISM as we know it*.

Such formation must be proceeding now; in fact astronomers identify regions of *current star-formation*.

There are *generations* of stars resulting from *'bursts'* of star formation.

Some of the material of the Galaxy must have been *processed through stars more than once*.

Astronomers proceeded to construct theories of *galactic evolution*. These are plausible so far as they may be justifiably carried at the present time. One result that has long emerged is that the age of the Galaxy since the oldest stars came into existence is between 10 and 20 x 10^9 years. A value near 16 x 10^9 has often been quoted as the best estimate, but some recent estimates tend to bring it somewhat lower to about 13 x 10^9 years. It is important to note that we are at this point dealing with ages derived from *astrophysical* considerations. *Cosmological* considerations related to the expansion of the Universe are not yet taken into account.

There is however one outstanding fundamental deficiency. Astronomers still *possess no accepted model for the actual process of forming a star out of ISM*. The process cannot be the simple contraction upon itself of any particular portion of ISM, of appropriate mass, as it normally exists. For such a portion would possess far too much *angular momentum* for it to be able to contract as a whole. This must be why stars are formed normally in clusters, presumably by some process of 'clotting' of the raw material that can leave angular momentum in the *relative motions* of condensing stars, rather than in spinning motion of those stars themselves. At present there is no consensus as to how nature achieves this. This is the distressing gap in the cosmogony of to-day.

Universe of Galaxies

The astronomers of about 1914 consigned most 'nebulae' to an outer fringe of the Galaxy so that these helped to form what they saw projected upon the sky as the *Milky Way*.

There had, however, been conjectures about 'island universes'. Eddington had been able to discuss as reasonable the suggestion that 'spiral nebulae' might actually be objects comparable to the Galaxy itself. In that case their apparent sizes and luminosities would imply *distances many times the linear extent of the Galaxy*. Eddington was favourably inclined to the idea. But at the time there was no independent estimate of distances that might check it.

By about 1925, however, things had changed drastically. Although distances were still underestimated by factors of up to about 10, the astronomers' Universe had been enlarged from

the *single Galaxy* with its indeterminate fringe

to a *system of 'galaxies'*, each separated from its nearest neighbour by distance, on average, of order 100 times its own diameter, apparently extending indefinitely in all directions.

Astronomers had contemplated a Universe in which practically all matter was in *one star or another* and, once formed, *any particular star evolved under almost no influence from outside itself*. Now they thought in terms of a Universe in which all observable matter belonged to *one galaxy or another* and, once formed, *a galaxy would evolve under no influence from outside itself*.

One direct consequence was, of course, that our own Galaxy became much better understood as a result of comparative studies with other galaxies. However, in this paper there can be no attempt to discuss the evolution of observational procedures. Having said that, it should be remarked that in more recent times the extension of observational capabilities into all parts of the electromagnetic spectrum has obviously been of immeasurable significance for all the developments to be noticed here.

Mass of a galaxy

Galaxies had become for astronomy the building-blocks of the Universe. So the crucial question became, *What amount of matter is in a galaxy?*

Astronomers could attempt to evaluate:-

(a) 'Luminosity mass' - based upon the inferred stellar composition and knowledge of the mass/luminosity ratio for the various sorts of star contributing to the composition.

(b) 'Rotation mass' - based upon the estimated size and the observed rate of rotation at various radial distances.

(c) 'Virial mass' - got by applying the virial theorem to a gravitationally bound cluster of galaxies using observed sight-line velocities of these galaxies, the result yielding a value of their mean mass.

Rotation masses and virial masses agree fairly well with each other. But they seem generally to be about 10 or more times the luminosity masses. For this reason, and because of the shape of the 'rotation curves', many galaxies are inferred to possess *dark halos*.

Most astronomers seem convinced that many spiral galaxies do have such a halo. However, it is difficult to obtain direct evidence in any particular case. It is disconcerting to think that 90 per cent of the material of a normal galaxy may be something we know nothing about! This is particularly the case if we proceed to ask how a galaxy is formed. How *what* is formed, you will ask. We shall ask this a little later.

Cosmology

Eddington's 'Universe' of 1914 was thought of as having diameter of
perhaps 25 kpc. He contemplated speeds of stars moving therein as
typically of order about 20 km/s. So the crossing-time of such a
star would be greater than about 10^9 years. Therefore the system
as a whole would not change greatly in a few times 10^9 years, or not
at all if the motions are mainly periodic.

   A. Einstein's celebrated paper 'Cosmological considerations
on the general theory of relativity' 1917 is rightly regarded as the
beginning of modern cosmology. In it he presented his *static
cosmological model* that required him to modify his original form of
the theory. That modification, he wrote, "is necessary only for the
purpose of making possible a quasi-static distribution of matter,
as required by the fact of the small velocities of the stars".

   Thus Einstein in 1917 had in mind a 'Universe' like Eddington's
of 1914, which as we have just seen would indeed be 'quasi-static'.
Einstein thought he was dealing with the Universe as it is; he had no
a priori conviction that it ought to be static or anything else.

   What astronomers actually discovered a decade later was an
expanding system of galaxies which evidently comprised the whole Universe.

   Theoretical work by Einstein (1917), W. de Sitter (1917), A.
Friedman (1922), G. Lemaitre (1927), Eddington (c.1930), when properly
understood all predicted an expanding system - maybe one should say
definitely expanding or definitely contracting.

   In our present context, the crucial feature was:- *The actual
Universe in the large was seen to be evolving; theory had indicated
this was to be expected.* The remarkable feature was that evolution
appears as a necessary feature *without reference to thermodynamics*.

Expansion of the Universe

To a good first approximation the expansion is specified by the single
parameter H_0, such that any two galaxies remote from each other are
separating at speed H_0 x distance apart. A time t_0 is defined by

Hubble time $t_0 \equiv H_0^{-1}$

From about 1930, E. Hubble claimed that his observations showed

$t_0 \overset{\sim}{\sim} 2 \times 10^9$ years.

Unfortunately a speculation of Eddington's happened to suggest a
'theoretical' value close to this.

   Geologists had already estimated the age of the Earth $\overset{\sim}{\sim} 4 \times 10^9$
years. Hubble's observations seemed to show that the Universe had
been in a very congested state about 2×10^9 years ago; evolution might
have been accelerated in that state, so perhaps the 4×10^9 years was
an overestimate. At that stage, therefore, the times coming out of
the work seemed not necessarily unacceptable. It is to be remembered
that this was nearly a decade before the investigation of thermonuclear

processes began to lead to 'astrophysical' ages.

In 1933 Lemaitre proposed what is now recognized to have been a first version of *hot big-bang cosmology*. Lemaitre identified the basic associated problems, but the significance of his ideas was not then recognised. Cosmology then produced few predictions of astronomical interest until after World War II.

G. Gamow was then the first to see that nuclear reactions of known sorts would be important in an early big bang universe. (Lemaitre had speculated on the subject before sufficient nuclear physics was known for it to be possible to reach useful conclusions.) He and collaborators about 1948 correctly predicted background radiation, but discovered difficulties in the way of predicting nuclear abundances.

Other cosmologists then became concerned about the *age-problem*. With Hubble's still current value of H_O, this problem had become acute for a simple big-bang cosmological model. Instead of querying the observations, they queried the postulates of the model.

H. Bondi and T. Gold, and F. Hoyle separately, invented *Steady-state cosmology* depending upon the hypothesized *continual creation* of new matter - presumably protons and electrons.

This theory consequently required nucleo-synthesis to proceed in *stars*. In a famous paper in 1957, G. & E.M. Burbidge, W.A. Fowler and F. Hoyle ('B^{2}FH') showed how this could occur. Later 'the helium problem' arose, but in the main B^{2}FH is still regarded as valid.

A great shake-up in the whole field followed from observational work by W. Baade announced in 1952. In the course of the next few years this led to a reduction of the accepted empirical value of H_O by a factor between 5 and 10. The time t_O was thus put at between 10 and 20 x 10^9 years. The age-problem had apparently become no longer insuperable for big-bang theory.

Radio astronomy proceeded to yield results of source-counts any one of which, if confirmed, would have invalidated simple steady-state theory. Unfortunately, at any rate at first, the results tended to invalidate each other.

But then in 1965 A.A. Penzias and R.W. Wilson discovered microwave background radiation. This had been predicted by simple big-bang cosmology. Astronomers forthwith took that theory seriously. They quickly, though somewhat regretfully, abandoned the steady-state model.

'Simple' big-bang cosmology

The specification of the simple model requires it to be *homogeneous*. In the early universe there is a phase in which the temperature T and density D are very great, and atomic nuclei interact so freely that the nuclear chemical composition depends only upon the instantaneous T, D values. As universe-expansion proceeds, T, D decrease in such a way that the parameter N *remains constant*, where N ≡ (number of photons in unit volume) ÷ (number of baryons in unit volume). Baryons are actual or potential hydrogen atomic nuclei. They compose nearly all the mass of 'ordinary' matter, which may be called *baryonic matter*. Unless otherwise stated, this is what we mean by 'matter'. Non-baryonic matter would be composed of neutrinos, or so-called 'exotic' particles whose existence is not yet established.

After T has fallen below about 10^8K, all nuclear reactions - in the homogeneous universe - die down. A particular chemical composition then remains frozen into the matter. This is all the material available to form galaxies and stars; for this purpose it is described as *primordial*.

Nuclear physics shows that the primordial composition depends almost entirely upon the value of N. The only nuclei of much interest at this stage are those of

hydrogen ^{1}H, deuterium ^{2}D, helium ^{4}He

If then we know the primordial abundance of ^{2}D, or of ^{4}He, nuclear physics yields a value of N.

About the most recent astrophysically inferred estimates of these abundances are still some made by B.E.J. Pagel in 1982. With the photon number-density for the temperature 2.7K of the microwave background at our cosmic epoch, the resulting N-values yield values of the present mean baryonic density ρ_b in the Universe. Pagel's error-bars for the estimates from the two nuclei overlap slightly so as to agree in allowing a value close to $\rho_b \sim 2 \times 10^{-31}$gcm^{-3}. This result does not depend upon the value of H_0.

A value of ρ_b can be got also from counts of galaxies and dynamical masses derived as already described. This value does depend upon H_0, which enters into distance estimates; in fact the derived density is proportional to $H_0{}^2$. If H_0 corresponds to $t_0 \sim 20 \times 10^9$years, this estimate is again $\rho_b \sim 2 \times 10^{-31}$gcm^{-3}.

Having regard to uncertainties at various points, the closeness of the agreement of the results must be somewhat fortuitous. Nevertheless the self consistency of the picture as a whole is impressive. Let us try to wrap it up, as they say:-

The abundances of H, D, He in the raw materials of galaxies is accounted for. Given these initial abundances we know that B^{2}FH can account for the rest. That is to say, the *chemical composition of the cosmos is explained*. This has depended upon assuming *the cosmic nature of the background radiation,* which may be claimed as thereby confirmed. The consistency is found to be best for the quoted cosmological value of t_0, which is in excellent accord with the 'astrophysical' age of the Galaxy. Thus cosmology and astrophysics have come into impressive agreement. It is even better than appears from these results alone, for in tracing back the history of the expansion of the universe we can locate approximately the *epoch at which the galaxy must have become isolated from other matter* in order for it to possess the mean density that we observe. Again the result is plausible. Also we note that all this in itself seems to confirm the *baryonic nature* of most of the material of the Universe. Incidentally if all this is mainly valid it may be claimed to confirm the inferred history back to within a few seconds after the big bang, which would be about the beginning of the phase in which helium formation took place.

Finally all this appears nowhere to produce a conflict between observation and theory.

So far as this goes it surely implies that the big-bang model is meaningful as giving a biography of what we have come to call the Universe, back to within seconds of its birth.

Problems outstanding

The foregoing account tells us very little compared with what it does not tell.

What determines the particles of physics and the laws of physics that shall constitute the Universe? What prescribes the very special properties of the simple big-bang model? How do the apparently simple initial properties produce the complex structure of the observed Universe?

Here we can say something about only the last topic.

Matter distribution

Observation shows that matter in the Universe is distributed in space with huge *non-uniformity* on the scales of stars, galaxies, clusters of galaxies, chains of clusters, and of enormous voids of order 10^8 parsecs across.

By contrast, on a still larger scale the Universe is inferred to be to a high degree homogeneous and isotropic – witness the astonishingly high degree of isotropy of background radiation. With the simple expansion of Friedman-Lemaitre models this would require the exactly similar behaviour of parts of the Universe before they have had time to 'communicate' with each other.

The *phenomenon of inflation* in the very early Universe is hypothesized to cope with this situation – supplemented by ideas on 'topological defects' and theories of 'strings' and 'superstrings'. Also the behaviour of the contents of the Universe in that phase would be governed by ultra *high-energy particle physics*.

Many cosmologists infer that this would result in the mean density of the Universe being equal to the 'closure density'.

This would be 10 or more times the derived baryonic density. It would bring back the age-problem. It presents the problem of *non-baryonic matter* ('dark' matter) and of *exotic particles*. (Note that dark galactic halos need not have anything to do with *this* dark matter.)

Universe of universes?

We began by recalling a time (c.1914) when the Galaxy in which we are located seemed to astronomers to be their whole universe. In due course they recognized the existence of other galaxies. They came to see our Galaxy as belonging to an (expanding) universe of galaxies. We now ask, might this belong to a UNIVERSE of such universes?

(a) This would be in keeping with the general historical development of astronomy – whatever astronomers have discovered, always they have gone on to discover other instances of the same thing.

(b) It would accord with the scientists' intuition that every experiment must be repeatable – even a big-bang.

(c) 'Simple' big-bang cosmology sees the whole universe as emerging from one singularity – a simple 'bang'. Simple steady-state cosmology saw every elementary particle as emerging from a singularity – its own 'micro' bang; this was the postulated continual creation. It is reasonable to explore the intermediate possibility of a set of little big (or big little) bangs.

(d) There is nothing essentially new in the suggestion. At one stage in the discussion of steady-state cosmology, Hoyle and J.V. Narlikar proffered a scheme of what looked like big-bang models all combining to produce a steady state system as a whole. More recently Hoyle sketched out another scheme of basically this character. But the present suggestion is *not* for an over-all steady state; this would not be of interest since it appears that the component 'universes' would not interact.

(e) The basic astronomical merit of the system here proposed is that it should be able to account for the *origin of galaxies* in our universe. There is the old problem. What caused the very first condensations in its contents? Apart from quantum effects, the only cause we can conceive must come from outside this universe itself – presumably from interaction with another such 'universe'. We should expect an interaction that causes shock waves of some sort. Obviously this begins to look like the start of an *infinite regress* – the UNIVERSE may be like that.

(f) As customarily presented, the concept of the big-bang implicitly requires something of the sort suggested here. For it looks upon the big-bang as operating according to some *pre-determined physical principles*. In itself it makes no provision for 'laws of physics' to emerge from the big-bang. That is to say, the concept evidently accepts the existence of something 'before' the big-bang.

(g) Some versions of an inflationary model do recognize the existence of something more than 'our universe' by regarding this as having been produced by the inflation of one tiny 'bubble' within something more extensive. However, in general no subsequent interaction with that 'something' seems to be invoked.

(h) Recent work on *large-scale streaming* in the observed universe (D. Lynden-Bell and collaborators) – it is tempting to suggest – might be taken as indicating some outside influence. But the interpretation of the observations as a whole appears to be still undecided.

(i) A model having the desired general character is in principle mathematically possible:– The simplest big-bang universe is the one known as the 'Milne universe'. The standard description is as the simplest Friedman-Lemaitre expanding universe. But it can be described alternatively by a pencil of world-lines through a single event in a Minkowksi space-time. In that description

there could be any number of other such systems in that same space-
time. Thus we should have a UNIVERSE of (Milne) universes. And
obviously they would be interacting systems. The model may be
sufficient to suggest some of the problems that would arise, but
it seems to be too crude to offer useful physical inferences.

The only safe inference is that ideas about evolution in astronomy
will go on evolving.

INDEX

506